Translocator Protein (TSPO)

Special Issue Editors

Giovanni Natile
Nunzio Denora

MDPI • Basel • Beijing • Wuhan • Barcelona • Belgrade

MDPI

Special Issue Editors

Giovanni Natile
University of Bari "Aldo Moro"
Italy

Nunzio Denora
University of Bari "Aldo Moro"
Italy

Editorial Office
MDPI AG
St. Alban-Anlage 66
Basel, Switzerland

This edition is a reprint of the Special Issue published online in the open access journal *International Journal of Molecular Sciences* (ISSN 1422-0067) from 2016–2017 (available at: http://www.mdpi.com/journal/ijms/special_issues/tspo).

For citation purposes, cite each article independently as indicated on the article page online and as indicated below:

Lastname, F.M., Lastname, F.M. Article title. *Journal Name*. **Year**. *Article number*, page range.

First Edition 2018

ISBN 978-3-03842-757-5 (Pbk)
ISBN 978-3-03842-758-2 (PDF)

Table of Contents

About the Special Issue Editors   v

Nunzio Denora and Giovanni Natile
An Updated View of Translocator Protein (TSPO)
doi: 10.3390/ijms18122640 . . . .   **1**

Anna M. Giudetti, Eleonora Stanca, Luisa Siculella, Gabriele V. Gnoni and Fabrizio Damiano
Nutritional and Hormonal Regulation of Citrate and Carnitine/Acylcarnitine Transporters:
Two Mitochondrial Carriers Involved in Fatty Acid Metabolism
doi: 10.3390/ijms17060817 . . . .   **6**

Leo Veenman, Alex Vainshtein, Nasra Yasin, Maya Azrad and Moshe Gavish
Tetrapyrroles as Endogenous TSPO Ligands in Eukaryotes and Prokaryotes: Comparisons with
Synthetic Ligands
doi: 10.3390/ijms17060880 . . . .   **21**

**Valentino Laquintana, Nunzio Denora, Annalisa Cutrignelli, Mara Perrone, Rosa Maria Iacobazzi,
Cosimo Annese, Antonio Lopalco, Angela Assunta Lopedota and Massimo Franco**
TSPO Ligand-Methotrexate Prodrug Conjugates: Design, Synthesis, and Biological Evaluation
doi: 10.3390/ijms17060967 . . . .   **47**

**Salvatore Savino, Nunzio Denora, Rosa Maria Iacobazzi, Letizia Porcelli, Amalia Azzariti,
Giovanni Natile and Nicola Margiotta**
Synthesis, Characterization, and Cytotoxicity of the First Oxaliplatin Pt(IV) Derivative
Having a TSPO Ligand in the Axial Position
doi: 10.3390/ijms17071010 . . . .   **60**

**Eleonora Da Pozzo, Chiara Giacomelli, Barbara Costa, Chiara Cavallini, Sabrina Taliani,
Elisabetta Barresi, Federico Da Settimo and Claudia Martini**
TSPO PIGA Ligands Promote Neurosteroidogenesis and Human Astrocyte Well-Being
doi: 10.3390/ijms17071028 . . . .   **78**

**Ji Young Choi, Rosa Maria Iacobazzi, Mara Perrone, Nicola Margiotta, Annalisa Cutrignelli,
Jae Ho Jung, Do Dam Park, Byung Seok Moon, Nunzio Denora, Sang Eun Kim and
Byung Chul Lee**
Synthesis and Evaluation of Tricarbonyl [99mTc]-Labeled 2-(4-Chloro)phenyl-imidazo[1,2-*a*]
pyridine Analogs as Novel SPECT Imaging Radiotracer for TSPO-Rich Cancer
doi: 10.3390/ijms17071085 . . . .   **94**

Gurpreet Manku and Martine Culty
Regulation of Translocator Protein 18 kDa (TSPO) Expression in Rat and Human Male
Germ Cells
doi: 10.3390/ijms17091486 . . . .   **104**

**Tamara Azarashvili, Olga Krestinina, Yulia Baburina, Irina Odinokova, Vladimir Akatov,
Igor Beletsky, John Lemasters and Vassilios Papadopoulos**
Effect of the CRAC Peptide, VLNYYVW, on mPTP Opening in Rat Brain and Liver Mitochondria
doi: 10.3390/ijms17122096 . . . .   **119**

Nasra Yasin, Leo Veenman, Sukhdev Singh, Maya Azrad, Julia Bode, Alex Vainshtein, Beatriz Caballero, Ilan Marek and Moshe Gavish
Classical and Novel TSPO Ligands for the Mitochondrial TSPO Can Modulate Nuclear Gene Expression: Implications for Mitochondrial Retrograde Signaling
doi: 10.3390/ijms18040786 134

About the Special Issue Editors

Giovanni Natile received the Laurea in Chemistry at the University of Padua in 1968. After postdoctoral positions at the University College London (1969–1970) and University of Padua (1971–1972), he was appointed Research Assistant and Lecturer at the University of Venice (1972–1980) and then Professor of Inorganic Chemistry at the University of Bari (since 1980). Professor Natile's main research interests are in the field of Organometallic and Medicinal Bioinorganic Chemistry. (i) In the field of Organometallic Chemistry, he has been interested in the activation of unsaturated molecules and formation of C–C, C–N, and C–O bonds. In this context, he initiated the synthesis of a large series of unprecedented five-coordinate complexes of platinum(II) containing olefins, which can evolve into 4-coordinate superelectrophilic species. (ii) In the field of Bioinorganic Chemistry, he has extensively investigated non-classical antitumor-active platinum compounds demonstrating that also compounds with trans-geometry can have high biological activity. (iii) More recently, his research has been devoted to the targeting and delivery of antitumoral platinum complexes and to the molecular understanding of their uptake and transport mechanisms. Giovanni Natile is co-author of more than 350 papers, international patents and books. He serves as an associate editor of the journal of Bioinorganic Chemistry and Applications and is member of the Editorial Board of the Journal of Inorganic Biochemistry and of the International Journal of Molecular Sciences. He has served as President of the European Association for Chemical and Molecular Sciences "EuCheMS" (2006–2008) and of the Italian Chemical Society (2002–2004) and has been a member of the Management Committee of the European COST ACTIONS D1, D8, D20, D39 and CM1105.

Nunzio Denora graduated in Chemistry/Pharmaceutical Chemistry at the University of Bari in 2001. After a PhD in Pharmaceutical Technology (2001–2004) at the University of Bari, he became postdoctoral researcher at the Pharmaceutical Chemistry Department, the University of Kansas (2005). From March 2006, he has been Assistant Professor (full time, permanent position) in Pharmaceutical Technology at the Department of Pharmacy—Drug Sciences, University of Bari Aldo Moro. His research interests concern the following: new prodrugs and conjugates of neurotransmitters useful to overcome the blood–brain barrier; biocompatible polymers and nanocarriers useful to improve the delivery of bioactive drugs; the role of translocator protein in the selective delivery of bioactive drugs to pathologies overexpressing the TSPO. Nunzio Denora is co-author of more than 80 papers, international patents and books.

International Journal of
Molecular Sciences

MDPI

Editorial

An Updated View of Translocator Protein (TSPO)

Nunzio Denora [1],* and Giovanni Natile [2],*

[1]   Department of Pharmacy-Drug Sciences, University of Bari "Aldo Moro", 70125 Bari, Italy
[2]   Department of Chemistry, University of Bari "Aldo Moro", 70125 Bari, Italy
*   Correspondence: nunzio.denora@uniba.it (N.D.); giovanni.natile@uniba.it (G.N.);
    Tel.: +39-080-544-2767 (N.D.); +39-080-544-2774 (G.N.)

Received: 13 November 2017; Accepted: 4 December 2017; Published: 6 December 2017

Abstract: Decades of study on the role of mitochondria in living cells have evidenced the importance of the 18 kDa mitochondrial translocator protein (TSPO), first discovered in the 1977 as an alternative binding site for the benzodiazepine diazepam in the kidneys. This protein participates in a variety of cellular functions, including cholesterol transport, steroid hormone synthesis, mitochondrial respiration, permeability transition pore opening, apoptosis, and cell proliferation. Thus, TSPO has become an extremely attractive subcellular target for the early detection of disease states that involve the overexpression of this protein and the selective mitochondrial drug delivery. This special issue was programmed with the aim of summarizing the latest findings about the role of TSPO in eukaryotic cells and as a potential subcellular target of diagnostics or therapeutics. A total of 9 papers have been accepted for publication in this issue, in particular, 2 reviews and 7 primary data manuscripts, overall describing the main advances in this field.

Keywords: translocator protein (TSPO); neuroinflammation; steroidogenesis; subcellular targeting

1. Introduction

In eukaryotic cells mitochondria play a vital role as they are involved in the control of oxidative phosphorylation and ATP synthesis and in the regulation of apoptosis and translocation of pro-apoptotic proteins from the mitochondrial intermembrane space to the cytosol [1]. Hence, it is evident that mitochondria can regulate cell survival and that mitochondrial dysfunctions are involved in the onset of several human pathological conditions. For this reason, mitochondria can be considered a potential target for the delivery of therapeutics, although its intracellular localization makes it difficult to reach. Decades of study on the role of mitochondria in living cells have evidenced, in particular, the importance of the 18 kDa mitochondrial translocator protein (TSPO), first discovered in the 1977 as an alternative binding site for the benzodiazepine diazepam in the kidneys. This protein participates in a variety of cellular functions, including cholesterol transport, steroid hormone synthesis, mitochondrial respiration, permeability transition pore opening, apoptosis, and cell proliferation [2–7]. In accordance with TSPO's diverse functions, changes in TSPO expression have been linked to multiple diseases, from cancer to endocrine and neurological diseases. Thus, TSPO has become an extremely attractive subcellular target for (1) early detection of disease states that involve the overexpression of this protein [8–10] and (2) selective mitochondrial drug delivery [11–14]. To date, several studies have been carried out on the synthesis of new structurally diverse TSPO ligands and on the preparation of nanosystems or metal complexes that can be directed to TSPO for diagnosis or therapy [15,16], thus highlighting the great interest of the scientific community in understanding the functions of this translocator protein in both normal and pathological conditions.

Investigation of the functions of this protein, both in vitro and in vivo, has been mainly carried out using high-affinity ligands, such as isoquinoline carboxamides (e.g., PK 11195) and benzodiazepines (e.g., Ro5-4864). For instance, PK 11195 and Ro5-4864 have been used to explore TSPO distribution and function in various tissues and pathologies, thus allowing the mapping of the "peripheral binding site" in almost every tissue examined [17].

As a result of the great interest in the role of TSPO and its potential use as subcellular target, this special issue, entitled, "Translocator Protein (TSPO)", was programmed to consolidate new knowledge focused specifically on this receptor. A total of 9 papers were accepted for publication in this issue: 2 reviews and 7 primary data manuscripts, focusing on (1) new functions attributable to TSPO; (2) new potent and selective TSPO ligands; (3) the use of ligands as imaging tools for the early diagnosis of diseases characterized by a high expression of TSPO, such as neuroinflammation and TSPO-rich cancers; (4) TSPO targeted nanocarriers that deliver therapeutics and diagnostics; (5) TSPO ligands that could be used to prepare coordination complexes of metallodrugs for use in diagnosis and therapy; (6) TSPO ligands as pro-apoptotic agents that are potentially useful for the treatment of cancers, and, finally; (7) in vitro and in vivo investigations of the ability of TSPO ligands to affect steroidogenesis.

2. Articles in This Special Issue

Mitochondria are involved in several metabolic processes, comprising the energy transduction mechanism which requires the transport of specific metabolites across the inner membrane which is achieved through mitochondrial carriers (MCs), a family of nuclear-encoded proteins sharing several structural features [18]. The MCs' function is to facilitate the exchange of metabolites through the inner mitochondrial membrane (IMM). In this regard, Damiano et al. contributed to this special issue with a review concerning two mitochondrial carriers, such as citrate and carnitine/acylcarnitine transporters, involved in fatty acid metabolism [19]. Interesting new research on the mechanisms involved in the regulation of lipid metabolism in the cell could be sparked from this study.

The study of Gavish and coworkers, also reported in this issue [20], provides a deeper understanding of the overall biological function of TSPO. By applying the microarray analysis of the gene expression in U118MG glioblastoma cells, they discovered that the classical TSPO ligand PK 11,195 can modulate gene expression in this tumor cell line and induce cell morphological changes. In particular, at exposure times of 15, 30, 45, and 60 min, as well as 3 and 24 h, to PK 11,195, changes in gene expression might be associated with several cellular functions including viability, proliferation, differentiation, adhesion, migration, tumorigenesis, and angiogenesis. This was supported microscopically by cell migration, cell accumulation, adhesion, and neuronal differentiation. The authors propose that the modulation in gene expression occurs via mitochondria-to-nucleus signaling; thus, TSPO modulates not only local mitochondrial functions but also nuclear gene expression. Furthermore, Gavish and coworkers identified a novel TSPO ligand, the 2-Cl-MGV-1, able to modulate gene expression of immediate early genes and transcription factors [20]. The results reported in this work highlight the possible effects on cellular and organismal functions induced by TSPO ligands possibly via modulation of nuclear genes expression promoted by mitochondrial TSPO. This type of modulation can influence several vital cell functions, with major implications on the whole organism in health and disease states.

The article of Culty and coworkers in this issue gives further insight into the role of TSPO in living cells [21]. They showed that TSPO is downregulated during gonocyte differentiation, which is indicative of a possible repressive role. Moreover, expression studies in human normal testes confirmed that TSPO is expressed in subsets of adult germ cells, suggesting a function in acrosome formation, while the analysis of tumor samples revealed the upregulation of its mRNA and protein localization in seminoma cells. The authors' prospect is to investigate the exact role of TSPO in normal spermatogenic cell development (from gonocyte to more mature germ cells) and in testicular cancer.

The investigation of the TSPO functions takes advantage of the use of synthetic TSPO ligands. Veenman and coworkers contributed to this issue with a review on the role of Tetrapyrroles as endogenous ligands for TSPO in comparison with synthetic ligands [22]. Interactions between the 18 kDa translocator protein (TSPO) and tetrapyrroles, including the tetrapyrrole protoporphyrin IX

(PPIX), have been studied for several decades in various species. Thus, Veenman et al. give an overview and the future perspectives for research regarding interactions between TSPO and tetrapyrroles. TSPO can be considered a receptor for PPIX, a transporter for tetrapyrroles, and a participant in the regulation of tetrapyrrole metabolism; vice versa, tetrapyrroles can modulate TSPO functions. A better understanding of the structure–function interactions between TSPO and its endogenous ligands such as tetrapyrroles, including PPIX, may aid in the development of new synthetic TSPO ligands as versatile drugs for the treatment of various diseases. In the review, apart from interactions between TSPO and tetrapyrroles, the effects of synthetic TSPO ligands in the context of TSPO–tetrapyrrole interactions are also presented.

Martini and coworkers [23] contributed here with 13 new high affinity TSPO ligands belonging to their previously described *N,N*-dialkyl-2-phenylindol-3-ylglyoxylamide (PIGA) class. The new ligands were evaluated for their potential ability to affect the cellular Oxidative Metabolism Activity/Proliferation index, which is used as a measure of astrocyte well-being. The relevance of neurosteroidogenesis in astrocyte well-being was investigated in a human astrocyte model and the positive effect of TSPO-stimulated neurosteroid release on astrocytes well-being was demonstrated. The development of molecules able to stimulate steroid release could represent a therapeutic strategy for central nervous system diseases characterized by astrocyte loss. Furthermore, these ligands may be exploited as pharmacological tools for investigating the autocrine/paracrine roles of neurosteroids in the control of astrocyte metabolism.

In order to develop highly selective and active TSPO ligands for cancer therapy and imaging, Lee and coworkers synthesized a new imidazopyridine-based TSPO ligand (CB256) for coordination to ^{99m}Tc and Re [24]. The ^{99m}Tc-labeled imidazopyridine-based bifunctional chelate ligand was prepared in one step with good radiochemical yield. The resulting complex showed high stability in vitro. The coordination to tricarbonyl rhenium did not alter the TSPO affinity of CB256. In vitro studies on TSPO-rich tumor cells suggested that the radiolabeled complex may have a potential as SPECT radiotracer for the evaluation of TSPO-overexpressing tissues, thus calling for further in vivo biological evaluation.

In this special issue Laquintana and coworkers present two TSPO ligand-methotrexate conjugates that are potentially useful for the treatment of TSPO-rich cancers, including brain tumors [25]. Methotrexate (MTX) is the drug of choice for the treatment of several cancers, but its permeability through the blood–brain barrier (BBB) is poor, making it unsuitable for the treatment of brain tumors. In contrast, the TSPO ligand-MTX conjugates prepared by these authors showed a high binding affinity and selectivity for TSPO, and a more marked toxicity toward glioma cells than MTX alone. These results confirm the ability of the selected TSPO ligand to transport a hydrophilic drug through the biological membranes and determine its accumulation in target cells overexpressing TSPO. The study of Laquintana and coworkers also demonstrates the effectiveness of the bio-conjugate strategy for bringing two agents with a distinct mechanism of action to cancer cells.

The use of TSPO ligands for preparing coordination complexes of metallodrugs with diagnostic and/or therapeutic potential has also been exploited by Margiotta and coworkers [26], who present here the first Pt(IV) derivative of oxaliplatin carrying a ligand for TSPO. This new Pt(IV) complex has been fully characterized from a chemical point of view and has been tested in vitro against human MCF7 breast carcinoma, U87 glioblastoma, and LoVo colon adenocarcinoma cell lines. The affinity for TSPO receptor, the cellular uptake, and the effect on cell cycle progression were also evaluated. The results obtained by these authors render this new coordination complex very promising in the context of a receptor-mediated drug targeting strategy toward TSPO-overexpressing tumors, in particular the colorectal cancer.

Finally, Papadopoulos and coworkers contributed to this issue with an article concerning the ability of the peptide VLNYYVW, designed on the TSPO's CRAC (cholesterol recognition/interaction amino acid consensus) domain, to prevent the opening of the mPTP and the release of apoptotic factors in rat brain mithocondria [27]. In addition, the authors showed that the TSPO specific drug ligand PK

Int. J. Mol. Sci. **2017**, *18*, 2640

11,195 modulates the effects of the CRAC peptide on the induction of mPTP opening and the release of apoptotic factors. These results suggest that TSPO via its C-terminal CRAC domain participates in mPTP function/regulation and apoptosis initiation and that TSPO drug ligands are regulators of this process.

3. Conclusions

The high number of papers submitted and ultimately accepted for publication in this special issue attests to the considerable amount of research being conducted on TSPO and TSPO's role in living cells. TSPO has become an extremely attractive subcellular target for the early detection of disease states (that involve the overexpression of this protein) and for the selective delivery to mitochondria of drugs for diagnostic and therapeutic purposes. Moreover, the effort in the design and synthesis of new, more specific and effective TSPO ligands has been valuable and cannot be neglected.

Acknowledgments: We would like to thank all the authors who submitted their work for this special issue. Special thanks also to all of the reviewers who participated and enhanced the quality of the articles by seeking clarification for arguments and requesting modifications where necessary.

Conflicts of Interest: The authors declare no conflict of interest.

References

1. Biswas, S.; Torchilin, V.P. Nanopreparations for organelle-specific delivery in cancer. *Adv. Drug Deliv. Rev.* **2014**, *66*, 26–41. [CrossRef] [PubMed]
2. Papadopoulos, V. Structure and function of the peripheral-type benzodiazepine receptor in steroidogenic cells. *Proc. Soc. Exp. Biol. Med.* **1998**, *217*, 130–142. [CrossRef] [PubMed]
3. Azarashvili, T.; Krestinina, O.; Yurkov, I.; Evtodienko, Y.; Reiser, G. High-affinity peripheral benzodiazepine receptor ligand, PK11195, regulates protein phosphorylation in rat brain mitochondria under control of Ca^{2+}. *J. Neurochem.* **2005**, *94*, 1054–1062. [CrossRef] [PubMed]
4. Hirsch, J.D.; Beyer, C.F.; Malkowitz, L.; Beer, B.; Blume, A.J. Mitochondrial benzodiazepine receptors mediate inhibition of mitochondrial respiratory control. *Mol. Pharmacol.* **1989**, *35*, 157–163. [PubMed]
5. Hirsch, T.; Decaudin, D.; Susin, S.A.; Marchetti, P.; Larochette, N.; Resche-Rigon, M.; Kroemer, G. PK11195, a ligand of the mitochondrial benzodiazepine receptor, facilitates the induction of apoptosis and reverses Bcl-2-mediated cytoprotection. *Exp. Cell Res.* **1998**, *241*, 426–434. [CrossRef] [PubMed]
6. Lee, D.H.; Kang, S.K.; Lee, R.H.; Ryu, J.M.; Park, H.Y.; Choi, H.S.; Bae, Y.C.; Suh, K.T.; Kim, Y.K.; Jung, J.S. Effects of peripheral benzodiazepine receptor ligands on proliferation and differentiation of human mesenchymal stem cells. *J. Cell. Physiol.* **2004**, *198*, 91–99. [CrossRef] [PubMed]
7. Veenman, L.; Papadopoulos, V.; Gavish, M. Channel like functions of the 18-kDa translocator protein (TSPO): Regulation of apoptosis and steroidogenesis as part of the host-defense response. *Curr. Pharm. Des.* **2007**, *13*, 2385–2405. [CrossRef] [PubMed]
8. Fanizza, E.; Iacobazzi, R.M.; Laquintana, V.; Valente, G.; Caliandro, G.; Striccoli, M.; Agostiano, A.; Cutrignelli, A.; Lopedota, A.; Curri, M.L.; et al. Highly selective luminescent nanostructures for mitochondrial imaging and targeting. *Nanoscale* **2016**, *14*, 3350–3361. [CrossRef] [PubMed]
9. Sekimata, K.; Hatano, K.; Ogawa, M.; Abe, J.; Magata, Y.; Biggio, G.; Serra, M.; Laquintana, V.; Denora, N.; Latrofa, A.; et al. Radiosynthesis and in vivo evaluation of N-[^{11}C]methylated imidazopyridineacetamides as PET tracers for peripheral benzodiazepine receptors. *Nucl. Med. Biol.* **2008**, *35*, 327–334. [CrossRef] [PubMed]
10. Denora, N.; Laquintana, V.; Lopalco, A.; Iacobazzi, R.M.; Lopedota, A.; Cutrignelli, A.; Iacobellis, G.; Annese, C.; Cascione, M.; Leporatti, S.; et al. In vitro targeting and imaging the translocator protein TSPO 18-kDa through G(4)-PAMAM-FITC labeled dendrimer. *J. Control. Release* **2013**, *172*, 1111–1125. [CrossRef] [PubMed]
11. Denora, N.; Cassano, T.; Laquintana, V.; Lopalco, A.; Trapani, A.; Cimmino, C.S.; Laconca, L.; Giuffrida, A.; Trapani, G. Novel codrugs with GABAergic activity for dopamine delivery in the brain. *Int. J. Pharm.* **2012**, *437*, 221–231. [CrossRef] [PubMed]

12. Denora, N.; Laquintana, V.; Trapani, A.; Lopedota, A.; Latrofa, A.; Gallo, J.M.; Trapani, G. Translocator protein (TSPO) ligand-Ara-C (cytarabine) conjugates as a strategy to deliver antineoplastic drugs and to enhance drug clinical potential. *Mol. Pharm.* **2010**, *7*, 2255–2269. [CrossRef] [PubMed]

13. Johnstone, T.C.; Suntharalingam, K.; Lippard, S.J. The Next Generation of Platinum Drugs: Targeted Pt(II) Agents, Nanoparticle Delivery, and Pt(IV) Prodrugs. *Chem. Rev.* **2016**, *116*, 3436–3486. [CrossRef] [PubMed]

14. Laquintana, V.; Denora, N.; Lopalco, A.; Lopedota, A.; Cutrignelli, A.; Lasorsa, F.M.; Agostino, G.; Franco, M. Translocator protein ligand-PLGA conjugated nanoparticles for 5-fluorouracil delivery to glioma cancer cells. *Mol. Pharm.* **2014**, *11*, 859–871. [CrossRef] [PubMed]

15. Denora, N.; Iacobazzi, R.M.; Natile, G.; Margiotta, N. Metal complexes targeting the Translocator Protein 18 kDa (TSPO). *Coord. Chem. Rev.* **2017**, *341*, 1–18. [CrossRef]

16. Iacobazzi, R.M.; Lopalco, A.; Cutrignelli, A.; Laquintana, V.; Lopedota, A.; Franco, M.; Denora, N. Bridging Pharmaceutical Chemistry with Drug and Nanoparticle Targeting to Investigate the Role of the 18-kDa Translocator Protein TSPO. *ChemMedChem* **2017**, *12*, 1261–1274. [CrossRef] [PubMed]

17. Awad, M.; Gavish, M. Binding of [^{3}H]Ro 5-4864 and [^{3}H]PK 11195 to cerebral cortex and peripheral tissues of various species: species differences and heterogeneity in peripheral benzodiazepine binding sites. *J. Neurochem.* **1987**, *49*, 1407–1414. [CrossRef] [PubMed]

18. Wohlrab, H. Transport proteins (carriers) of mitochondria. *IUBMB Life* **2009**, *61*, 40–46. [CrossRef] [PubMed]

19. Giudetti, A.M.; Stanca, E.; Siculella, L.; Gnoni, G.V.; Damiano, F. Nutritional and Hormonal Regulation of Citrate and Carnitine/Acylcarnitine Transporters: Two Mitochondrial Carriers Involved in Fatty Acid Metabolism. *Int. J. Mol. Sci.* **2016**, *17*, 817. [CrossRef] [PubMed]

20. Yasin, N.; Veenman, L.; Singh, S.; Azrad, M.; Bode, B.; Vainshtein, A.; Caballero, B.; Marek, I.; Gavish, M. Classical and Novel TSPO Ligands for the Mitochondrial TSPO Can Modulate Nuclear Gene Expression: Implications for Mitochondrial Retrograde Signaling. *Int. J. Mol. Sci.* **2017**, *18*, 786. [CrossRef] [PubMed]

21. Manku, G.; Culty, M. Regulation of Translocator Protein 18 kDa (TSPO)Expression in Rat and Human Male Germ Cells. *Int. J. Mol. Sci.* **2016**, *17*, 1486. [CrossRef] [PubMed]

22. Veenman, L.; Vainshtein, A.; Yasin, N.; Azrad, M.; Gavish, M. Tetrapyrroles as Endogenous TSPO Ligands in Eukaryotes and Prokaryotes: Comparisons with Synthetic Ligands. *Int. J. Mol. Sci.* **2016**, *17*, 880. [CrossRef] [PubMed]

23. Da Pozzo, E.; Giacomelli, C.; Costa, B.; Cavallini, C.; Taliani, S.; Barresi, E.; Da Settimo, F.; Martini, C. TSPO PIGA Ligands Promote Neurosteroidogenesis and Human AstrocyteWell-Being. *Int. J. Mol. Sci.* **2016**, *17*, 1028. [CrossRef] [PubMed]

24. Choi, J.Y.; Iacobazzi, R.M.; Perrone, M.; Margiotta, N.; Cutrignelli, A.; Jung, J.H.; Park, D.D.; Moon, B.S.; Denora, N.; Kim, S.E.; et al. Synthesis and Evaluation of Tricarbonyl ^{99m}Tc-Labeled 2-(4-Chloro)phenyl-imidazo[1,2-a] pyridine Analogs as Novel SPECT Imaging Radiotracer for TSPO-Rich Cancer. *Int. J. Mol. Sci.* **2016**, *17*, 1085. [CrossRef] [PubMed]

25. Valentino Laquintana, V.; Denora, N.; Cutrignelli, A.; Perrone, M.; Iacobazzi, R.M.; Annese, C.; Lopalco, A.; Lopedota, A.A.; Franco, M. TSPO Ligand-Methotrexate Prodrug Conjugates: Design, Synthesis, and Biological Evaluation. *Int. J. Mol. Sci.* **2016**, *17*, 967. [CrossRef] [PubMed]

26. Savino, S.; Denora, N.; Iacobazzi, R.M.; Porcelli, L.; Azzariti, A.; Natile, G.; Margiotta, N. Synthesis, Characterization, and Cytotoxicity of the First Oxaliplatin Pt(IV) Derivative Having a TSPO Ligand in the Axial Position. *Int. J. Mol. Sci.* **2016**, *17*, 1010. [CrossRef] [PubMed]

27. Azarashvili, T.; Krestinina, O.; Baburina, Y.; Odinokova, I.; Akatov, V.; Beletsky, I.; Lemasters, J.; Papadopoulos, V. Effect of the CRAC Peptide, VLNYYVW, on mPTP Opening in Rat Brain and Liver Mitochondria. *Int. J. Mol. Sci.* **2016**, *17*, 2096. [CrossRef] [PubMed]

International Journal of
Molecular Sciences

MDPI

Review

Nutritional and Hormonal Regulation of Citrate and Carnitine/Acylcarnitine Transporters: Two Mitochondrial Carriers Involved in Fatty Acid Metabolism

Anna M. Giudetti, Eleonora Stanca, Luisa Siculella *, Gabriele V. Gnoni and Fabrizio Damiano

Laboratory of Biochemistry and Molecular Biology, Department of Biological and Environmental Sciences and Technologies, University of Salento, Lecce 73100, Italy; anna.giudetti@unisalento.it (A.M.G.); eleonora.stanca@unisalento.it (E.S.); gabriele.gnoni@unisalento.it (G.V.G.); fabrizio.damiano@unisalento.it (F.D.)
* Correspondence: luisa.siculella@unisalento.it; Tel.: +39-8-3229-8696

Academic Editors: Giovanni Natile and Nunzio Denora
Received: 13 April 2016; Accepted: 19 May 2016; Published: 25 May 2016

Abstract: The transport of solutes across the inner mitochondrial membrane is catalyzed by a family of nuclear-encoded membrane-embedded proteins called mitochondrial carriers (MCs). The citrate carrier (CiC) and the carnitine/acylcarnitine transporter (CACT) are two members of the MCs family involved in fatty acid metabolism. By conveying acetyl-coenzyme A, in the form of citrate, from the mitochondria to the cytosol, CiC contributes to fatty acid and cholesterol synthesis; CACT allows fatty acid oxidation, transporting cytosolic fatty acids, in the form of acylcarnitines, into the mitochondrial matrix. Fatty acid synthesis and oxidation are inversely regulated so that when fatty acid synthesis is activated, the catabolism of fatty acids is turned-off. Malonyl-CoA, produced by acetyl-coenzyme A carboxylase, a key enzyme of cytosolic fatty acid synthesis, represents a regulator of both metabolic pathways. CiC and CACT activity and expression are regulated by different nutritional and hormonal conditions. Defects in the corresponding genes have been directly linked to various human diseases. This review will assess the current understanding of CiC and CACT regulation; underlining their roles in physio-pathological conditions. Emphasis will be placed on the molecular basis of the regulation of CiC and CACT associated with fatty acid metabolism.

Keywords: β-oxidation; carnitine/acylcarnitine translocase; citrate carrier; fatty acid synthesis; hormonal regulation; nutritional regulation

---

1. Introduction

Mitochondria are well-defined cytoplasmic organelles, which undertake multiple critical functions in the cell. In addition to oxidative phosphorylation (OXPHOS), a pathway in which nutrients are oxidized to form adenosine triphosphate (ATP), mitochondria are involved in several pathways including citric acid cycle, gluconeogenesis, fatty acid oxidation and lipogenesis, amino acid degradation and heme biosynthesis. They produce most of the cellular reactive oxygen species (ROS), buffer cellular Ca^{2+} and they initiate cellular apoptosis [1–3]. Moreover, mitochondria participate in cell communication and inflammation, and play an important role in aging, drug toxicity, and pathogenesis [4].

Mitochondria and cytosol are engaged in numerous metabolic processes which, due to enzyme compartmentalization, involve the exchange of metabolites among them.

Energy transduction in mitochondria requires the transport of specific metabolites across the inner membrane, achieved through mitochondrial carriers (MCs), a family of nuclear-encoded proteins sharing several structural features. Their common function is to provide a link between mitochondria and cytosol by facilitating the flux of a high number of metabolites through the permeability barrier of the inner mitochondrial membrane (IMM).

In humans, MCs are encoded by the *SLC25* genes, and some of them have isoforms encoded by different genes [4]. Until now, 53 mitochondrial carriers have been identified on the human genome and more than half have been functionally characterized [4].

Members of SLC25 family, mainly located in mitochondria, have been found in all eukaryotes; few of them are present in peroxisomes and chloroplasts [5]. *SLC25* genes are highly variable in size and organization, whereas their products are very similar sharing a tripartite structure composed by 100 amino acid repeats [4].

The complete sequence analysis of some MCs during the late 1990s showed that each domain contains two transmembrane α-helices, separated by hydrophilic regions, and a common signature motif which can be divided into a first part, P-X-D/E-X-X-K/R, and a second part as [D/E]GXXXX[W/Y/F][R/K]G [4]. The common structure is reflected in a similar function. To transport solutes across IMM, many MCs catalyze an exchange reaction: they have only one binding site, which is alternately exposed to the two opposite sides of the membrane. The substrate-induced conformational changes occur during the transition from cytosol to matrix and vice versa [4]. However, MCs do not adopt only antiport as transport; the carnitine-acylcarnitine translocase (CACT) catalyzes both unidirectional transport of carnitine and the carnitine/acylcarnitine exchange [6], whereas the uncoupling protein catalyzes uniport as the exclusive transport mode [7].

Functional characterization of MCs was carried out through their purification and reconstitution in artificial membranes, such as proteoliposomes. Studies on the transport activity performed with the liposomal systems show a dependence of the kinetic parameters on the lipid composition of the mitochondrial membrane, particularly on the cardiolipin (CL) levels [8]. CL interacts with a number of proteins and enzymes involved in fundamental mitochondrial bioenergetic processes [9]. Thus, CL is crucial for mitochondrial OXPHOS and for correct structure and function of IMM. It has been proposed that CL creates an environment protecting and stabilizing MCs in a functionally intact state [8].

MCs play a crucial role in intermediary metabolism. In this respect, citrate carrier (CiC) and CACT are two MCs mainly involved in fatty acid metabolism. CiC, encoded by *SLC25A1*, promotes the efflux of citrate from the mitochondria to the cytosol where citrate is cleaved by ATP-citrate lyase to oxaloacetate (OAA) and acetyl-coenzyme A (acetyl-CoA), which is used for fatty acid and sterol syntheses. CACT catalyzes the transport of fatty acids, in the form of acylcarnitine, into mitochondria, where they are oxidized by the enzymes of β-oxidation pathway (Figure 1). These MCs are mutually regulated to allow the synthesis and oxidation of fatty acids to take place at different times. The aim of this review is to summarize biochemical, molecular, and physio-pathological aspects of CiC and CACT.

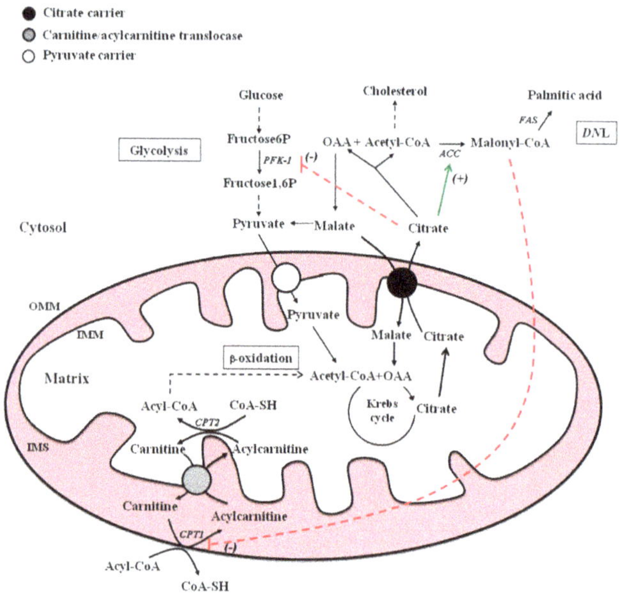

Figure 1. Schematic model of citrate carrier (CiC) and carnitine-acylcarnitine translocase (CACT) in lipogenesis and β-oxidation, and their metabolic interrelationship. Abbreviations: ACC, acetyl-CoA carboxylase; CoA-SH, coenzyme-A; CPT1, carnitine palmitoyltransferase 1; CPT2, carnitine palmitoyltransferase 2; *DNL*, *de novo* lipogenesis; FAS, fatty acid synthase; IMM, inner mitochondrial membrane; IMS, intermembrane space; OAA, oxaloacetate; OMM, outer mitochondrial membrane; PFK-1, phosphofructokinase-1. The green arrow and dashed red lines represent, respectively, positive (+) and negative (-) allosteric modulation of the indicated target enzymes.

2. Citrate Carrier (CiC) and Carnitine-Acylcarnitine Translocase (CACT): Mitochondrial Carriers in Fatty Acid Metabolism

Fatty acids perform several functions in cells; as components of triacylglycerols, they represent the main form of stored energy, and as constituents of phospholipids they play important structural roles, while some of them are also involved in intracellular signaling.

Fatty acid metabolism requires the involvement of both cytosolic and mitochondrial reactions (Figure 1).

The *de novo* lipogenesis (*DNL*) (*i.e.*, *de novo* fatty acid synthesis) takes place in the cytosol. The initiation of *DNL* occurs in the presence of high levels of blood glucose, indicating a sufficient energy intake. In this condition, the pancreas secretes insulin, which not only promotes the uptake of glucose from blood into the cells but also stimulates the synthesis of two enzymes of the *DNL*, acetyl-CoA carboxylase (ACC) and fatty acid synthase (FAS) [10]. These enzymes work in sequence to convert first acetyl-CoA to malonyl-CoA, in a reaction catalyzed by ACC, and then, by a series of reactions catalyzed by FAS, to produce palmitate, a saturated fatty acid with 16 carbon atoms. The condensation of malonyl, bound to acyl carrier protein (ACP), with acetyl-CoA by the ketoacyl-ACP synthase, is the first reaction catalyzed by FAS.

Acetyl-CoA utilized for *DNL* is mainly derived from carbohydrate metabolism. In this respect, after glucose conversion into pyruvate, the latter enters through its specific transporter into mitochondria, where it is converted in acetyl-CoA in a reaction catalyzed by pyruvate dehydrogenase. In the Krebs cycle, acetyl-CoA is then converted into citrate (tricarboxylate) after condensation with OAA. In a good energetic state, citrate is transported into the cytosol via CiC in exchange for malate (dicarboxylate) (the citrate/malate antiporter). The exchange is electroneutral being citrate efflux compensated by a contemporary efflux of a proton [4]. CiC transport activity is particularly high in the liver where active fatty acid synthesis occurs, and it is virtually absent in other tissues. CiC mRNA and/or protein levels are high in the liver, pancreas, and kidney, but are low or absent in the brain, heart, skeletal muscle, placenta, and lungs [11].

By the action of ATP-citrate lyase, cytosolic citrate is converted into OAA and acetyl-CoA, and this latter is used for the synthesis of fatty acids and cholesterol. OAA produced in the cytosol by ATP-citrate lyase is reduced to malate, which is converted to pyruvate via the malic enzyme with production of cytosolic NADPH plus H^+ necessary for fatty acid and sterol syntheses. Moreover, citrate in the cytoplasm blunts glycolysis by inhibiting phosphofructokinase-1 (PFK-1), and positively modulates ACC, a key enzyme of the *DNL* pathway [12]. Additionally, the entry of malate into mitochondria in exchange for citrate stimulates OXPHOS [13].

Differently from fatty acid synthesis, fatty acid oxidation occurs in mitochondria. The signal for fatty acid oxidation begins with the secretion of glucagon or, in some cases, epinephrine [14]. These hormones stimulate enzymes that clip off fatty acids from triacylglycerol molecules. During fatty acid oxidation, two-carbon units are sequentially cleaved from the fatty acid chain, as acetyl-CoA, which then enters the Krebs cycle.

β-Oxidation can occur when fatty acids cross the IMM. This is achieved through CACT encoded by *SLC25A20* [15]. The *CACT* gene is differently expressed in human tissues. High levels of transcripts are found in liver, heart and skeletal muscle, where β-oxidation is essential for energy production; much lower levels are observed in other tissues, such as brain, placenta, kidney, pancreas and lung [16].

This carrier belongs to the carnitine palmitoyl transferase (CPT) system, the major site of control of fatty acid β-oxidation, which transports cytosolic long chain fatty acids (LCFA) in the form of esters of CoA-SH (LCFA-CoA) into the mitochondrial matrix for their oxidation [17]. Three different proteins are involved in the carnitine-dependent transport: carnitine palmitoyl transferase 1 (CPT1), CACT and carnitine palmitoyl transferase 2 (CPT2). The acyl-CoAs, products of the cytosolic activation of fatty acids, are transformed into the corresponding carnitine esters by CPT1 localized in the outer mitochondrial membrane (OMM) [18]. Then, acyl-carnitine permeates the IMM by CACT and reacts with a matrix pool of CoA-SH in a reaction catalyzed by CPT2 on the inner face of the IMM. The reformed acyl-CoA then enters the β-oxidation pathway, while the released carnitine returns to the extramitochondrial compartment.

This pathway is the major source of energy for heart and skeletal muscles during fasting and physical exercise. Besides the exchange, CACT performs unidirectional transport of carnitine across IMM but to a much lower rate (about one tenth of the exchange); uniport of carnitine balances the matrix carnitine pool, a prerequisite for optimal carnitine/acylcarnitine activity [6].

Malonyl-CoA, the first committed intermediate in the pathway of fatty acid synthesis, represents an allosteric modulator of fatty acid oxidation. In the fed state, when insulin/glucagon ratio is high, hepatic lipogenesis is active, the concentration of malonyl-CoA rises and becomes sufficient to inhibit CPT1 [19], the enzyme catalyzing the rate-limiting step in fatty acid oxidation. In this condition, while fatty acid oxidation is low or absent, lipogenesis is up-regulated, being malonyl-CoA a substrate for FAS.

Conversely, in ketotic states (low insulin/glucagon ratio) carbon flow through glycolysis and ACC diminishes, the malonyl-CoA level falls. In this setting, CPT1 is disinhibited, and incoming fatty acids readily undergo β-oxidation with accelerated production of ketone bodies.

It follows that fatty acid oxidation and fatty acid synthesis fluctuate reciprocally with changes in malonyl-CoA levels [20].

3. CiC and CACT Involvement in Pathological States

The human *SLC25A1* gene, encoding for CiC, maps on the chromosome 22.q11.2. Micro deletions involving 22q11.2 have been associated with developmental disorders known as DiGeorge syndrome, velo-cardio-facial syndrome, and a subtype of schizophrenia [21]. However, direct evidence of the relevance of *SLC25A1* haploinsufficiency in these syndromes so far has not been found.

Recessive mutations in *SLC25A1* with impaired mitochondrial citrate efflux are found in patients with combined D-2- and L-2-hydroxyglutaric aciduria (D,L-2-HGA), a disease characterized by epileptic encephalopathy, respiratory insufficiency, developmental arrest and early death [22]. Recently, it has been reported that null/missense *SLC25A1* mutations, besides classic clinical features of D,L-2-HGA, showed marked facial dysmorphism and prominent lactic acidosis [23]. Moreover, *SLC25A1* knockdown has been related to pre-synaptic nerve terminal abnormalities and neuromuscular junction impairment [24].

Increased levels of CiC have been found in human cancers, while inhibition of CiC activity showed anti-tumor activity, and *SLC25A1* gene has been implied in epigenetic regulation or cancer biology [22]. Citrate exported from mitochondria via CiC and its downstream metabolic intermediate, acetyl-CoA, are necessary for cytokine induced inflammatory signals [25] and CiC acetylation plays a key role in the production of inflammatory mediators in activated immune cells [26].

Low levels of CiC activity and expression were measured in primary biliary cirrhosis, together with a low synthesis of fatty acids. The impaired CiC activity and expression was almost completely prevented by treatment with Silybin, an extract of silymarin with antioxidant and anti-inflammatory properties [27].

Several lines of evidence suggest the involvement of CiC in the development of liver steatosis, which is characterized by accumulation of lipid droplets in hepatocytes. This is partially due to the increased lipogenic gene expression, linked to the Unfolded Protein Response (UPR) pathway [28,29]. Indeed, in HepG2 and in BRL-3A cells, CiC expression was increased upon the induction of endoplasmic reticulum (ER) stress and the consequent activation of UPR pathway [29].

SLC25A20 gene maps on the chromosome 3p21.31 [30], spread over 42 kb, consist of nine exon and eight intron and encode for CACT, a protein of 301 amino acids [31]. CACT deficiency, described for the first time in 1992 [32], may present two different phenotypes: the most common with an early onset in the neonatal period and a milder form with onset in infancy or, less frequently, in childhood.

A mutation in the *SLC25A20* gene was firstly individuated by Huizing [15]. Since then, 35 others mutations have been identified in the *SLC25A20* gene [33]. Usually, disease-triggering mutations affect residues of the carrier signature motif. Deficiency of CACT activity can also be a consequence of an elongation in the C-terminal portion of the protein [15]. In any case, defective CACT results in decreased carnitine/acylcarnitine transport and impaired fatty acid β-oxidation. The clinical consequences of such alterations may involve hypoglycaemia, hyperammonaemia, cardiomyopathy, liver failure and encephalopathy [33].

4. Hormonal Regulation of CiC and CACT Activity and Expression

4.1. Thyroid Hormones

Thyroid hormones influence synthesis, mobilization and degradation of lipids. In this regards, thyroid hormones have been demonstrated to affect the activities of MCs directly involved in lipid metabolism [34–38].

Modulation of CACT activity by thyroid hormones has been reported by Paradies *et al.* [35]. In this study, an increased rate of palmitoylcarnitine/carnitine exchange in heart mitochondria from hyperthyroid rats has been demonstrated. Conversely, the hypothyroid state reduced fatty acid oxidation in rat heart mitochondria due to a decreased CACT activity which was restored to normal levels after 3,3′,5-triiodo-L-thyronine (T3) administration [36]. Analysis of kinetic parameters of CACT demonstrated that both hyper- and hypothyroidism significantly affected Vmax without changing Km value and the changes in heart CACT activity were ascribed to changes in CL level [36] (Table 1).

Moreover, T3 stimulated in the liver the transcription of *CPT1* gene in coordination with other genes involved in fatty acid oxidation [39,40], binding to a thyroid hormone response element (TRE), present in CPT1 promoter [41].

Liver CiC activity was demonstrated to be significantly stimulated in hyperthyroid with respect to euthyroid rats [34]. Changes in the mitochondrial membrane lipid composition and in the amount of CL strictly associated to CiC were reported to be involved in the T3-induced increase of CiC activity [34] (Table 1).

Conversely, hepatic CiC activity and expression were decreased, together with the expression of lipogenic genes *ACC* and *FAS*, in hypothyroid rats [37]. A decrease of mRNA abundance and protein level, due to a lower transcription rate and splicing of CiC pre-mRNA, were responsible for the impaired CiC activity in the hypothyroid state [38].

It has been demonstrated that T3 is able to directly increase ACC and FAS mRNA transcription through interaction with TRE located on the respective gene promoters [42,43]. To date, a TRE has not been found in the CiC promoter. Thus, the mechanism of T3 effect on *CiC* gene expression has not been yet fully clarified. It is plausible that T3 may regulate *CiC* gene expression through Sterol Regulatory Element-Binding Protein-1 (SREBP-1), which is considered the master transcription factor involved in the regulation of lipogenic gene expression [44]. Indeed, T3 affecting SREBP-1 expression [45] could regulate *CiC* gene expression through the SREBP-1 binding site found on human [46] and rat [47] CiC promoters.

Thus, the T3 activation of liver fatty acid synthesis and oxidation can support the establishment of a futile cycle.

4.2. Diabetes and Insulin

4.2.1. Type 1 Diabetes

A decrease in CiC activity was measured in experimental type 1 diabetic rats [48,49] and kinetic studies showed a reduction in Vmax and almost unchanged Km of CiC protein [48] (Table 1). The observed reduction of CiC activity was mainly ascribed to a reduced level of both CiC mRNA and translated protein. The reduction of CiC expression in diabetic rats is attributed either to transcriptional and post-transcriptional gene regulation. Indeed, both the transcriptional rate of *CiC* gene and the splicing reaction of CiC pre-mRNA decreased in nuclei from diabetic rats [50]. Injection of insulin to diabetic rats increased hepatic CiC activity and protein level to values higher than those measured in control animals [48].

Table 1. Effect of the hormonal status on activity, kinetics, protein and mRNA levels of CACT and CiC.

Hormonal Status	Carrier	Activity	Km	Vmax	Protein	mRNA	References
Hyperthyroidism	CACT	+43%	↔	↑			[35]
Hypothyroidism	CACT	−41%	↔	↓			[36]
Hypothyroidism	CiC	−60%	↔	↓	−35%	−30%	[37]
Hyperthyroidism	CiC	+43%	↔	↑			[34]
Streptozotocin-induced diabetes (1–8 weeks)	CiC	−31% ÷ −51%					[49]
Streptozotocin-induced diabetes (3 weeks)	CiC	−35%	↔	↑	−37%	−35%	[48]

↔: no change; ↓: decrease; ↑: increase.

Several lines of evidence suggest that insulin regulation of CiC is mediated by SREBP-1 [50]. Sites for SREBP-1a transactivation of *CiC* gene have been characterized on human and rat *CiC* promoter, at −1696 bp [46] and −67 bp [47], respectively. In rat hepatocytes cultured in the absence of insulin, a reduction of *CiC* promoter activity was observed as a consequence of a decrease of SREBP-1 expression [50]. Furthermore, the binding of SREBP-1 to the *CiC* promoter was reduced in diabetic rats with respect to control ones and it was restored to the control values after insulin treatment [50]. Overall, these results emphasize that insulin participates in the regulation of the functional levels of CiC in rat liver.

The activity and the expression of CiC were increased in liver of rats fed on a high carbohydrate diet [51]. In liver from diabetic rats, correction of hyperglycemia with phlorizin restored CiC activity and protein level to values measured in control animals [48] (Table 1). However, the molecular mechanism/s by which hyperglycemia affects CiC expression is not so far understood.

4.2.2. Type 2 Diabetes

Increased plasma levels of acylcarnitines have been associated with type 2 diabetes and insulin resistance [52,53]. Acute carnitine administration is able to improve peripheral insulin sensitivity in non-insulin-dependent diabetic patients [54] and to relieve glucose intolerance in obese rodents [55]. CACT expression was up-regulated in pancreatic islets of diabetic obese mice [56]. Moreover, CACT levels decreased in the kidney of diabetic rats [57] and in the muscle of insulin-resistant patients [58]. These data highlight the relevance of fatty acid accumulation in the muscle for the etiology of insulin resistance. No effect of type 2 diabetes on rat liver CiC activity has been observed [59].

4.2.3. Insulin Secretion

Studies have reported that both CiC and CACT may be involved in insulin responses.

CACT down-regulation by RNA interference enhances the insulin secretion in murine pancreatic Langerhans islets [60]. On the contrary, reduction in CiC expression by CiC-specific siRNA inhibited glucose-stimulated insulin secretion in normal rat pancreatic islets [61].

Recently, a study provided evidence that the inhibition of CiC by the specific substrate analog 1,2,3-benzenetricarboxylate resulted in a reduction of glucose-stimulated insulin secretion and autocrine insulin secretion by sperm [62]. These data furnish a new site of action for CiC in the regulation of sperm energetic metabolism to sustain capacitation process and acrosome reaction.

5. Nutritional Regulation of CiC and CACT Activity and Expression

5.1. Starvation

During fasting, an up-regulation of genes involved in fatty acid catabolism is observed [63,64]. This metabolic response is mediated by Peroxisome Proliferator-Activated Receptor α (PPARα), a transcriptional factor activated by free fatty acids. Different hypotheses have been proposed on the origins of the free fatty acids activating PPARα in the fasting state [65,66]. In response to fasting, PPARα stimulates the transcription of a large number of genes encoding for proteins involved in fatty acid catabolism, such as fatty acid transporters, fatty acid binding proteins, acyl-CoA synthase, CPT1 and CPT2 [63,64]. PPARα is also a strong activator of *CACT* gene expression [67]. CACT [68] as well as CACL, a CACT-like protein expressed in the brain [69], are up-regulated during fasting (Table 2). Anti-hyperlipidemic drugs, such as statins and fibrates, up-regulate expression of CACT [67,68]. This would be expected to enhance fatty acid oxidation and, therefore, may contribute to the lipid-lowering effects of these agents. Fibrates exert their effect by binding to PPARα [70], statins by inhibiting the Rho-signaling pathway [71], and retinoic acid by activating PPARα-RXRα heterodimer when bound to the PPAR responsive element (PPRE) [72].

On the other hand, a noticeable reduction of CiC activity has been observed in starved rats [73,74] (Table 2). No change in the membrane lipid composition and fluidity was detected in mitochondria from liver of starved rats suggesting that the reduction of CiC activity could be ascribed to modifications of CiC expression [73]. When compared to fed animals, in starved rats a considerable reduction of CiC mRNA abundance was observed. The reported reduction was caused by an increment in mRNA turnover, suggesting that starvation accelerates the degradation of CiC mRNA [74]. Altogether, the up-regulation of CACT and the down-regulation of CiC are aimed at increasing fatty acid oxidation, the major metabolic pathway for energy supply in the fasting state.

5.2. Saturated and Unsaturated Fatty Acids

It is well known that fatty acids, and in particular polyunsaturated fatty acids (PUFA), are potent regulators of cell metabolism. Fatty acids can influence hormonal signaling events by modifying membrane lipid composition, but they also have a direct effect on the regulation of genes mainly involved in carbohydrate metabolism and lipogenesis [75] as well as in fatty acid β-oxidation [76].

CACT activity and expression were affected by diets supplemented with different fat types [77,78] (Table 2). Fish oil (FO)-enriched diets, rich in ω-3 PUFA, increased rat liver CACT mRNA level and activity, whereas olive oil (OO)-enriched diets, rich in monounsaturated fatty acids (MUFA), reduced CACT mRNA level when compared to beef tallow (BT) rich in saturated fatty acids (SFA). ω-6 PUFA-supplemented diets did not affect CACT activity and expression [78] analogously to what was reported for CPT1 and CPT2 [77,78]. Recently, an up-regulation of CACT mRNA in FO-fed grass carp has also been reported [79].

FO treatment increased the transcriptional rate of CACT mRNA. On the other hand, OO modulated the splicing of the last intron of CACT pre-mRNA, and the rate of 3′-end formation [78]. It has been demonstrated that ω-3 PUFA are important positive regulators of PPARα [80], a transcriptional factor which positively regulates *CACT* gene expression. Notably, a functional PPRE in the CACT promoter has been identified [67,68].

Differently from CACT, several studies (Table 2) reported that CiC activity and expression were inhibited in liver of rats fed on a diet supplemented (15%) with FO [81,82] or safflower oil [83,84], rich in ω-3 or ω-6 PUFA, respectively. When compared with control rats, $C_{18:2}$ conjugated linoleic acids did not affect CiC activity [85], whereas CiC activity was lower in rats fed on diets enriched in oleic ($C_{18:1}$ cis) or elaidic ($C_{18:1}$ trans) acid [86]. A significant decrease in citrate transport was also observed when rats were fed on a diet containing a small percentage (2.5%) of FO for a relatively short period of treatment (2–3 weeks) [87]. The decrease was more pronounced when ω-3 PUFA were administered in the form of krill oil [87]. Corn oil and pine nut oil, rich in ω-6 PUFA, were able to reduce the activities of hepatic CiC and the cytosolic lipogenic enzymes in mice [88] (Table 2).

Table 2. Effect of the nutritional status on activity, kinetics, protein and mRNA levels of CACT and CiC.

Treatment	Carrier	Activity	Km	Vmax	Protein	mRNA	References
	CACT					+60%	[68]
Fasting	CACT				↑	↑	[69]
	CiC	−40%				−35%	[73,74]
ω-6 PUFA							
15% safflower oil for 3 weeks	CACT	↔	↔	↔	↔	↔	[78]
15% safflower oil for 3 weeks	CiC	−40%	↔	↓	−30%	−35%	[83]
15% safflower oil for 1–4 weeks	CiC	−50%				−35%	[84]
7.5% corn oil for 8 weeks	CiC	−60%			−70%		[88]
7.5% pine nut oil for 8 weeks	CiC	−40%					
ω-3 PUFA							
15% fish oil for 3 weeks	CACT	+50%	↔	↑	+60%	+70%	[78]
15% fish oil for 3 weeks	CiC	−60%	↔	↓	−50%	−40%	[81,82]
2.5% fish oil for 3 weeks	CiC	−30%	↔	↓	−30%	↔	[87]
2.5% fish oil for 6 weeks	CiC	−65%	↔	↓	−70%	−30%	[87]
2.5% krill oil for 6 weeks	CiC	−65%	↔	↔	−70%	−30%	[87]
CLA							
2.25% CLA for 2 weeks	CiC	↔					[85]
MUFA							
15% olive oil for 3 weeks	CACT	−10%	↔	↓	−20%	−20%	[78]
15% olive oil for 3 weeks	CiC	↔			↔	↔	[82]
14% oleic acid for 2 weeks	CiC	−22%					[85]
9.5% elaidic acid for 2 weeks	CiC	−36%					[85]
SFA							
20.2% SFA for 1 week	CiC	−54%	↔	↓	−40%	−30%	[58]
35.2% SFA for 1 week	CiC	−80%	↔	↓	−60%	−70%	[58]
Carbohydrate							
70% carbohydrate for 1 week	CiC	+20%	↔	↑	+20%	+45%	[58]

↔: no change; ↓: decrease; ↑: increase. CLA = conjugated linoleic acids; MUFA = monounsaturated fatty acids; PUFA = polyunsaturated fatty acids SFA = saturated fatty acids.

Feeding rats for three weeks on a diet supplemented with FO significantly decreased liver CiC activity when compared to SFA [81]. A study in which rats were fed on diets enriched in OO, FO or BT showed that CiC transcription rate, mRNA turnover and RNA processing were decreased only upon FO-feeding [82]. These data indicate that FO administration regulates *CiC* gene at transcriptional and post-transcriptional levels, whereas BT- and OO-feeding affects neither CiC activity nor *CiC* gene expression [82]. Inhibition of activity and expression of CiC, as well as of lipogenic enzymes ACC and FAS, were found in the liver of rats fed on a high fat diet (HFD) enriched with two different doses of SFA for one week [58]. However, this inhibition was progressively attenuated in long feeding experiments suggesting that the effects of HFD enriched in SFA on CiC activity and expression was time- and dose-dependent [58].

Diets enriched with PUFA of the ω-6 or ω-3 series strongly reduce hepatic lipogenesis both in human and in animal models [75]. Although changes in the lipid composition have been found in the membrane of hepatic mitochondria from PUFA-fed rats, the decreased CiC activity was mainly ascribed to a reduction of *CiC* gene expression, as shown by *in vivo* [81–83] and *in vitro* studies [47].

Several molecular events were reported for the PUFA-mediated reduction of CiC expression. Dietary PUFA administration reduces both the transcriptional rate and the splicing process of CiC pre-mRNA, whereas no change in the estimated half-life of the transcript was found [82,84]. Taken together, the aforementioned reports indicate that ω-3 and ω-6 PUFA-supplemented diets down-regulate hepatic *CiC* gene expression by both transcriptional and post-transcriptional mechanisms [82,84].

Two distinct transcription factors, SREBP-1 and PPARs, mediate the regulation of *CiC* gene expression by PUFA.

It has been demonstrated that PUFA of the ω-3 or ω-6 series reduce the transcriptional activity of SREBP-1. In HepG2 [46,47] and in H4IIE hepatoma cell lines [47] SREBP-1 mRNA and protein levels are reduced by the ω-3 PUFA docosahexaenoic acid (DHA, $C_{22:6}$). Taking into account that CiC promoter is activated by SREBP-1 [47], the reduction of CiC mRNA level observed in DHA-treated cells was ascribed to the PUFA-mediated inhibition of *CiC* promoter transactivation by SREBP-1 [47].

Different results have been obtained in the non tumoral BRL-3A hepatic cell line treated with DHA [89]. A strong decrement of CiC expression and *CiC* promoter activity was observed in BRL-3A treated with 50 μM DHA. However, both CiC expression and CiC promoter activity increased in hepatocytes treated with concentrations of DHA higher than 50 μM. As PUFA are natural ligands of PPARα, these findings were ascribed to the transactivation of CiC promoter by this nuclear receptor [89]. This hypothesis was supported by the induction of *CiC* promoter activity in BRL-3A cells upon PPARα/RXRα overexpression or treatment with WY-14,643, a specific PPARα agonist [89]. Moreover, a functional PPRE has been identified at −625 bp of the *CiC* promoter [89]. Since PPARα is the master regulator of genes for β-oxidation enzymes, the physiological role of *CiC* gene transactivation by this transcriptional factor is not yet fully understood. However, an implication of CiC transcriptional activation by PPARα in gluconeogenesis has been suggested [89]. An increase in CiC mRNA abundance and protein level was observed during the induction of murine 3T3-L1 cell differentiation into mature adipocytes, as well as in cells treated with rosiglitazone, a PPARγ agonist, suggesting the involvement of CiC in adipogenesis [89,90].

6. Conclusions

In the last years, the molecular studies on the expression of carrier genes and the functional characterization of their promoters have provided information about specific functions of different MCs. *CACT* and *CiC*, members of *SLC25* family of MCs, are involved in fatty acid metabolism. They have been functionally characterized and their regulation at the transcriptional level has been investigated.

Although much progress has recently been made in the study of the regulation of *CiC* and *CACT* gene expression, underlying mechanisms in different species, tissues, metabolic and hormonal states are not completely understood.

While the effects of different nutritional and hormonal states on the activity and expression of the cytosolic lipogenic enzymes as well as of the mitochondrial enzymes involved in fatty acid oxidation have been deeply investigated, to date a similar work has not been done on CiC and CACT. Considering that the rate of efflux or influx of metabolites through IMM could regulate cellular pathways, the knowledge of the mechanisms by which different nutritional and hormonal factors can control CiC and CACT functions is crucial. These studies may also provide insight into the interconnection existing between catabolic and anabolic pathways inside the cells.

In the future, it could be interesting to investigate whether transcriptional regulation of *CiC* and *CACT* is relevant in diseases associated with insulin signal deregulation, such as obesity and metabolic syndrome. In this respect, in light of the role played by acylcarnitine accumulation in metabolic syndrome [60], the development of therapeutic strategies to regulate CACT activity might furnish valid approaches to the management of syndromes associated with altered fatty acid oxidation.

It is important to note that cytoplasm citrate, conveyed by CiC, is cleaved to acetyl-CoA which is not only the precursor for fatty acid and sterol biosynthesis but it is also the universal donor for protein and histone acetylation [12]. It is worth underlining that CiC expression is increased in cancer cells, in which high levels of acetyl-CoA are required for both lipid synthesis and histone acetylation [12].

Furthermore, the N-terminal (Nt) acetylation of most cellular proteins plays a crucial role in different cellular pathways including apoptosis, regulation of protein degradation through recruitment of ubiquitin ligases [91], prevention of protein translocation from the cytosol to the endoplasmic reticulum (ER), protein complex formation and membrane attachment of small GTPases involved in organelle trafficking [91]. Since it has been demonstrated that the level of acetyl-CoA can regulate the abundance of acetylated proteins, it might be interesting to study: (i) the potential role of CiC and CACT in these cellular processes and (ii) if CACT activity, similarly to CiC, can be regulated by acetylation reactions. These studies could open up interesting new fields on the mechanisms involved in the regulation of lipid metabolism in the cell.

Conflicts of Interest: The authors declare no conflict of interest.

References

1. Duchen, M.R. Roles of mitochondria in health and disease. *Diabetes* **2004**, *53*, S96–S102. [CrossRef] [PubMed]
2. Palmieri, F. Diseases caused by defects of mitochondrial carriers: A review. *Biochim. Biophys. Acta* **2008**, *1777*, 564–578. [CrossRef] [PubMed]
3. Clémençon, B.; Babot, M.; Trézéguet, V. The mitochondrial ADP/ATP carrier (SLC25 family): Pathological implications of its dysfunction. *Mol. Asp. Med.* **2013**, *34*, 485–493. [CrossRef] [PubMed]
4. Palmieri, F. The mitochondrial transporter family SLC25: Identification, properties and physiopathology. *Mol. Asp. Med.* **2013**, *34*, 465–484. [CrossRef] [PubMed]
5. Agrimi, G.; Russo, A.; Pierri, C.L.; Palmieri, F. The peroxisomal NAD$^+$ carrier of *Arabidopsis thaliana* transports coenzyme A and its derivatives. *J. Bioenerg. Biomembr.* **2012**, *44*, 333–340. [CrossRef] [PubMed]
6. Indiveri, C.; Tonazzi, A.; Palmieri, F. The reconstituted carnitine carrier from rat liver mitochondria: Evidence for a transport mechanism different from that of the other mitochondrial translocators. *Biochim. Biophys. Acta* **1994**, *1189*, 65–73. [CrossRef]
7. Klingenberg, M.; Winkler, E. The reconstituted isolated uncoupling protein is a membrane potential driven H$^+$ translocator. *EMBO J.* **1985**, *4*, 3087–3092. [PubMed]
8. Klingenberg, M. Cardiolipin and mitochondrial carriers. *Biochim. Biophys. Acta* **2009**, *1788*, 2048–2058. [CrossRef] [PubMed]
9. Paradies, G.; Paradies, V.; de Benedictis, V.; Ruggiero, F.M.; Petrosillo, G. Functional role of cardiolipin in mitochondrial bioenergetics. *Biochim. Biophys Acta* **2014**, *837*, 408–417. [CrossRef] [PubMed]
10. Sanders, F.W.; Griffin, J.L. *De novo* lipogenesis in the liver in health and disease: More than just a shunting yard for glucose. *Biol. Rev. Camb. Philos. Soc.* **2016**, *91*, 452–468. [CrossRef] [PubMed]
11. Gnoni, G.V.; Priore, P.; Geelen, M.J.; Siculella, L. The mitochondrial citrate carrier: Metabolic role and regulation of its activity and expression. *IUBMB Life* **2009**, *61*, 987–994. [CrossRef] [PubMed]

12. Icard, P.; Poulain, L.; Lincet, H. Understanding the central role of citrate in the metabolism of cancer cells. *Biochim. Biophys. Acta* **2012**, *1825*, 111–116. [CrossRef] [PubMed]

13. Owen, O.E.; Kalhan, S.C.; Hanson, R.W. The key role of anaplerosis and cataplerosis for citric acid cycle function. *J. Biol. Chem.* **2002**, *277*, 30409–30412. [CrossRef] [PubMed]

14. Mora-Rodriguez, R.; Coyle, E.F. Effects of plasma epinephrine on fat metabolism during exercise: Interactions with exercise intensity. *Am. J. Physiol. Endocrinol. Metab.* **2000**, *278*, E669–E676. [PubMed]

15. Huizing, M.; Iacobazzi, V.; Ijlst, L.; Savelkoul, P.; Ruitenbeek, W.; van den Heuvel, L.P.; Indiveri, C.; Smeitink, J.; Trijbels, F.J.M.; Wanders, R.J.A.; *et al.* Cloning of the human carnitine-acylcarnitine carrier cDNA, and identification of the molecular defect in a patient. *Am. J. Hum. Genet.* **1997**, *61*, 1239–1245. [CrossRef] [PubMed]

16. Huizing, M.; Wendel, U.; Ruitenbeek, W.; Iacobazzi, V.; IJlst, L.; Veenhuizen, P.; Savelkoul, P.; van den Heuvel, L.P.; Smeitink, J.A.; Wanders, R.J.; *et al.* Carnitine-acylcarnitine carrier deficiency: Identification of the molecular defect in a patient. *J. Inherit. Metab. Dis.* **1998**, *21*, 262–267. [CrossRef] [PubMed]

17. Kerner, J.; Hoppel, C. Fatty acid import into mitochondria. *Biochim. Biophys. Acta* **2000**, *1486*, 1–17. [CrossRef]

18. Murthy, M.S.; Pande, S.V. Characterization of a solubilized malonyl-CoA-sensitive carnitine palmitoyltransferase from the mitochondrial outer membrane as a protein distinct from the malonyl-CoA-insensitive carnitine palmitoyltransferase of the inner membrane. *Biochem. J.* **1990**, *268*, 599–604. [CrossRef] [PubMed]

19. McGarry, J.D. Malonyl-CoA and carnitine palmitoyltransferase I: An expanding partnership. *Biochem. Soc. Trans.* **1995**, *23*, 481–485. [CrossRef] [PubMed]

20. Wakil, S.J.; Abu-Elheiga, L.A. Fatty acid metabolism: Target for metabolic syndrome. *J. Lipid Res.* **2009**, *50*, S138–S143. [CrossRef] [PubMed]

21. Stoffel, M.; Karayiorgou, M.; Espinosa, R.; Beau, M.M. The human mitochondrial citrate transporter gene SLC20A3 maps to chromosome band 22q11 within a region implicated in DiGeorge syndrome, velo-cardio-facial syndrome and schizophrenia. *Hum. Genet.* **1996**, *98*, 113–115. [CrossRef] [PubMed]

22. Nota, B.; Struys, E.A.; Pop, A.; Jansen, E.E.; Fernandez Ojeda, M.R.; Kanhai, W.A.; Kranendijk, M.; van Dooren, S.J.M.; Bevova, M.R.; Sistermans, E.A.; *et al.* Deficiency in SLC25A1, encoding the mitochondrial citrate carrier, causes combined D-2- and L-2-hydroxyglutaric aciduria. *Am. J. Hum. Genet.* **2013**, *92*, 627–631. [CrossRef] [PubMed]

23. Prasun, P.; Young, S.; Salomons, G.; Werneke, A.; Jiang, Y.H.; Struys, E.; Paige, M.; Avantaggiati, M.L.; McDonald, M. Expanding the clinical spectrum of mitochondrial citrate carrier (SLC25A1) deficiency: Facial dysmorphism in siblings with epileptic encephalopathy and combined D,L-2-hydroxyglutaric aciduria. *JIMD Rep.* **2015**, *19*, 111–115. [PubMed]

24. Chaouch, A.; Porcelli, V.; Cox, D.; Edvardson, S.; Scarcia, P.; de Grassi, A.; Pierri, C.L.; Cossins, J.; Laval, S.H.; Griffin, H.; *et al.* Mutations in the mitochondrial citrate carrier SLC25A1 are associated with impaired neuromuscular transmission. *J. Neuromuscul. Dis.* **2014**, *1*, 75–90. [PubMed]

25. Infantino, V.; Iacobazzi, V.; Menga, A.; Avantaggiati, M.L.; Palmieri, F. A key role of the mitochondrial citrate carrier (SLC25A1) in TNFα- and IFNγ-triggered inflammation. *Biochim. Biophys. Acta* **2014**, *1839*, 1217–1225. [CrossRef] [PubMed]

26. Palmieri, E.M.; Spera, I.; Menga, A.; Infantino, V.; Porcelli, V.; Iacobazzi, V.; Pierri, C.L.; Hooper, D.C.; Palmieri, F.; Castegna, A. Acetylation of human mitochondrial citrate carrier modulates mitochondrial citrate/malate exchange activity to sustain NADPH production during macrophage activation. *Biochim. Biophys. Acta* **2015**, *1847*, 729–738. [CrossRef] [PubMed]

27. Serviddio, G.; Bellanti, F.; Stanca, E.; Lunetti, P.; Blonda, M.; Tamborra, R.; Siculella, L.; Vendemiale, G.; Capobianco, L.; Giudetti, A.M. Silybin exerts antioxidant effects and induces mitochondrial biogenesis in liver of rat with secondary biliary cirrhosis. *Free Radic. Biol. Med.* **2014**, *73*, 117–126. [CrossRef] [PubMed]

28. Pagliassotti, M.J. Endoplasmic reticulum stress in nonalcoholic fatty liver disease. *Annu. Rev. Nutr.* **2012**, *32*, 17–33. [CrossRef] [PubMed]

29. Damiano, F.; Tocci, R.; Gnoni, G.V.; Siculella, L. Expression of citrate carrier gene is activated by ER stress effectors XBP1 and ATF6α, binding to an UPRE in its promoter. *Biochim. Biophys. Acta* **2015**, *1849*, 23–31. [CrossRef] [PubMed]

30. Viggiano, L.; Iacobazzi, V.; Marzella, R.; Cassano, C.; Rocchi, M.; Palmieri, F. Assignment of the carnitine/ acylcarnitine translocase gene (CACT) to human chromosome band 3p21.31 by *in situ* hybridization. *Cytogenet. Cell. Genet.* **1997**, *79*, 62–63. [CrossRef] [PubMed]

31. Indiveri, C.; Iacobazzi, V.; Tonazzi, A.; Giangregorio, N.; Infantino, V.; Convertini, P.; Console, L.; Palmieri, F. The mitochondrial carnitine/acylcarnitine carrier: Function, structure and physiopathology. *Mol. Aspects Med.* **2011**, *32*, 223–233. [CrossRef] [PubMed]

32. Stanley, C.A.; Hale, D.E.; Berry, G.T.; Deleeuw, S.; Boxer, J.; Bonnefont, J.P. Brief report: A deficiency of carnitine-acylcarnitine translocase in the inner mitochondrial membrane. *N. Engl. J. Med.* **1992**, *327*, 19–23. [CrossRef] [PubMed]

33. Palmieri, F. Mitochondrial transporters of the SLC25 family and associated diseases: A review. *J. Inherit. Metab. Dis.* **2014**, *37*, 565–575. [CrossRef] [PubMed]

34. Paradies, G.; Ruggiero, F.M. Enhanced activity of the tricarboxylate carrier and modification of lipids in hepatic mitochondria from hyperthyroid rats. *Arch. Biochem. Biophys.* **1990**, *278*, 425–430. [CrossRef]

35. Paradies, G.; Ruggiero, F.M.; Petrosillo, G.; Quagliariello, E. Stimulation of carnitine acylcarnitine translocase activity in heart mitochondria from hyperthyroid rats. *FEBS Lett.* **1996**, *397*, 260–262. [CrossRef]

36. Paradies, G.; Ruggiero, F.M.; Petrosillo, G.; Quagliariello, E. Alterations in carnitine-acylcarnitine translocase activity and in phospholipid composition in heart mitochondria from hypothyroid rats. *Biochim. Biophys. Acta* **1997**, *1362*, 193–200. [CrossRef]

37. Giudetti, A.M.; Leo, M.; Siculella, L.; Gnoni, G.V. Hypothyroidism down-regulates mitochondrial citrate carrier activity and expression in rat liver. *Biochim. Biophys. Acta* **2006**, *1761*, 484–491. [CrossRef] [PubMed]

38. Siculella, L.; Sabetta, S.; Giudetti, A.M.; Gnoni, G.V. Hypothyroidism reduces tricarboxylate carrier activity and expression in rat liver mitochondria by reducing nuclear transcription rate and splicing efficiency. *J. Biol. Chem.* **2006**, *281*, 19072–19080. [CrossRef] [PubMed]

39. Flores-Morales, A.; Gullberg, H.; Fernandez, L.; Ståhlberg, N.; Lee, N.H.; Vennström, B.; Norstedt, G. Patterns of liver gene expression governed by TRβ. *Mol. Endocrinol.* **2002**, *16*, 1257–1260. [CrossRef] [PubMed]

40. Santillo, A.; Burrone, L.; Falvo, S.; Senese, R.; Lanni, A.; Chieffi Baccari, G. Triiodothyronine induces lipid oxidation and mitochondrial biogenesis in rat Harderian gland. *J. Endocrinol.* **2013**, *219*, 69–78. [CrossRef] [PubMed]

41. Jackson-Hayes, L.; Song, S.; Lavrentyev, E.N.; Jansen, M.S.; Hillgartner, F.B.; Tian, L.; Wood, P.A.; Cook, G.A.; Park, E.A. A thyroid hormone response unit formed between the promoter and first intron of the carnitine palmitoyltransferase-Iα gene mediates the liver-specific induction by thyroid hormone. *J. Biol. Chem.* **2003**, *278*, 7964–7972. [CrossRef] [PubMed]

42. Huang, C.; Freake, H. C. Thyroid hormone regulates the acetyl-CoA carboxylase PI promoter. *Biochem. Biophys. Res. Commun.* **1998**, *249*, 704–708. [CrossRef] [PubMed]

43. Radenne, A.; Akpa, M.; Martel, C.; Sawadogo, S.; Mauvoisin, D.; Mounier, C. Hepatic regulation of fatty acid synthase by insulin and T3: Evidence for T3 genomic and nongenomic actions. *Am. J. Physiol. Endocrinol. Metab.* **2008**, *295*, E884–E894. [CrossRef] [PubMed]

44. Shao, W.; Espenshade, P.J. Expanding roles for SREBP in metabolism. *Cell Metab.* **2012**, *16*, 414–419. [CrossRef] [PubMed]

45. Gnoni, G.V.; Rochira, A.; Leone, A.; Damiano, F.; Marsigliante, S.; Siculella, L. 3,5,3′ Triiodo-L-thyronine induces SREBP-1 expression by non-genomic actions in human HEP G2 cells. *J. Cell. Physiol.* **2012**, *227*, 2388–2397. [CrossRef] [PubMed]

46. Infantino, V.; Iacobazzi, V.; de Santis, F.; Mastrapasqua, M.; Palmieri, F. Transcription of the mitochondrial citrate carrier gene: Role of SREBP-1, upregulation by insulin and downregulation by PUFA. *Biochem. Biophys. Res. Commun.* **2007**, *356*, 249–254. [CrossRef] [PubMed]

47. Damiano, F.; Gnoni, G.V.; Siculella, L. Functional analysis of rat liver citrate carrier promoter: Differential responsiveness to polyunsaturated fatty acids. *Biochem. J.* **2009**, *417*, 561–571. [CrossRef] [PubMed]

48. Gnoni, G.V.; Giudetti, A.M.; Mercuri, E.; Damiano, F.; Stanca, E.; Priore, P.; Siculella, L. Reduced activity and expression of mitochondrial citrate carrier in streptozotocin-induced diabetic rats. *Endocrinology* **2010**, *151*, 1551–1559. [CrossRef] [PubMed]

49. Kaplan, R.S.; Oliveira, D.L.; Wilson, G.L. Streptozotocin induced alterations in the levels of functional mitochondrial anion transport proteins. *Arch. Biochem. Biophys.* **1990**, *280*, 181–191. [CrossRef]

50. Damiano, F.; Mercuri, E.; Stanca, E.; Gnoni, G.V.; Siculella, L. Streptozotocin-induced diabetes affects in rat liver citrate carrier gene expression by transcriptional and posttranscriptional mechanisms. *Int. J. Biochem. Cell Biol.* **2011**, *43*, 1621–1629. [CrossRef] [PubMed]

51. Ferramosca, A.; Conte, A.; Damiano, F.; Siculella, L.; Zara, V. Differential effects of high-carbohydrate and high-fat diets on hepatic lipogenesis in rats. *Eur. J. Nutr.* **2014**, *53*, 1103–1114. [CrossRef] [PubMed]

52. Mihalik, S.J.; Goodpaster, B.H.; Kelley, D.E.; Chace, D.H.; Vockley, J.; Toledo, F.G.; DeLany, J.P. Increased levels of plasma acylcarnitines in obesity and type 2 diabetes and identification of a marker of glucolipotoxicity. *Obesity* **2010**, *18*, 1695–1700. [CrossRef] [PubMed]

53. Schooneman, M.G.; Vaz, F.M.; Houten, S.M.; Soeters, M.R. Acylcarnitines, reflecting or inflicting insulin resistance? *Diabetes* **2013**, *62*, 1–8. [CrossRef] [PubMed]

54. Capaldo, B.; Napoli, R.; Di Bonito, P.; Albano, G.; Saccà, L. Carnitine improves peripheral glucose disposal in non-insulin-dependent diabetic patients. *Diabetes Res. Clin. Pract.* **1991**, *14*, 191–195. [PubMed]

55. Power, R.A.; Hulver, M.W.; Zhang, J.Y.; Dubois, J.; Marchand, R.M.; Ilkayeva, O.; Muoio, D.M.; Mynatt, R.L. Carnitine revisited: Potential use as adjunctive treatment in diabetes. *Diabetologia* **2007**, *50*, 824–832. [CrossRef] [PubMed]

56. Keller, M.P.; Choi, Y.; Wang, P.; Davis, D.B.; Rabaglia, M.E.; Oler, A.T.; Stapleton, D.S.; Argmann, C.; Schueler, K.L.; Edwards, S.; *et al.* A gene expression network model of type 2 diabetes links cell cycle regulation in islets with diabetes susceptibility. *Genome Res.* **2008**, *18*, 706–716. [CrossRef] [PubMed]

57. Di Noia, M.A.; van Driesche, S.; Palmieri, F.; Yang, L.M.; Quan, S.; Goodman, A.I.; Abraham, N.G. Heme oxygenase-1 enhances renal mitochondrial transport carriers and cytochrome *c* oxidase activity in experimental diabetes. *J. Biol. Chem.* **2006**, *281*, 15687–15693. [CrossRef] [PubMed]

58. Peluso, G.; Petillo, O.; Margarucci, S.; Mingrone, G.; Greco, A.V.; Indiveri, C.; Palmieri, F.; Melone, M.A.; Reda, E.; Calvani, M. Decreased mitochondrial carnitine translocase in skeletal muscles impairs utilization of fatty acids in insulin-resistant patients. *Front. Biosci.* **2002**, *7*, a109–a116. [CrossRef] [PubMed]

59. Kaplan, R.S.; Mayor, J.A.; Blackwell, R.; Wilson, G.L.; Schaffer, S.W. Functional levels of mitochondrial anion transport proteins in non-insulin-dependent diabetes mellitus. *Mol. Cell. Biochem.* **1991**, *107*, 79–86. [CrossRef] [PubMed]

60. Soni, M.S.; Rabaglia, M.E.; Bhatnagar, S.; Shang, J.; Ilkayeva, O.; Mynatt, R.; Zhou, Y.; Schadt, E.E.; Thornberry, N.A.; Muoio, D.M.; *et al.* Downregulation of carnitine acyl-carnitine translocase by miRNAs 132 and 212 amplifies glucose-stimulated insulin secretion. *Diabetes* **2014**, *63*, 3805–3814. [CrossRef] [PubMed]

61. Joseph, J.W.; Jensen, M.V.; Ilkayeva, O.; Palmieri, F.; Alárcon, C.; Rhodes, C.J.; Newgard, C.B. The mitochondrial citrate/isocitrate carrier plays a regulatory role in glucose-stimulated insulin secretion. *J. Biol. Chem.* **2006**, *281*, 35624–35632. [CrossRef] [PubMed]

62. Cappello, A.R.; Guido, C.; Santoro, A.; Santoro, M.; Capobianco, L.; Montanaro, D.; Madeo, M.; Andò, S.; Dolce, V.; Aquila, S. The mitochondrial citrate carrier (CIC) is present and regulates insulin secretion by human male gamete. *Endocrinology* **2012**, *153*, 1743–1754. [CrossRef] [PubMed]

63. Kersten, S.; Seydoux, J.; Peters, J.M.; Gonzalez, F.J.; Desvergne, B.; Wahli, W. Peroxisome proliferator-activated receptor α mediates the adaptive response to fasting. *J. Clin. Investig.* **1999**, *103*, 1489–1498. [CrossRef] [PubMed]

64. Mandard, S.; Zandbergen, F.; Tan, N.S.; Escher, P.; Patsouris, D.; Koenig, W.; Kleemann, R.; Bakker, A.; Veenman, F.; Wahli, W.; *et al.* The direct peroxisome proliferator-activated receptor target fasting-induced adipose factor (FIAF/PGAR/ANGPTL4) is present in blood plasma as a truncated protein that is increased by fenofibrate treatment. *J. Biol. Chem.* **2004**, *279*, 34411–34420. [CrossRef] [PubMed]

65. Chakravarthy, M.V.; Pan, Z.; Zhu, Y.; Tordjman, K.; Schneider, J.G.; Coleman, T.; Turk, J.; Semenkovich, C.F. "New" hepatic fat activates PPARα to maintain glucose, lipid, and cholesterol homeostasis. *Cell Metab.* **2005**, *1*, 309–322. [CrossRef] [PubMed]

66. Zechner, R.; Zimmermann, R.; Eichmann, T.O.; Kohlwein, S.D.; Haemmerle, G.; Lass, A.; Madeo, F. FAT SIGNALS—lipases and lipolysis in lipid metabolism and signaling. *Cell Metab.* **2012**, *15*, 279–291. [CrossRef] [PubMed]

67. Iacobazzi, V.; Convertini, P.; Infantino, V.; Scarcia, P.; Todisco, S.; Palmieri, F. Statins, fibrates and retinoic acid upregulate mitochondrial acylcarnitine carrier gene expression. *Biochem. Biophys. Res. Commun.* **2009**, *388*, 643–647. [CrossRef] [PubMed]

68. Gutgesell, A.; Wen, G.; König, B.; Koch, A.; Spielmann, J.; Stangl, G.I.; Eder, K.; Ringseis, R. Mouse carnitine-acylcarnitine translocase (CACT) is transcriptionally regulated by PPARα and PPARδ in liver cells. *Biochim. Biophys. Acta* **2009**, *1790*, 1206–1216. [CrossRef] [PubMed]

69. Sekoguchi, E.; Sato, N.; Yasui, A.; Fukada, S.; Nimura, Y.; Aburatani, H.; Ikeda, K.; Matsuura, A. A novel mitochondrial carnitine-acylcarnitine translocase induced by partial hepatectomy and fasting. *J. Biol. Chem.* **2003**, *278*, 38796–38802. [CrossRef] [PubMed]

70. Peters, J.M.; Hennuyer, N.; Staels, B.; Fruchart, J.C.; Fievet, C.; Gonzalez, F.J.; Auwerx, J. Alterations in lipoprotein metabolism in peroxisome proliferator-activated receptor α deficient mice. *J. Biol. Chem.* **1997**, *272*, 27307–27312. [CrossRef] [PubMed]

71. Martin, G.; Duez, H.; Blanquart, C.; Berezowski, V.; Poulain, P.; Fruchart, J.C.; Najib-Fruchart, J.; Glineur, C.; Staels, B. Statin-induced inhibition of the Rho-signaling pathway activates PPARα and induces HDL apoA-I. *J. Clin. Investig.* **2001**, *107*, 1423–1432. [CrossRef] [PubMed]

72. Kliewer, S.A.; Lehmann, J.M.; Willson, T.M. Orphan nuclear receptors: Shifting endocrinology into reverse. *Science* **1999**, *284*, 757–760. [CrossRef] [PubMed]

73. Zara, V.; Gnoni, G.V. Effect of starvation on the activity of the mitochondrial tricarboxylate carrier. *Biochim. Biophys. Acta* **1995**, *1239*, 33–38. [CrossRef]

74. Siculella, L.; Sabetta, S.; di Summa, R.; Leo, M.; Giudetti, A.M.; Palmieri, F.; Gnoni, G.V. Starvation-induced posttranscriptional control of rat liver mitochondrial citrate carrier expression. *Biochem. Biophys. Res. Commun.* **2002**, *299*, 418–423. [CrossRef]

75. Jump, D.B. Fatty acid regulation of hepatic lipid metabolism. *Curr. Opin. Clin. Nutr. Metab. Care* **2011**, *14*, 115–120. [CrossRef] [PubMed]

76. Takeuchi, H.; Nakamoto, T.; Mori, Y.; Kawakami, M.; Mabuchi, H.; Ohishi, Y.; Ichikawa, N.; Koike, A.; Masuda, K. Comparative effects of dietary fat types on hepatic enzyme activities related to the synthesis and oxidation of fatty acid and to lipogenesis in rats. *Biosci. Biotechnol. Biochem.* **2001**, *65*, 1748–1754. [CrossRef] [PubMed]

77. Ide, T.; Kobayashi, H.; Ashakumary, L.; Rouyer, I.A.; Takahashi, Y.; Aoyama, T.; Hashimoto, T.; Mizugaki, M. Comparative effects of perilla and fish oils on the activity and gene expression of fatty acid oxidation enzymes in rat liver. *Biochim. Biophys. Acta* **2000**, *1485*, 23–35. [CrossRef]

78. Priore, P.; Stanca, E.; Gnoni, G.V.; Siculella, L. Dietary fat types differently modulate the activity and expression of mitochondrial carnitine/acylcarnitine translocase in rat liver. *Biochim. Biophys. Acta* **2012**, *1821*, 1341–1349. [CrossRef] [PubMed]

79. Tian, J.J.; Lu, R.H.; Ji, H.; Sun, J.; Li, C.; Liu, P.; Lei, C.X.; Chen, L.Q.; Du, Z.Y. Comparative analysis of the hepatopancreas transcriptome of grass carp (*Ctenopharyngodon idellus*) fed with lard oil and fish oil diets. *Gene* **2015**, *565*, 192–200. [CrossRef] [PubMed]

80. Jump, D.B. *n*-3 Polyunsaturated fatty acid regulation of hepatic gene transcription. *Curr. Opin. Lipidol.* **2008**, *19*, 242–247. [CrossRef] [PubMed]

81. Giudetti, A.M.; Sabetta, S.; Di Summa, R.; Leo, M.; Damiano, F.; Siculella, L.; Gnoni, G.V. Differential effects of coconut oil and fish oil-enriched diets on tricarboxylate carrier in rat liver mitochondria. *J. Lipid Res.* **2003**, *44*, 2135–2141. [CrossRef] [PubMed]

82. Siculella, L.; Sabetta, S.; Damiano, F.; Giudetti, A.M.; Gnoni, G.V. Different dietary fatty acids have dissimilar effects on activity and gene expression of mitochondrial tricarboxylate carrier in rat liver. *FEBS Lett.* **2004**, *578*, 280–284. [CrossRef] [PubMed]

83. Zara, V.; Giudetti, A.M.; Siculella, L.; Palmieri, F.; Gnoni, G.V. Covariance of tricarboxylate carrier activity and lipogenesis in liver of polyunsaturated fatty acid (*n*-6) fed rats. *Eur. J. Biochem.* **2001**, *268*, 5734–5739. [CrossRef] [PubMed]

84. Siculella, L.; Damiano, F.; Sabetta, S.; Gnoni, G.V. *n*-6 PUFAs downregulate expression of the tricarboxylate carrier in rat liver by transcriptional and posttranscriptional mechanisms. *J. Lipid Res.* **2004**, *45*, 1333–1340. [CrossRef] [PubMed]

85. Giudetti, A.M.; Beynen, A.C.; Lemmens, A.G.; Gnoni, G.V.; Geelen, M.J. Hepatic lipid and carbohydrate metabolism in rats fed a commercial mixture of conjugated linoleic acids (Clarinol G-80). *Eur. J. Nutr.* **2005**, *44*, 33–39. [CrossRef] [PubMed]

86. Giudetti, A.M.; Beynen, A.C.; Lemmens, A.G.; Gnoni, G.V.; Geelen, M.J. Hepatic fatty acid metabolism in rats fed diets with different contents of C18:0, C18:1 cis and C18:1 trans isomers. *Br. J. Nutr.* **2003**, *90*, 887–893. [CrossRef] [PubMed]

87. Ferramosca, A.; Conte, L.; Zara, V. A krill oil supplemented diet reduces the activities of the mitochondrial tricarboxylate carrier and of the cytosolic lipogenic enzymes in rats. *J. Anim. Physiol. Anim. Nutr.* **2012**, *96*, 295–306. [CrossRef] [PubMed]

88. Ferramosca, A.; Zara, V. Dietary fat and hepatic lipogenesis: Mitochondrial citrate carrier as a sensor of metabolic changes. *Adv. Nutr.* **2014**, *5*, 217–225. [CrossRef] [PubMed]

89. Damiano, F.; Gnoni, G.V.; Siculella, L. Citrate carrier promoter is target of peroxisome proliferator-activated receptor α and γ in hepatocytes and adipocytes. *Int. J. Biochem. Cell Biol.* **2012**, *44*, 659–668. [CrossRef] [PubMed]

90. Bonofiglio, D.; Santoro, A.; Martello, E.; Vizza, D.; Rovito, D.; Cappello, A.R.; Barone, I.; Giordano, C.; Panza, S.; Catalano, S.; *et al.* Mechanisms of divergent effects of activated peroxisome proliferator-activated receptor-γ on mitochondrial citrate carrier expression in 3T3-L1 fibroblasts and mature adipocytes. *Biochim. Biophys. Acta* **2013**, *1831*, 1027–1036. [CrossRef] [PubMed]

91. Starheim, K.K.; Gevaert, K.; Arnesen, T. Protein N-terminal acetyltransferases: When the start matters. *Trends Biochem. Sci.* **2012**, *37*, 152–161. [CrossRef] [PubMed]

International Journal of
Molecular Sciences

MDPI

Review

Tetrapyrroles as Endogenous TSPO Ligands in Eukaryotes and Prokaryotes: Comparisons with Synthetic Ligands

Leo Veenman *, Alex Vainshtein, Nasra Yasin, Maya Azrad and Moshe Gavish *

Department of Neuroscience, Faculty of Medicine, Rappaport Family Institute for Research in the Medical Sciences, Technion-Israel Institute of Technology, Ephron Street, P.O.B. 9649, Bat-Galim, Haifa 31096, Israel; alexanderv21184@gmail.com (A.V.); blackpearl.black2@gmail.com (N.Y.); mayabz@gmail.com (M.A.)
* Correspondence: veenmanl@techunix.technion.ac.il (L.V.); mgavish@tx.technion.ac.il (M.G.);
  Tel.: +972-4-829-5276 (L.V.); +972-4-829-5275 (M.G.)

Academic Editor: Giovanni Natile
Received: 2 May 2016; Accepted: 19 May 2016; Published: 4 June 2016

Abstract: The 18 kDa translocator protein (TSPO) is highly 0conserved in eukaryotes and prokaryotes. Since its discovery in 1977, numerous studies established the TSPO's importance for life essential functions. For these studies, synthetic TSPO ligands typically are applied. Tetrapyrroles present endogenous ligands for the TSPO. Tetrapyrroles are also evolutionarily conserved and regulate multiple functions. TSPO and tetrapyrroles regulate each other. In animals TSPO-tetrapyrrole interactions range from effects on embryonic development to metabolism, programmed cell death, response to stress, injury and disease, and even to life span extension. In animals TSPOs are primarily located in mitochondria. In plants TSPOs are also present in plastids, the nuclear fraction, the endoplasmic reticulum, and Golgi stacks. This may contribute to translocation of tetrapyrrole intermediates across organelles' membranes. As in animals, plant TSPO binds heme and protoporphyrin IX. TSPO-tetrapyrrole interactions in plants appear to relate to development as well as stress conditions, including salt tolerance, abscisic acid-induced stress, reactive oxygen species homeostasis, and finally cell death regulation. In bacteria, TSPO is important for switching from aerobic to anaerobic metabolism, including the regulation of photosynthesis. As in mitochondria, in bacteria TSPO is located in the outer membrane. TSPO-tetrapyrrole interactions may be part of the establishment of the bacterial-eukaryote relationships, *i.e.*, mitochondrial-eukaryote and plastid-plant endosymbiotic relationships.

Keywords: TSPO; tetrapyrrole; eukaryotes; prokaryotes; TSPO ligand binding; TSPO binding site structures; cell function; stress; homeostasis; life expectancy

1. Concise Introduction to the Theme

Interactions between the 18 kDa translocator protein (TSPO) and tetrapyrroles, including the tetrapyrrole protoporphyrin IX (PPIX), have been studied for several decades, in various species. These species cover animals, plants, fungi, bacteria, and archea, *i.e.*, eukaryotes as well as prokaryotes. TSPOs as well as tetrapyrroles are known to be involved in various vital functions regarding molecular cell biology as well as organismal functions (Figure 1). Thus, it is worthwhile to present an overview of the knowledge that has been gathered so far, and mention the future perspectives for research regarding interactions between TSPO and tetrapyrroles. In the end such research will reveal further potential implications regarding health and disease, and even plant cultivation. In this review, apart from interactions between TSPO and tetrapyrroles, including PPIX as an endogenous ligand, the effects of synthetic TSPO ligands in the context of TSPO-tetrapyrrole interactions are also presented (summarily indicated in Figure 1).

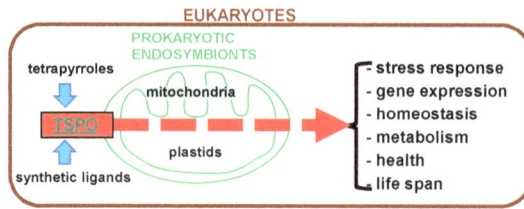

Figure 1. Overview of the subject of this review. Endogenous ligands (tetrapyrroles) as well as synthetic ligands for translocator protein (TSPO) affect functions of free living prokaryotes as well as the derived endosymbionts present as mitochondria and plastids in eukaryotes. A few of these functions modulated by TSPO and its ligands are listed on the right-hand side.

2. The 18 kDa Translocator Protein (TSPO)

2.1. General TSPO Characteristics and Functions

The 18 kDa translocator protein (TSPO) is a highly conserved protein in eukaryotic as well as prokaryotic species and is involved in various life essential functions (Tables 1 and 2) [1,2]. Previously, TSPO was known as peripheral-type benzodiazepine receptor (PBR) because of its ability to bind benzodiazepines in various peripheral tissues in mammals [1,3,4]. A major intracellular location of this five-α-helices membrane-spanning protein in eukaryotes is the outer mitochondrial membrane at the contact sites with the inner mitochondrial membrane [4–6]. TSPO can form a complex with the voltage-dependent anion channel (VDAC, 32 kDa) and the adenine nucleotide translocator (ANT, 30 kDa), which are located at the outer and inner mitochondrial membrane, respectively [1,6]. The isoquinoline carboxamide PK 11195, the benzodiazepine Ro5-4864, and the indole derivative FGIN-1-27 are the classical synthetic ligands of TSPO [1,7]. Their full names are, respectively, 1-(2-chlorophenyl)-*N*-methyl-*N*-(1-methylpropyl)-3-isoquinoline carboxamide; 7-chloro-5-(4-chlorophenyl)-1-methyl-3*H*-1,4-benzodiazepin-2-one; and *N,N*-di-n-hexyl 2-(4-fluorophenyl)indole-3-acetamide.

A well-known endogenous TSPO ligand is the tetrapyrrole protoporphyrin IX (PPIX) [8,9]. The present review deals with the potential of tetrapyrroles to modulate TSPO functions in animals, plants, and bacteria. It has been found that TSPO has various functions in mammals, such as programmed cell death induction, regulation of mitochondrial membrane potential transition (MPT) including mitochondrial membrane potential ($\Delta\Psi$m) collapse, respiratory chain regulation, cholesterol transportation, regulation of steroidogenesis, heme metabolism (heme is PPIX containing a Fe^2 ion in its center), anion transportation, modulation of voltage-dependent anion channel (VDAC) opening, immune response, glial activation related to brain damage, cell growth and differentiation, and cancer cell proliferation [1,4,10–14] (see also Table 2). At organismal levels, the effects include modulation of endocrine, reproductive, and cardiovascular functions, local responses to brain damage due to injury and disease, and other neuropathological, emotional, and mental disorders, in particular including responses to stress [1,12,15–22], as well as life span enhancement [19,21,22] (see also Table 2). At cellular levels, programmed cell death regulation by the TSPO may include changes in TSPO expression levels, and typically includes mitochondrial reactive oxygen species (ROS) generation, cardiolipin oxidation, and collapse of the $\Delta\Psi$m, all under the control of the TSPO and its ligands [10,14,23–27] (see also Table 2). Previous studies have shown that knockdown of TSPO by genetic manipulation as well as application of its synthetic ligands Ro5-4864, PK 11195, and FGIN-1-27 can protect various human and animal cell lines against apoptotic cell death [14,18,24,26,28]. Indicated in Table 2, TSPO-associated functions studied in plants and bacteria are reminiscent of several TSPO-associated functions in animals. This is also discussed in Sections 6–8 of this review, respectively dealing with: "Bacteria, TSPO and tetrapyrroles"; "Plants, TSPO and tetrapyrroles"; and "TSPO-tetrapyrrole interactions from an evolutionary perspective".

Table 1. Translocator protein (TSPO) gene length in base pairs (bp) and protein length in amino acids (aa) in different species that are discussed in this review, plus a few additional ones, to obtain a representative view of what is known regarding TSPO in living organisms in general. The left column gives the names of the species. The species are organized according to: human, mammals, insects, archea, bacteria, plants, and fungi. The middle column shows the TSPO gene lengths in base pairs (bp) for each species. The right column shows the protein lengths, which are between 151 and 211 aa. TSPO protein molecular weight for all species typically is 18 kDa. Interestingly, while protein size does not differ essentially from species to species, as shown here, gene length varies from 11,729 bp in humans to as low as 456 bp in *Bacillus anthracis str. Ames*. (After the resources "Gene" and "Protein" from the National Center for Biotechnology Information, National Library of Medicine, 8600 Rockville Pike, Bethesda, MD, USA).

Various Species Expressing TSPO	TSPO Gene Length (bp)	TSPO Protein Length (aa)
Human		
Homo sapiens	11,729 bp	169 aa
Mammals		
Rattus norvegicus	10,253 bp	169 aa
Mus musculus	10,631 bp	169 aa
Amphibians		
Xenopus tropicalis	5970 bp	211 aa
Insects		
Drosophila melanogaster	6569 bp	185 aa
Aedes aegypti	1236 bp	176 aa
Archea		
Haloferax mediterranei	486 bp	161 aa
Bacteria		
Bacillus anthracis str. Ames	456 bp	151 aa
Rhodobacter sphaeroides	480 bp	158 aa
Rhodobacter capsulatus	483 bp	160 aa
Plants		
Arabidopsis thaliana	1044 bp	196 aa
Solanum tuberosum	895 bp	203 aa
Ricinus communis	1530 bp	196 aa
Vitis vinifera	1055 bp	185 aa
Fungi		
Cryptococcus gattii	865 bp	174 aa
Schizosaccharomyces cryophilus	480 bp	164 aa
Aspergillus fumigatus	632 bp	177 aa
Kluyveromyces	486 bp	160 aa

Also, TSPO involvement in steroid production, including neurosteroids, has attracted a lot of attention [18,29–31] (see also Table 2). These studies emphasize that aberrant regulation of steroid production via TSPO activity can be linked to cancer, neurodegeneration, neuropsychiatric disorders, and primary hypogonadism. TSPO ligands have been proposed as therapeutic agents to regulate steroid levels in the brain and reproductive system. *Vice versa*, it is also well known that at system levels, various types of steroids modulate TSPO expression [1,15,18]. It is also becoming more and more appreciated that TSPO is involved in the differentiation of various cell types [22,32,33] (see also Table 2). Synthetic TSPO ligands promoting neuronal differentiation have been proposed as therapeutic agents for the repair of brain and spinal cord damage due to injury and disease [21]. Recent studies have also revealed the involvement of TSPO in the modulation of nuclear gene expression, giving some explanation as to how TSPO can be involved in so many and such diverse functions [34–36] (see also Table 2). As this regulation of gene expression most likely takes place via the retrograde mitochondrial-nuclear signaling pathway for the regulation of nuclear gene expression, it probably is a TSPO function that can be found in all eukaryotes.

Table 2. TSPO is involved in various functions in animals, plants, and bacteria. As described in this review, and summarized in this table, TSPO performs, regulates, and modulates a rich spectrum of life essential functions. These TSPO functions have been studied extensively in mammals, ranging from molecular biological mechanisms to various stress responses, at cellular and organismal levels, and even to enhancement of life expectancy. (Note: TSPO functions uncovered in insects show great similarity to those in mammals.) While research on plant and bacterial TSPO thus far has been more restricted, the TSPO functions described in various plant and bacteria species are reminiscent of several TSPO functions described in animals. In this table, comparable functions of animals, plants and bacteria are placed in one row.

TSPO-Associated Functions in Animals, Plants, and Bacteria		
Animals	Plants	Bacteria
Mitochondrial membrane potential transition		Interactions with large membrane channels
Transport of porphyrin intermediates	Translocation of tetrapyrrole intermediates	Transport of porphyrin intermediates
Heme metabolism	Tetrapyrrole metabolism	Heme metabolism
ROS generation	Oxidative stress	ROS generation
Programmed cell death	Cell death	Induction of apoptosis in eukaryotes
Mitochondrial protein transport		
Mitochondrial metabolism		Anaerobic and aerobic metabolism
Mitochondrial cholesterol transport		Cholesterol binding
Steroidogenesis		
Nuclear gene expression		Gene expression
Cell cycle	Cell cycle	Cell cycle
Cell growth		Cell growth
Cell proliferation		
Cell migration		
Cell adhesion		Adhesion
Cell differentiation		
Embryonic development	Seed and plant development	
Endocrinological function		
Reproduction		
Stress response	Stress response	Stress response
Immune response	Response to pathogens	
Inflammatory response		
Glial activation		
Response to brain disease and injury		
Emotional health		
Mental health		
Cardiovascular health		
Homeostasis	Homeostasis	Homeostasis
Life span of multicellular organisms		

2.2. TSPO Ligands, Endogenous and Synthetic

Several endogenous TSPO ligands have been identified. PPIX, which was first reported in 1987 as a TSPO ligand [8,37], is the most studied tetrapyrrole in this respect [9]. Other endogenous TSPO ligands include: phospholipase A2 (PLA2) (*Naja naja*) [38], and diazepam binding inhibitor (DBI) and its post-translational products [39–42]. Apart from PPIX no other small endogenous molecules are known that display ligands binding to the TSPO in animals. It would be worthwhile to endeavor targeted research to detect small endogenous molecules binding specifically to TSPO in animal species, a phenomenon as, for example, suggested by plant TSPO research (see Section 7 dealing with: "Plants, TSPO and tetrapyrroles").

Since the discovery of TSPO in rats by Braestrup and Squires [3], an increasing number of synthetic TSPO ligands are incessantly being developed [7,21,27,43]. As mentioned, PK 11195, Ro5-4864, and FGIN-1-27 can be considered classical TSPO ligands [44,45] (presented in Figure 2). Their full names are, respectively, 1-(2-chlorophenyl)-*N*-methyl-(1-methylpropyl)-3 isoquinolinecarboxamide; 7-chloro-5-(4-chlorophenyl)-1-methyl-3*H*-1,4-benzodiazepin-2-one; and *N,N*-di-n-hexyl 2-(4-fluorophenyl)indole-3-acetamide. Later synthetic TSPO ligands encompass: (i) derivatives of Ro5-4864; (ii) derivatives of the 2 aryl-3-indoleacetamide FGIN-1, including FGIN-1-27; and several other types of molecules [7,21,27,43]. In general, the most common structure of a synthetic TSPO ligand includes a backbone of three carbocycles, typically including heteroatoms such as O or N [7,21,27,43–45]. In addition, carbon-based side chains including an acetamide component are frequently part of these TSPO ligands. Furthermore, halogenations at several locations, and/or additional carbocycles linked to the basic structure, are often part of the synthetic TSPO ligands. A few examples of synthetic TSPO ligands, together with the endogenous TSPO ligand PPIX are given in Figure 2. Apart from these small molecules, based on carbocycles, synthetic peptides with the motif STXXXXP can also act as TSPO ligands.

Comparisons of effects and structures of endogenous ligands, such as PPIX, with those of synthetic ligands (as illustrated by Figure 2) may lead to insights into the functional aspects of their interactions with the TSPO. This may include their potential structural interactions with ligand binding sites present on the TSPO. In turn, these elucidations may inform us what is essential for the design of efficacious TSPO ligands, for example for treatment of diseases while avoiding undesired concomitant side effects. As the scope of functional capabilities of the TSPO is broad (Table 2), we are dealing with a double-edged sword. While specific TSPO ligands may display beneficial characteristics regarding various diseases, at the same time such ligands targeting a specific disease may induce effects not related to the disease in question, and may even show adverse additional effects. Knowledge of endogenous ligands may teach us which aspects, affinity-wise and functional, one should focus on for the development of TSPO ligands targeting specific diseases.

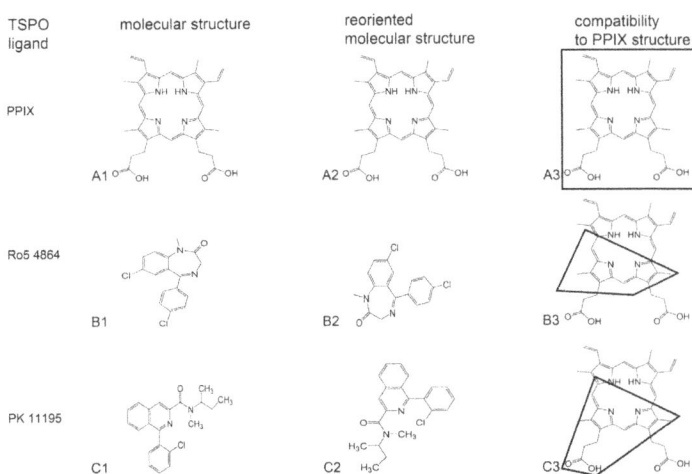

Figure 2. *Cont.*

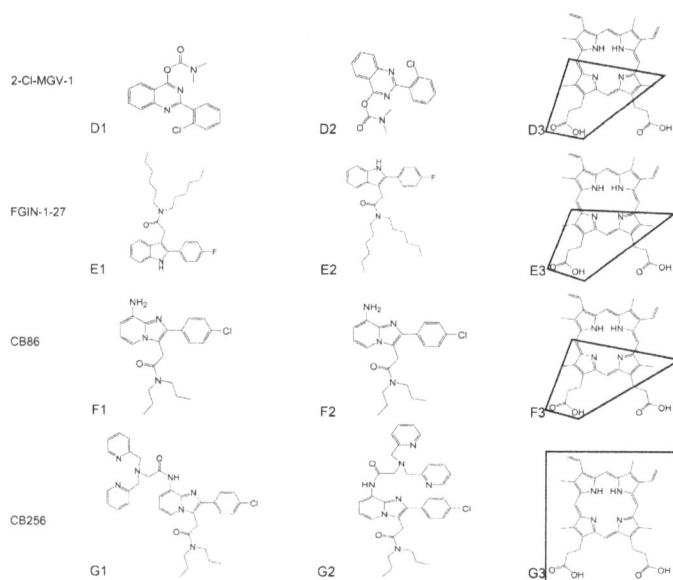

Figure 2. This figure presents line drawings of the molecular structures of seven known TSPO ligands (listed in the most left hand column), to visualize for each one the structural compatibilities of the synthetic ligands with the molecular structure of other synthetic ligands and the endogenous TSPO ligand PPIX. For orientation in the figure, adjacent to each molecular structure the letter refers to the row (lined up with the compound's name) and the numbers refer to the columns related to the molecular structure characteristics (*i.e.*, 1 relates to Molecular Structure, 2 relates to Reoriented Molecular Structure, and 3 relates to Compatibility to PPIX structure). These TSPO ligands were first described as such by: Verma *et al.* [8] (PPIX), Le Fur *et al.* [44] (Ro5-4864 and PK 11195), Vainshtein *et al.* [22] (2-Cl-MGV-1), Romeo *et al.* [45] (FGIN-1-27), Denora *et al.* [46] (CB86 and CB256). Their full names at the left hand beginnings or their rows are, respectively : 3-[18-(2-carboxyethyl)-8,13-bis(ethenyl)-3,7,12,17-tetramethyl-22,23-dihydroporphyrin-2-yl]propanoic acid (protoporphyrin IX ; abbreviation PPIX in row A), 7-chloro-5-(4-chlorophenyl)-1-methyl-3*H*-1,4-benzodiazepin-2-one (Ro5-4864 in row B), 1-(2-chlorophenyl)-*N*-methyl-*N*-(1-methylpropyl)-3-isoquinoline carboxamide (PK 11195 in row C); [2-(2-chlorophenyl)quinazolin-4-yl dimethylcarbamate] (2-Cl-MGV-1 in row D); *N,N*-di-n-hexyl 2-(4-fluorophenyl)indole-3-acetamide (FGIN-1-27 in row E); 2-(8-amino-2-(4-chlorophenyl)*H*-imidazo[1,2-*a*]pyridin-3-yl)-*N,N*-dipropylacetamide (CB86 in row F) 2-(8-(2-(bis(pyridin-2-yl)methyl)amino)acetamido)-2-(4-chlorophenyl)*H*-imidazo[1,2-*a*]pyridin-3-yl)-*N,N*-dipropylacetamide (CB256 in row G). In the first, left hand column the names of the ligands are given as they are generally used in the scientific community (not numbered here). In the second column (indicated with #1) the molecular structures are given as they are typically presented in the literature. In the third column (indicated with #2) the molecular structures are reoriented to facilitate visualization of a potential match with a corresponding part of PPIX. This reorientation typically is no more than flipping and rotating the original drawing, if required at all. Regarding the drawing of "CB256", rotations of several bonds are applied (using ChemBioDraw™) to achieve a configuration that matches the structure of PPIX. In the fourth, most right handed column (indicated with #3), in each row, the PPIX molecular structure is presented. In this fourth column, for each row, angular shapes are drafted, outlining the parts of PPIX that may potentially correspond to the full molecular structures of the ligands in the rows in question. Thus, this figure presents structural characteristics common to various TSPO ligands. One can assume that the structural commonalities are related to shared functions (as well as affinity for the TSPO), while the structural differences may be related to differences in effects (as well as differences in affinity for the TSPO). The molecular structures were drawn with the aid of ChemBioDraw ™ of PerkinElmer, 940 Winter Street, Waltham, MA, USA.

3. Tetrapyrroles, including Protoporphyrins such as PPIX, as Ligands for TSPO

3.1. Tetrapyrroles Binding to TSPO

Interestingly, as TSPO is an evolutionarily conserved protein in archea, bacteria, fungi, plants, and animals (Table 1), found throughout all tissues studied [1,2,21], tetrapyrroles, in a similar vein, are probably one of the most ancient prosthetic groups in all kingdoms of living organisms and comprise the most abundant pigment molecules on earth [47]. The multifunctionality of tetrapyrroles, in particular in association with their interactions with TSPO in eukaryotes as well as prokaryotes in this respect, suggests a relationship between TSPO and tetrapyrroles, at least from the geological time point that endosymbiotic relationships between specific bacterial species and eukaryotes were established (see also Section 8 dealing with: "TSPO-tetrapyrrole interactions from an evolutionary perspective"). Tetrapyrroles are a class of chemical compounds that contain four pyrrole rings held together by one-carbon bridges (=(CH)– or –CH: two units) or by direct covalent bonds, in either a linear or a cyclic fashion. Tetrapyrroles are involved in metabolism in all kingdoms of living organisms. In the animal kingdom, well-known tetrapyrroles are porphyrins that are part of the heme synthesis pathway [48]. In plants, tetrapyrroles are part of the pathways leading to heme formation and chlorophyll [49]. In cyanobacteria and red algae, such pathways lead to the formation of different phycobilines [49]. Below, we will give an overview of what is known thus far regarding interactions between tetrapyrroles and TSPO.

In 1987, it was reported that porphyrins (cyclic tetrapyrroles with prominent physiological functions), extracted from rat as well as human tissue, in particular PPIX (Figure 2), present themselves as endogenous ligands for TSPO [8,37]. In this respect, the endogenous PPIX and the synthetic PK 11195 showed relative constancy in affinity (in the nM range). In contrast, the affinity of the synthetic benzodiazepine Ro5-4864 was shown to vary several orders of magnitude in competing for receptors in different organs and species [8,37,50,51]. One typical aspect of the Ro5-4864 molecule is that it lacks an elongated side chain (Figure 2). In contrast, both PK 11195 and PPIX display elongated side chains containing a number of carbon atoms (Figure 2). Based on present knowledge, one can suggest that the side chain presenting a number of carbon atoms attached to the part of the molecule binding to the TSPO in question is important for constancy in affinity [22]. In Table 3, TSPO functions are listed that were found to be modulated by tetrapyrroles in animals, plant, and bacteria, as presented in this review.

Table 3. TSPO functions affected by tetrapyrroles in animals, plants and bacteria. While it is known that tetrapyrroles can bind TSPO in animals, plants, and bacteria, TSPO functions affected by porphyrins have been described in particular for animals and humans. Although not studied as extensively as in animals and humans, tetrapyrrole effects on TSPO functions in plants and bacteria are reminiscent of those described for animals and humans.

TSPO Functions Affected by Tetrapyrroles		
Animals	**Plants**	**Bacteria**
TSPO expression	TSPO expression	
Mitochondrial membrane potential transition		
ROS generation	Stress response	
Mitochondrial protein transport		
Mitochondrial cholesterol transport		
Regulation of steroidogenesis		
Heme metabolism		
Transport of porphyrin intermediates		
Modulation of nuclear gene expression	Seed and plant development	Photosynthetic gene expression
Cell migration		
Programmed cell death		
Mitochondrial metabolism		Switch between anaerobic and aerobic metabolism
Life span		

3.2. Implications of Tetrapyrrole-TSPO Interactions

It was suggested that ligands specific to TSPO could be applied for the treatment of porphyrias [52]. Porphyrias are diseases in which porphyrins ccumulate [53]. Hepatic protoporphyria induced by DDC (3,5-diethoxycarbonyl-1,4-dihydrocollidine) includes increased levels of PPIX and N-methylprotoporphyrin IX (N-MePPIX), which is associated with a decrease in TSPO ligand binding in the liver, as measured in rats' liver homogenates [54]. This included decreased affinity for Ro5-4864 and PK 11195, as well as a 55% decrease in the maximum number of binding sites (Bmax). Further studies on cultured hepatocytes suggested that, depending on the energetic states of the mitochondria, TSPO-PPIX interactions may have multifunctional effects, including membrane permeability transition (MPT) and transport of PPIX across the mitochondrial membranes [55]. In this study [55], in hepatocyte cultures de-energized by rotenone, nanomolar concentrations of PPIX potentiated the induction of the MPT, *i.e.*, induced ΔΨm collapse, and enhanced the extent of cell killing. In short, PPIX enhanced the effects of rotenone. This appears to be the opposite of what happens with nanomolar concentrations of PK 11195, Ro5-4864, and FGIN-1-27 in U118MG cells, which at these concentrations counter ΔΨm collapse and cell death otherwise induced by ammonia [28]. This suggests that at least in this respect, these synthetic TSPO ligands can counter the functions of the endogenous PPIX. As a specific example, PK 11195 as well as TSPO knockdown by siRNA can counteract the cytotoxic effects of hemin (PPIX containing a ferric iron ion with a chloride ligand) in colonic epithelial (Caco-2) cells [56]. Already early on it was found that the TSPO ligands PK 11195, Ro5-4864, and PPIX had different functional effects. For example, Ro5-4864 and PK 11195 could modulate prolactin-stimulated mitogenesis in the Nb2 lymphoma cell, while PPIX had no such effect [57]. Rats suffering from porphyria induced by DDC showed interesting effects of PK 11195 administration (15 mg/kg/day) [58]. This PK 11195 administration aggravates PPIX accumulation and cellular damage in the liver. It was suggested that in this paradigm, PK 11195 blocks the binding of PPIX to TSPO, thereby elevating the content of PPIX in the liver [59], *i.e.*, it then already was assumed that TSPO contributes to metabolization of PPIX, which was corroborated by later studies [13].

In rat liver mitochondria, PK 11195 and N-MePPIX dose-dependently stimulate cholesterol translocation and incorporation into inner membranes, while PPIX and Ro5-4864 are ineffective in this respect [59]. In another study, time- and dose-dependent effects of synthetic and endogenous TSPO ligands were seen. Briefly, PK 11195, N-MePPIX, and PPIX either stimulated mitochondrial 27-hydroxylation of [4-14C] cholesterol *in vitro* (PK 11195 and N-MePPIX being more effective than PPIX), or at relatively long-time exposures and increased doses of these TSPO ligands, mitochondrial 27-hydroxylation of [4-14C] cholesterol was decreased [60]. Thus, bimodal effects on the acidic pathway to bile acids, depending on the concentrations of these endogenous and synthetic TSPO ligands and exposure times, were reported. In contrast to PK 11195, N-MePPIX, and PPIX, mentioned effects were not seen for Ro5-4864 and hemin [61]. Various studies have suggested that the concentration-dependent bimodal effects of TSPO ligands may be associated with the presence of high affinity and low affinity binding sites [10,28,60,62]. For example, bimodal effects of TSPO ligands on ammonia-induced toxicity showed that nanomolar concentrations of PK 11195, Ro5-4864, and FGIN-1-27 protected U118MG cells from ammonia-induced cell death while these same TSPO ligands applied at μM concentrations enhanced ammonia-induced cell death of U118MG cells [28].

In human tissues and cells, applying quantitative evaluation of the positron emission tomography (PET) signal of the ^{11}C-PBR28 TSPO ligand revealed high affinity binding sites (HAB) and low affinity binding sites (LAB) [63]. In particular, PBR28 can bind the TSPO with high affinity (binding affinity as indicated by the dissociation constant K_i ~4 nM), low affinity (K_i, ~200 nM), or mixed affinity (two sites with K_i, ~4 and ~300 nM). Other TSPO ligands, including DAA1106, DPA713, PBR06, PBR111, and XBD173 also bind with different affinities to TSPO binding sites. However, PK 11195 does not present such a distinction in affinity [63,64]. The differences in affinity appear to relate to two types of human TSPOs differing at just one amino acid site. In this common polymorphism, an alanine at position 147 of the wild-type TSPO is replaced by threonine [65]. An 18 kDa translocator protein (TSPO) polymorphism explains differences in binding affinity of the PET radioligand PBR28. It has been suggested that in the cases of PBR28 and other second-generation ligands, A147T mTSPO might no longer be able to retain the same structural and dynamic profile as the wild-type protein and thus binds these ligands with lower affinity [66]. Nonetheless, this polymorphism does not affect affinity for PK 11195 [67].

Regarding interactions of synthetic TSPO ligands with functions of the endogenous TSPO ligand PPIX, in one paradigm the effects of PK 11195 and Ro5-4864 on the metabolism and function of PPIX were studied. For this study, accumulation of photoactive PPIX was achieved by application of the exogenous PPIX precursor δ-aminolaevulinic acid (ALA) to rat pancreatoma AR4-2J cells in culture [68]. Under these conditions, exposure to light ($\lambda > 400$ nm) at an intensity of 0.2 mW·cm^2 for 8 min resulted in cytolysis. PPIX generates singlet oxygen upon illumination, *i.e.*, it generates ROS. PK 11195 and Ro5-4864 (at 10 and 20 µM for both ligands) exerted a photoprotective effect in this paradigm [68]. In another study it was found that TSPO can serve to reduce PPIX levels, most likely in association with ROS generation, as determined with TSPO knockdown [13]. A similar effect was induced by 25 µM of PK 11195 [13]. The suggestion that TSPO can serve to catalytically metabolize PPIX to tetrapyrrole products other than hemin in human U118MG glioblastoma cells [13] was recently corroborated for TSPO from *Bacillus cereus*, *Xenopus*, and mammals, in addition to humans [69,70]. Thus, one type of interaction between PPIX and TSPO may simply serve to catabolize excess PPIX in conjugation with ROS generation [13].

Thus, it was found that interactions between TSPO and PPIX cover various aspects, ranging from receptor ligand interactions, modulation of the mitochondrial MPT, and even including collapse of the $\Delta\Psi m$, ROS generation, initiation of programmed cell death, gene expression regulation, cholesterol transport, and heme transport and metabolism. From these studies it can be concluded that it is worthwhile to apply further studies to deepen our insights into TSPO-PPIX interactions. For example, applying recombinant mouse TSPO expressed in *Escherichia coli* showed that PPIX could displace PK 11195 binding in *E. coli* expressing this recombinant mouse gene product [71]. Moreover, induced TSPO protein expression in *E. coli* protoplasts caused an uptake of PPIX that could be completely inhibited by cholesterol and, to a lesser extent, inhibited by PK 11195 and Ro5-4864 [71]. In another study it was found that enhanced TSPO levels in glioma cells were associated with enhanced PPIX production [72]. Also, in rat *in vivo*, PPIX binding to the TSPO could be demonstrated with positron emission tomography, giving further indication of TSPO-tetrapyrrole interactions [73].

3.3. Implications of Tetrapyrrole-TSPO Interactions for Brain Disease

Focusing on TSPO-tetrapyrrole interactions, in a model for acute hepatic encephalopathy, ammonia-induced astrocyte swelling in culture could be attenuated by PK 11195 and PPIX, while Ro5-4864, diazepam binding inhibitor (DBI51-70), and octadecaneuropeptide exacerbated the swelling [74]. To gain a better understanding of PPIX-TSPO interactions in relation to inflammatory processes in the brain, effects on free radical generation by TSPO ligands were studied in cultured neural cells, including primary cultures of rat brain astrocytes and neurons as well as cells of the murine BV-2 microglial cell line [75]. For this purpose, the fluorescent dye dichlorofluorescein-diacetate was used. Free radical production was measured at the time points of 2, 30, 60, and 120 min of treatment with the TSPO ligands PK 11195, Ro5-4864, and PPIX (all at 10 nM). In astrocytes, all ligands showed a significant increase in free radical production at 2 min. The increase was short-lived with PK 11195, whereas with Ro5-4864 it persisted for at least 2 h. PPIX caused an increase at 2 and 30 min, but not at 2 h. Similar results were observed in microglial cells [75]. In this same study, the application of PK 11195 and PPIX to neurons showed an increase in free radical production only at 2 min, while Ro5-4864 had no effect. All in all, even though differences from cell type to cell type and ligand to ligand can be discerned, TSPO ligands can induce free radical production virtually instantly when applied to cells. Cyclosporin A (CsA), an inhibitor of the MPT, could prevent free radical formation by these TSPO ligands. CsA (1 µM) completely blocked free radical production following PK 11195 and

Ro5-4864 treatment in all the cell types of this study [75]. CsA was also effective in blocking free radical production in astrocytes following PPIX treatment, but it failed to do so in neurons and microglia. These studies indicated that exposure of neural cells to TSPO ligands generates free radicals, and that the MPT may be involved in this process [75]. Again, the study by the group of Norenberg [74,75] makes it clear that different TSPO ligands have different effects. Furthermore, the effects of the TSPO ligands were different from cell type to cell type.

Later studies applying various agents typically inducing programmed cell death showed that ROS generation at mitochondrial levels, measured with acridine orange 10-nonyl bromide (NAO) in U118MG cells, could be attenuated by TSPO knockdown, as well as by the TSPO ligands PK 11195, Ro5-4864, and FGIN-1-27 (applied for 24 h, optimal concentrations to achieve these effects typically range around 25 and 50 µM) [14,26,28]. Studies such as these substantiated that TSPO serves to initiate programmed cell death, including ROS generation, MPT, collapse of the $\Delta\Psi$m, and mitochondrial cytochrome C release [11,27,28]. Such ROS generation and MPT can also be part of the retrograde mitochondrial nuclear signaling expression pathway for regulation of nuclear gene expression [34–36]. An important aspect of these studies discussed above is that TSPO ligands, including PPIX, modulate ROS generation. This modulation of ROS generation typically is bimodal. In particular, with short durations of TSPO ligand exposures as well as with the application of low concentrations of ligands, ROS generation is enhanced. This is in contrast with the reduction of ROS generation induced by long durations of TSPO ligand exposures as well as by the application of high concentrations of TSPO ligands. This gives some indication as to why and how TSPO ligands, in a time- and dose-dependent way, can have opposite functional effects, *i.e.*, reductions and enhancements of ROS generation may serve as signals to induce particular functions, and may even be integral to such functional effects. Thus, while each effect in its own paradigm is very reproducible and TSPO-dependent, there is very high, context-associated variability in response, including the effects of the duration of exposure to TSPO ligands, the concentration of the TSPO ligands, and the cell types used. Importantly, PPIX is not an exception regarding these phenomena. These time-dependent and dose-dependent effects of TSPO ligands may reflect the TSPO's function to maintain homeostasis of various cell types, and moreover the health of the complete organism, including the brain in animals [7,76,77].

3.4. Tetrapyrrole-TSPO Interactions in Insects

Apart from vertebrates, TSPO is also found in insects [2]. A few studies regarding TSPO in association with tetrapyrroles have been performed in insects. First of all, the *Drosophila* homolog for TSPO, CG2789/dTSPO, has been identified [19]. This dTSPO was then inactivated by P-element insertion, RNAi knockdown, and inhibition by ligands (PK 11195 and Ro5-4864). Such inhibition of dTSPO in turn inhibited wing disk apoptosis in response to γ-irradiation or H_2O_2 exposure [19]. In the whole animal, dTSPO inhibition enhanced the male fly life span and inhibited Aβ42-induced neurodegeneration. These effects found in insects [19] are reminiscent of the control of TSPO on apoptosis, life span, and neurodegeneration in mammals [11,21,78]. Interestingly, PPIX can also enhance the lifespan of *Drosophila*, both female and male [79,80]. Data regarding *Drosophila* reported by Curtis *et al.* [81] suggest that upregulation of the mitochondrial antioxidant manganese superoxide dismutase (Mn-SOD) and a retrograde signal of ROS from the mitochondria can serve as intermediate steps in life span extension of *Drosophila*. PPIX, PK 11195, and Ro5-4864 were all found to enhance mitochondrial processing of the hMn-SOD precursor protein, suggesting a role for the TSPO in the regulation of mitochondrial transport of proteins, as well as life span extension [82]. It was also found in *Drosophila* that TSPO serves to counteract infection and promote wound healing [83]. In addition, PPIX can reduce genetic damage in *Drosophila* [81]. In embryonic grasshoppers, PPIX promotes the migration of neurons [84]. In adult mosquitoes, PPIX inhibits heme metabolism [85]. It appears that *Drosophila* presents an interesting model to study TSPO function in a whole animal in association with PPIX effects, in particular regarding health issues.

4. Functional Effects of PPIX and Synthetic TSPO Ligands in Mammals, including Human

4.1. Differences between PPIX and Synthetic Ligands in Interactions with TSPO

Studies regarding TSPO-PPIX interactions have also been applied to human primary cells [86–88]. Moreover, these studies have been done in comparison to synthetic TSPO ligands. First of all, it was shown that TSPO is abundant in primary human osteoblasts in cell culture showing ligand affinity in the nM range as assayed with [^{3}H]PK 11195 [86]. In primary human osteoblasts in culture, following exposure to PPIX (10 µM), cellular [^{18}F]-FDG incorporation, mitochondrial mass, and ATP content were suppressed, indicative of reduced metabolism [87]. Cellular proliferation was not affected. The ΔΨm collapsed, *i.e.*, MPT was enhanced, while no increase in apoptotic cell death was observed. Nonetheless, lactate dehydrogenase activity, indicative of overall cell death, was enhanced in culture media. Accordingly, cell numbers decreased. Protein expression of TSPO, VDAC1, and hexokinase 2 decreased [87]. It was found that the TSPO ligands PK 11195, Ro5-4864, and FGIN-1-27 applied at a 10 µM concentration did not exert the exact same effects as PPIX in the primary osteoblast cell culture [86–88]. For example, PK 11195 did not significantly affect cell death, maturation, [^{18}F]-FDG incorporation, and hexokinase 2 protein expression. There was an increase in mitochondrial mass and mitochondrial ATP content, and a reduction in ΔΨm collapse, *i.e.*, a decrease of MPT. The differences and similarities are discussed in more detail in a recent study submitted for publication [89]. It is considered that the functional differences may due to the structural differences between PPIX, Ro5-4864, PK 11195, and FGIN-1-27 (Figure 2) [89].

By studies on homogenates *in vitro*, as well as on cell culture and animals *in situ*, we have investigated which molecular parts are important for functional effects of tricyclic TSPO ligands based on the quinazoline scaffold [21,22,90,91]. One of these tricyclic TSPO ligands based on the quinazoline scaffold, named 2-Cl-MGV-1 [2-(2-chlorophenyl)quinazolin-4-yl dimethylcarbamate], protected cells of astroglial origin from glutamate-induced cell death, induced differentiation of neuronal progenitor cells, ameliorated behavioral abnormalities of R6-2 mice (a transgenic mouse model for Huntington Disease), and enhanced the life span of R6-2 mice. The ligand 2-Cl-MGV-1 is given as one example for structural characteristics of TSPO ligands in Figure 2. These studies have indicated that the side chains of TSPO ligands affect affinity [22]. The N atoms in the central carbocycles (the second carbocycle of the three carbocycles) affect affinity as well as function, *i.e.*, regulation of programmed cell death. This, for example, has been determined by comparisons of derivatives of phthalazines, quinoxalines, and quinazolines [90]. As 2-Cl-MGV-1 and MGV-1 (2-phenylquinazolin-4-yl dimethylcarbamate) better protect against cell death induction than PK 11195, and also are more effective in the induction of cell differentiation than PK 11195 [21,22], it appears that the location of an N atom on the side of the central carbocycle facing away from the side chain is elementary for proper functioning. This location of N atoms can be also recognized in PPIX (Figure 2), *i.e.*, the N atoms surrounding the center of the PPIX molecule that face away from the side chains at the outside of the cyclic-shaped PPIX. The N atoms at the center of porphyrins are well known to be essential for function, including the holding of metals. Halogenation by a single halogen of the third, rotatable carbocycle in tricyclic TSPO ligands based on a quinazoline scaffold affects function not affinity. In particular, this halogenation prevents cell death induction at higher concentrations in contrast to the typical lethal effects of "normal" TSPO ligands at such concentrations [10,22,26]. Further enhancement of halogenation, from one to more halogens at this part of the compound, reduces affinity in addition to enhanced functional beneficial effects on programmed cell death [85]. This could imply that modifications of cyclic tetrapyrroles that can interact with TSPO could have beneficial effects, for example by halogenation at specific locations and/or by variations of the length and structure of the side chains. A potentially illustrative example of this in Figure 2 is CB256 ((2-(8-(2-(bis(pyridin-2-yl)methyl)amino)acetamido)-2-(4-chlorophenyl) *H*-imidazo[1,2-*a*]pyridin-3-yl)-*N*,*N*-dipropylacetamide). CB256 is a synthetic TSPO ligand based on a tricyclic backbone, with the addition of a side chain presenting two additional carbocycles with heteroatoms (Ns) [46]. One could consider it as a hybrid form of a tricyclic synthetic TSPO ligand

and a tetracyclic TSPO ligand (reminiscent of a tetrapyrrole). *Vice versa*, one could also test whether selected segments of the tetrapyrroles by themselves could function as TSPO ligands.

4.2. Effects of Synthetic TSPO Ligands and the Endogenous TSPO Ligand PPIX in Blood Cells

In addition to human primary osteoblasts, TSPO-specific binding by PK 11195 and PPIX was also found in human mononuclear cells drawn from the blood circulation, a first indication that TSPO may be important for the host defense system [7,92]. Displacement studies applying [^{3}H]PK 11195 indicated that TSPO ligand binding sites recognized by PK 11195 in human lymphocytes are also recognized by Ro5-4864 and endogenous ligands, including PPIX and diazepam binding inhibitor (DBI) [93]. PPIX appears to be specifically targeting mitochondrial TSPO in these cells [93]. Interestingly, while TSPO is an outer mitochondrial membrane protein, Ro5-4864 and PPIX at nanomolar concentrations do inhibit the activity of inner membrane ion channels, namely the multiple conductance channel (MCC) and the mitochondrial centum-picosiemen (mCtS) [94]. PK 11195 inhibits mCtS activity at similar concentrations. Higher concentrations of PPIX induced MCC activity [94], a first indication that TSPO ligands including the endogenous TSPO ligand PPIX can have concentration-dependent bimodal effects, as also found later for protection against and induction of programmed cell death [27]. In addition to the regulation of transport of ions and small molecules over the mitochondrial membranes, the TSPO also appears to regulate the import of proteins into the mitochondria, including the matrix [82]. For example, apart from its function as a channel for ions, MCC can serve as the pore of the import complex present in the inner mitochondrial membrane [95]. In short, MCC is also considered to be a protein import channel. One can ask the question: by which mechanisms can TSPO regulate functions of interacting outer membrane proteins, including specific protein complexes? The question then still remains: by which mechanism can TSPO regulate functions of proteins located at the inner mitochondrial membrane?

In mouse erythroleukemia (MEL) cells it was found that TSPO may be involved in cell differentiation and heme biosynthesis [96]. In this study, RNA blot analysis revealed that treatment of MEL cells with dimethyl sulfoxide to induce differentiation led to an increase in TSPO mRNA levels for up to 72 h, with a concomitant induction of mRNAs for the heme biosynthetic enzymes, coproporphyrinogen oxidase, and ferrochelatase, *i.e.*, modulation of their gene expression. These results suggested that TSPO may be involved in porphyrin transport and may even be a critical factor in erythroid-specific induction of heme biosynthesis. Later studies in human glioblastoma cells suggested that TSPO may be involved in the degradation of excess levels of PPIX, by means of ROS generation [13]. PPIX was also found to induce erythroid differentiation of the human leukemia cell line K562, which was dose-dependently inhibited by PK 11195 [97]. Exposing platelets to PPIX (10 µM) caused a significant decrease in affinity for TSPO ligand binding, whereas the number of TSPO ligand binding sites remained unaltered [98]. Thus, PPIX can interact with TSPO in all types of human blood cells [7]. Studies in other human cells have shown that TSPO is a major participant in the modulation of gene expression in human glioblastoma cells [34–36].

5. Structural Relationships between Synthetic and Endogenous TSPO Ligands and Their Binding Sites

5.1. Differences between PPIX and Synthetic Ligands in Interactions with TSPO

As stipulated at various points above, the synthetic TSPO ligands Ro5-4864, PK 11195, and FGIN-1-27 and the endogenous TSPO ligand PPIX have common but not identical effects. Historically, it was first established by displacement studies (typically applying [^{3}H]PK 11195) that PPIX showed affinity for TSPO, as discussed above. A question that emerged from these studies was whether the binding sites on the TSPO for PPIX were identical or not. Subsequent studies dealt with this question and its ramifications by investigating which parts of the TSPO bound to its ligands. These studies are presented in this Section 5: "Structural relationships between synthetic and endogenous TSPO ligands and their binding sites". These studies may also answer the question why these TSPO ligands affect similar functions, but in different ways (such apparent TSPO functions are discussed above).

In Figure 2, the structural differences between these TSPO ligands are shown. Regarding their binding sites on the TSPOs of various species, the structural relations with their TSPO ligand binding sites of PK 11195 (in mouse and *Bacillus cereus*) and PPIX (in *Rhodobacter sphaeroides*) were experimentally determined by crystallography and electron microscopy [9,69,99–101]. In a first study, Li *et al.* [99] expressed and purified the homologue of mammalian TSPO from *Rhodobacter sphaeroides* (*Rs*TSPO). They constructed a computational model of the *Rs*TSPO dimer using EM-Fold, Rosetta, and a cryo-electron microscopy density map. In this computational model it appeared that *Rs*TSPO binding sites for PK 11195 and PPIX were not identical, *i.e.*, not in the same location and without the same structure [9,99]. Furthermore, they reported that the equilibrium dissociation constant (Kd) for PK 11195 in *Rhodobacter sphaeroides* is 10 μM and for PPIX it is 0.3 μM [99], while for eukaryotes it is in the nM range [1,37]. Regarding mammals, Jaremko *et al.* [100] presented a three-dimensional high-resolution structure of mouse TSPO reconstituted in detergent micelles in complex with PK 11195. This mouse TSPO-PK 11195 structure is described by a tight bundle of TSPO's five transmembrane α helices that form a hydrophobic pocket accepting PK 11195. The TSPO ligand binding site in question completely encapsulates PK 11195 [100]. As PPIX is larger than PK 11195, PPIX would not fit into this pocket. *Vice versa*, theoretically, by nature the larger binding pocket for PPIX would be large enough to accommodate the typical tricycle synthetic TSPO ligand. However, the location and structure of a mammalian PPIX binding site on TSPO has not been described. Interestingly, the mammalian gene for TSPO has 25 times more base pairs than the bacterial gene. Nonetheless, the bacterial and mammalian forms of TSPO by themselves present about the same number of amino acids and molecular weight (Table 1). Thus, one would expect that a PPIX binding site could be on the mammalian TSPO, even though its location and structure thus far have not been established. Thus, it is obvious that more studies dealing with structural ligand binding site interactions must be awaited to resolve such questions.

Studying *Rhodobacter sphaeroides*, Hinsen *et al.* [101] applied cryo-electron microscopy to tubular crystals of TSPO with lipids and analyzed possible ligand binding sites for protoporphyrin, PK 11195, and cholesterol, which appeared to not be identical, neither among each other nor with their mammalian counterparts. Guo *et al.* [69] reported on crystal structures for *Bacillus cereus* TSPO (*Bc*TSPO) down to a 1.7 Å resolution, including a complex with PK 11195. They also described *Bc*TSPO-mediated catalytic degradation of PPIX and showed that TSPO from *Bacillus cereus*, *Xenopus*, and human have similar PPIX-directed activities. In addition it was shown in *Rhodobacter sphaeroides* that TSPO appears to present a pocket that would be able to completely encapsulate an endogenous tetrapyrrole (porphyrin ligand) [9]. Keeping this in mind, it would be highly desirable to establish structural relations between PPIX and TSPO also in mammals, including explanations of the causes of binding competition between the endogenous PPIX and synthetic TSPO ligands. The recent studies on structural TSPO-ligand interactions provide very interesting data and interpretations. Nonetheless, it is also recognized that interpretation of the structure of a TSPO ligand binding site, including the interaction with the ligand, includes the bias of the observer [102]. In this respect, more studies are needed to address points of contention. As the list of extant TSPO ligands has become very long, it would be worthwhile to establish an effective computational approach to rigorously establish structural and functional commonalities and differences between all ligands, including their binding interactions with the TSPO.

Another essential question is how concentration-dependent bimodal effects of TSPO ligands come about. This is important because, for example, for drug development it is desirable that no adverse effects are induced at high doses, *i.e.*, accidental as well as intentional overdoses need to be precluded. In this context it was found that halogenation of the third, rotatable carbocycle of specific tricyclic TSPO ligands prevented lethal effects of otherwise identical ligands [22]. Also, moderation of the affinity of the ligands by specific modifications contributed to a reduction of adverse effects, while beneficial effects were promoted [22]. This may be important information for the selection of effective TSPO ligand–based drugs for treatments of various diseases.

5.2. Accommodation of Various Types of TSPO Ligands by Binding Sites on the TSPO

As mentioned, when looking at Figure 2, one can see that all tricyclic synthetic TSPO ligands shown have slightly different shapes. Regarding the tricyclic compounds, their ground form varies from hooked to straight, with acute angles, right angles, obtuse angles and straight angles at their connection points between the third carbocycle and the first two carbocycles. Furthermore, TSPO ligands can range from single carbocycle forms to multicarbocycle forms, including but not restricted to tetrapyrroles [7,46,103–105]. Looking at the PK 11195 binding pocket of mouse TSPO (as described by Jaremko *et al.* [101]), it would seem that none of these other TSPO ligands would comfortably fit into this PK 11195 binding site. This would, for example, be particularly true for the multicarbocycle forms of CB256 [92] and ZBD-2 [105], which are larger than PK 11195. Nonetheless, for example, ZBD-2 presents an affinity similar to PK 11195 [106].

One possibility is that the TSPO binding site can conform its shape to the particular shape of at least the tricyclic TSPO ligands. It is known that TSPO exchanges between multiple conformations in the absence of ligands [107], so it theoretically is possible that TSPO can conform its shape to various ligands. Alternatively, each ligand has a separate site on the TSPO. It is known that there are several binding sites for various molecules on the TSPO, including retinoic acid, curcumin, and a known Bcl-2 inhibitor, gossypol, apart from cholesterol, PPIX, and PK 11195, at least in *Rhodobacter sphaeroides* [99]. So the latter assumption that each ligand has its own docking site on the TSPO is also theoretically possible. In particular, bacterial studies regarding TSPO and its interactions with PPIX have provided us with titillating bits of information, but as of yet are not at a stage where they can solve all the questions regarding binding sites for TSPO ligands and interactions with tetrapyrroles including PPIX. Furthermore, it has been suggested that ligand binding sites on mammalian TSPO with high affinity for specific TSPO ligands are not presented by TSPO of all bacterial species [1]. More studies in prokaryotes as well as eukaryotes are needed to attain a better understanding of interactions between TSPO and its numerous binding agents.

In extensive studies, where the effects on affinity and function following small modifications of specific tricyclic cores, side chains, and halogenations of potential TSPO ligands were analyzed (as part of a project to design TSPO ligands with curative properties), it was extrapolated which components of TSPO ligands contributed to their affinity and which ones to function [21,22,90,91]. The side chain and the N atoms in the second carbocycle appeared to be important for affinity. The third carbocycle, including its halogenation, and also the N atoms in the second carbocycle were found to be important for function. We initially assumed that the first two carbocycles inserted into a binding pocket in the TSPO, while the third carbocycle would remain outside. In this way, the third carbocycle would present an element that could interact with the milieu surrounding the TSPO. The recent structural studies, however, show that while the first two carbocycles of PK 11195 indeed insert deep into a binding pocket in the TSPO, the remainder of the ligand is also encapsulated by the TSPO [9,69,99–101]. The question remains as to how the N atoms of the second carbocycle and the halogenation of the third carbocycle actually contribute to the functional characteristics of the ligands at hand. As PPIX also presents N atoms in its carbocycles and side chains, we assume that answers to the questions may also be relevant for understanding PPIX-TSPO interactions and their effects.

The observations of various binding sites for various factors on the TSPO may present some explanation for the context-dependent responses of TSPO to various stimuli. While this on the one hand defines the complexity of TSPO characteristics, making it enigmatic for our understanding [1], on the other hand it exemplifies TSPO's functional versatility against the various life-threatening challenges for animals as well as plants and bacteria [7]. This evolutionarily conserved protein and its endogenous ligands, including tetrapyrroles such as PPIX, allow organisms from prokaryotes to eukaryotes to respond not just to toxic environmental insults presented by chemicals to unicellar organisms, but also to challenges to multicellular organisms, even to physical and mental injuries emanating from human social interactions (ranging from inflammation and traumatic brain injury to maladaptive responses to stress, including anxiety).

6. Bacteria, TSPO, and Tetrapyrroles

TSPO has also been identified in the genome of bacteria (Table 1). As reviewed by Li *et al.* [99], TSPO was discovered in the carotenoid gene cluster known as CrtK in purple non-sulfur bacteria [108]. Purple non-sulfur bacteria are considered to be closely related to the free living bacterial ancestors of mitochondria [109]. TSPO was discovered first in *Rhodobacter capsulatus* and *Rhodobacter sphaeroides* [108,110,111]. The acronym TSPO is originally derived from the name tryptophan-rich sensory protein (TspO) which was first applied to these bacteria [4,110,111]. In 2009, Chapalain *et al.* [112] identified 98 bacteria presenting TSPO in their genome. Bacterial TSPO typically appears to be organized as a dimer, and in general the affinity for PK 11195 is in the same nM range as for eukaryotic TSPO. Treating *Pseudomonas fluorescens* MF37 with PK 11195 (10^{-5} M) increased adhesion to living or artificial surfaces and biofilm formation activity; at the same time, the apoptotic potential of bacteria on eukaryotic cells was significantly reduced [112]. As several structural and functional characteristics are shared with the mammalian TSPO, they may present an original source of TSPO function (see also Table 2). It appears that sometime during evolution, relationships between eukaryotes and particular prokaryotes changed from lethal to symbiotic (Figures 1 and 3). For example, eukaryotes ingesting prokaryotes did not digest them or prokaryotes originally clinging to eukaryotes and inducing programmed cell death become internalized endosymbionts and typically no longer induce programmed cell death. Assuming that PPIX and TSPO coexisted at the time point of this evolutionary transition, they may have been important participants in the establishment of this endosymbiotic relationship between bacteria and eukaryotes.

As TSPO of Archeobacteria is homologous to TSPO of species from the other kingdoms (Table 1), TSPO indeed is an evolutionarily conserved protein [1,2]. Moreover, rat TSPO has been shown to substitute for TSPO in *Rhodobacter sphaeroides* (*Rs*TSPO), also indicating conserved functional characteristics [113]. Similarly, PPIX is also present from archea to eukaryotes [114]. It has been recognized that the location of TSPO in bacteria and eukaryotes is similar. As in mitochondria, in bacteria including *Rhodobacter* TSPO is located in the outer membrane [113,115]. *Rhodobacter* TSPO is involved in regulating photosynthetic gene expression in response to oxygen and light conditions [113,115]. It has been suggested that TSPO-porphyrin interactions underlie the regulation of this gene expression, allowing *Rhodobacter* to switch from oxygen respiration to photosynthesis and back [116]. A question is whether tetrapyrroles act on TSPO as ligands to induce TSPO to activate its function as a nuclear gene expression regulator, or whether tetrapyrroles present intermediate steps for the mechanisms whereby TSPO regulates nuclear gene expression. Similar to its mammalian and plant orthologs, in *Rhodobacter sphaeroides* TSPO appears to be involved in the transport of small molecules such as porphyrin intermediates of the heme and in chlorophyll biosynthesis-degradation pathways [115,117]. *Rs*TSPO shares considerable sequence homology with human TSPO (*Hs*TSPO). Apart from an overall significant level of sequence identity (30%), *Rs*TSPO presents particular sequence similarity with *Hs*TSPO in the first extra-membrane loop (loop 1) considered to participate in porphyrin binding [117], synthetic TSPO ligand binding [118], and cholesterol binding [119]. Regarding *Rs*TSPO, it has been suggested that PK 11195 and PPIX interact with different sets of tryptophans [99]. Furthermore, the Kd in *Rhodobacter sphaeroides* for PK 11195 is 10 μM and for PPIX it is 0.3 μM, *i.e.*, not the same as for eukaryotes, where these Kds are in the nM range [99]. *E. coli* does not even present appreciable TSPO ligand binding [1]. Nonetheless, as mentioned above, Kd for TSPO in bacteria typically appears to be in the nM range [112]. As discussed above, high affinity and low affinity TSPO ligand binding sites can also be found in mammals [10,28,60,62–67]. Thus, physiological functions of TSPO and tetrapyrroles in bacteria are reminiscent of those in animals (see also Table 2). It indeed is tempting to correlate this with the bacterial origin of mitochondria, including their TSPO [99] (see also Section 8 dealing with "TSPO-tetrapyrrole interactions from an evolutionary perspective").

7. Plants, TSPO, and Tetrapyrroles

Plant TSPO homologs, with a molecular size of 18–20 kDa, show 15%–22% identical residues with bacterial and mammalian TSPO [120]. Interestingly, the various plant TSPO homologs present a distal segment of 40–50 amino acids, which is not identical to the bacterial and mammalian counterparts [120,121]. Similar to the mammalian and bacterial TSPO, plant TSPO homologs have a high affinity binding site, and a second binding site with a low affinity (μM range) has also been reported [120]. Several endogenous TSPO ligands were found in plants, such as benzodiazepines (including delorazepam and temazepam), porphyrins, cholesterol, and diazepam binding inhibitor (DBI) [120]. Regarding animal TSPO research, no endogenous, diazepam-like TSPO ligands have been identified in animals to date. TSPO cellular localization in plants can be in the mitochondria and plastids, as well as in the nuclear fraction, the endoplasmatic reticulum, and the Golgi stacks [120,122,123]. These various localizations may relate to TSPO's function in the translocation of tetrapyrrole intermediates across organelle membranes [124]. In this context, it was shown, both *in vitro* and *in vivo*, that the plant TSPO is able to bind heme and PPIX [125].

Four major tetrapyrroles are biosynthesized in the plastids of higher plants: chlorophyll, heme, siroheme, and phytochromobilin [124]. Among the plant tetrapyrroles, chlorophyll and siroheme act in plastids, heme is universally distributed to all cellular compartments, and phytochromobiline occurs in the cytoplasm [47,126]. Regarding *Arabidopsis thaliana* TSPO (*At*TSPO), several genes related to tetrapyrrole biosynthesis were downregulated in an *At*TSPO knockdown cell line, indicative of TSPO regulation of the tetrapyrrole metabolism [124]. *Vice versa*, treatment of the wild-type plants with tetrapyrrole biosynthesis inhibitors increased TSPO mRNA levels. Furthermore, mutations in different genes for tetrapyrrole metabolism also affect TSPO expression levels. For example, mutations in *gun4* (protoporphyrin IX- and Mg-Protoporphyrin IX-binding protein) lead to increased TSPO levels [124]. *At*TSPO levels are regulated at the transcriptional, post-transcriptional and post-translational levels in response to abiotic stress conditions [124]. Feeding the PPIX precursor 5-aminolevulinic acid (ALA) to *Arabidopsis thaliana* seeds enhanced porphyrin biosynthesis as well as downregulation of AtTSPO, and improved salt tolerance [123,124]. Heme synthesis was suggested to be responsible for TSPO downregulation [125]. Treatment of *Arabidopsis* cell culture with inhibitors of porphyrin biosynthesis significantly increased *At*TSPO expression, compared to the control [125].

It was proposed that TSPO serves to protect seed germination from the toxic effects of tetrapyrroles [123]. Plant TSPO is able to bind heme, as for example is required for TSPO degradation through autophagy [125,126]. In an *Arabidopsis* transgenic cell line over-expressing TSPO, ROS levels were found to be higher than in the wild type. This suggests that the plant TSPO, which is considered a heme scavenger, may mediate ROS homeostasis [127]. Moreover, TSPO levels are decreased 48 h after abscisic acid (ABA)-induced stress [125]. It is possible that heme levels are regulated by the over-expressed TSPO during ABA stress [128]. Further evidence for TSPO-tetrapyrrole interactions related to stress came from studies on the moss *Physcomitrella patens*, which has three TSPO homologs [129]. Under stress conditions, the *Pp*TSPO1 null mutants show elevated H_2O_2 levels, enhanced lipid peroxidation and cell death, indicating an important role of *Pp*TSPO1 in redox homeostasis. Furthermore, in *Physcomitrella patens*, knockout of one of its TSPOs led to increased levels of heme and PPIX [129]. These knockdown mutants had higher activity of class III peroxidase (PRX34), which is produced as a defense response to pathogens and is responsible for the oxidative burst response. Over-expression of other oxidative stress-related genes was seen in these mutants [130]. Overall, it appears that stress reduction leads to decreased levels of TSPO. These various studies in plants suggest that, possibly, increased levels of TSPO serve to reduce the stress response, which can be prevented by TSPO knockout and TSPO knockdown.

Furthermore, TSPO expression appears to be related to plant development. For example, dry seeds have high levels of TSPO. In contrast, TSPO levels in plantlets and leaves are undetectable [121]. This is reminiscent of neurodevelopment observed for mouse cells, where TSPO is abundant in neural progenitor cells but not detected in the derived healthy, mature neurons [32]. Tetrapyrroles also may be inseparable from development. For example, siroheme deficiency affects plant growth and development [131], reminiscent of PPIX promoting migration of neurons in embryonic insects [84].

Thus, physiological functions of TSPO in plants, including its interactions with tetrapyrroles, appear to be comparable to what is observed in animals and bacteria. Such functions range from embryonic development to the stress response in adult life (see also Table 2). Obviously, more studies in plants regarding TSPO-tetrapyrrole interactions, by themselves and in comparison to such interactions in animals and bacteria, will give very interesting insights in the life-supporting functions of TSPO. This will have implications in future approaches to human health issues, as well as agricultural production, including plant cultivation and animal breeding. Such research comparing bacterial, plant, and animal TSPO characteristics will also have implications for our understanding of TSPO evolution, including its interactions with tetrapyrroles.

8. TSPO-Tetrapyrrole Interactions from an Evolutionary Perspective

As valid for animal mitochondria endosymbiosis, plastids and mitochondria of plants also derive from prokaryotic symbionts [132,133]. Without wanting to go too much into detail of what is known regarding the organelle evolution of eukaryotes, we mention here that plants present the oldest eukaryote fossils. Microfossils of algae have been found in *ca.* 1.5-billion-year-old rocks in northern Australia [132,133]. Moreover, steroid molecules preserved as steranes have been identified in samples of sediments from 2.5 to 2.8 billion years old [134]. Thus, these studies suggest that eukaryotic algae already existed during the Late Archean period, implying that the endosymbiotic relation of eukaryotes and prokaryotes already had been established at that time. Interesting for our perspective from TSPO research, in particular regarding TSPO functions associated with steroidogenesis, is the early occurrence of steroid production in the geological record. It has been appreciated for some time that evolutionary aspects of TSPO can have implications for our understanding of TSPO function [1,135]. Regarding prokaryotes, present day cyanobacteria possess TSPOs that appear to be associated with stress, ROS generation, cell cycle regulation, heme metabolism, and homeostasis [136–138], reminiscent of TSPO functions in eukaryotes (see also Table 2, Figures 1 and 3). It appears that iron ores have been deposited by cyanobacteria already around four billion years ago, as a consequence of typical metabolic characteristics of cyanobacteria [139]. So, even as no direct evidence is available, it can be suggested that TSPO has been present for about four billion years in the evolution of Earth's living organisms [139,140].

These studies, suggesting the early existence of the bacterial ancestors of the endosymbiotic mitochondria and plastids, allow for the assumption that eukaryote evolution, including endosymbiosis of mitochondria and plastids with their TSPO, may indeed have been initiated early in the Earth's existence. The same appears to be true for PPIX. For example, the phylogenetic distribution of enzymes for the tetrapyrrole biosynthesis pathway, as described by Kobayashi *et al.* [141], parallels well-acknowledged views of evolution on mitochondrial and plastid endosymbiosis [132,133]. Briefly, in the tetrapyrrole biosynthesis pathway, protoporphyrinogen IX oxidase (Protox), present in mitochondria and plastids, catalyzes the formation of PPIX, the last common intermediate for the biosynthesis of heme and chlorophyll. In particular the phylogenetic distribution of HemY, one of the three nonhomologous isofunctional Protox forms (HemG, HemJ, and HemY), reflects endosymbiosis with bacteria in the evolution of eukaryotes. HemY is ubiquitous in prokaryotes, including cyanobacteria and purple bacteria, and is the only Protox in eukaryotes. These studies [132,133,141] suggest that HemY may be the ancestral bacterial Protox form nowadays present in the mitochondria and plastids of eukaryotes. Furthermore, regarding genes for the synthesis of bacteriochlorophyll from PPIX, by phylogenetic analysis of available genomic data it can be postulated that the transfer of such genes took place from cyanobacteria to purple non-sulfur phototropic bacteria (*Rhodobacter* and *Rhodopseudomonas*), at least as early as the Proterozoic era [142], which ranges from 2500 to 542.0 ± 1.0 million years ago, presenting in the most recent part of the Precambrian. Also from a phylogenetic perspective, studies on *Rhodobacter sphaeroides* have suggested that TSPO's association with larger membrane channels, such as VDAC in eukaryotes and porin in prokaryotes, can be considered a conserved TSPO characteristic [140]. This is particularly so as purple non-sulfur phototropic bacteria are considered the free living ancestral form of mitochondria [109].

Thus, we see the emergence of a body of data indicating that vital functions of TSPO such as modulation of membrane potential, steroidogenesis, and tetrapyrrole synthesis appear to already have been present very early in the geological history. As in present day organisms, these functions are unequivocally associated with the TSPO in bacteria, as well as in the mitochondria and plastids of eukaryotes, and it can be assumed that TSPO was present early on in the geological history of the Earth. A time point of 3.5 billion years ago has been postulated [140]. The same can be said for the existence of tetrapyrroles such as PPIX. Thus, TSPO-tetrapyrrole interactions indeed may have originated very early on in the evolution of living organisms.

9. Conclusions and Perspectives

In conclusion, TSPO and tetrapyrroles including PPIX are universally distributed, apparently in all six kingdoms of living organisms. Importantly, PPIX and TSPO demonstrate structural and functional interactions. Their life-supporting functions are of a wide variety, including gene expression, membrane functions, and programmed cell death, that translate to the regulation of homeostasis including adequate responses to environmental challenges. In this context, TSPO can be considered a receptor for PPIX, a transporter for tetrapyrroles, and a participant in the regulation of tetrapyrrole metabolism; *vice versa*, tetrapyrroles can modulate TSPO functions. Figure 3 presents a schematic overview of the evolution of endosymbiosis, from free living bacteria to mitochondria and plastids in animals and plants, including associations with TSPO functions.

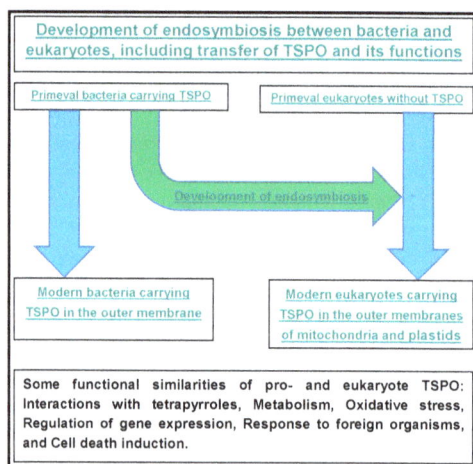

Figure 3. This scheme very concisely summarizes evolutionary relations between bacteria and eukaryotes regarding the presence of TSPO in these life forms, as well as the associated functions. In particular, it appears that during evolution, originally free living bacteria with TSPO became part of eukaryotes in the form of organelles, such as mitochondria and plastids with TSPO. In eukaryotes the TSPO functions that can be found in bacteria appear to basically have been maintained in cellular organelles. Beyond this, in prokaryotes as well eukaryotes, including multicellular organisms, TSPO serves to maintain homeostasis and viability.

Within this perspective it appears that further research, basic as well as applied, may give relevant information on how these functional relations between tetrapyrroles and TSPO can give insights into various biological mechanisms relevant for human health issues as well as agricultural advances. As TSPO-tetrapyrrole interactions appear to relate to eukaryote-prokaryote endosymbiotic relations (Figure 1), such research may also give insights into basic biology questions ranging from evolution to ecology. A better understanding of structure-function interactions between TSPO and its endogenous ligands such as tetrapyrroles, including PPIX, may aid in the development of new synthetic TSPO ligands as adequate drugs for treatments of various diseases. Some basic elements for such structural characteristics are indicated in Figure 2. It is noteworthy that diseases that have been associated with TSPO and its functions present a very broad range, for example including but not restricted to: porphyrias, developmental disorders, inflammatory diseases, cancer, and neuropathological disorders, including maladaptive responses to stressors. Finally, the application of TSPO ligands with efficacious curative effects may enhance the life span of patients as well as healthy individuals.

Acknowledgments: This work is supported in part by a joint grant from the Center for Absorption in Science of the Ministry of Immigrant Absorption and the Committee for Planning and Budgeting of the Council for Higher Education under the framework of the KAMEA program (Leo Veenman, Moshe Gavish). The Israel Science Foundation is thankfully acknowledged for their support for this project (Leo Veenman, Moshe Gavish).

Author Contributions: Leo Veenman conceived the subject, did the literature research and analysis, and designed its presentation. Alex Vainshtein dealt with general TSPO characteristic and functions. Nasra Yasin focused on studies pertaining to TSPO homology in phylogeny, including bacteria. Maya Azrad studied TSPO-tetrapyrrole relations in plants. Leo Veenman and Alex Vainshtein prepared the figures. Leo Veenman and Nasra Yasin prepared the tables. Leo Veenman, Maya Azrad, Alex Vainshtein, and Nasra Yasin wrote the paper. Moshe Gavish contributed tools for the research, design and presentation, and critically read and commented.

Conflicts of Interest: The authors declare no conflict of interest.

References

1. Gavish, M.; Bachman, I.; Shoukrun, R.; Katz, Y.; Veenman, L.; Weisinger, G.; Weizman, A. Enigma of the peripheral benzodiazepine receptor. *Pharmacol. Rev.* **1999**, *51*, 629–650. [PubMed]
2. Fan, J.; Lindemann, P.; Feuilloley, M.G.; Papadopoulos, V. Structural and functional evolution of the translocator protein (18 kDa). *Curr. Mol. Med.* **2012**, *12*, 369–386. [CrossRef] [PubMed]
3. Braestrup, C.; Squires, R.F. Specific benzodiazepine receptors in rat brain characterized by high-affinity (3H)diazepam binding. *Proc. Natl. Acad. Sci. USA* **1977**, *74*, 3805–3809. [CrossRef] [PubMed]
4. Papadopoulos, V.; Baraldi, M.; Guilarte, T.R.; Knudsen, T.B.; Lacapère, J.J.; Lindemann, P.; Norenberg, M.D.; Nutt, D.; Weizman, A.; Zhang, M.R.; *et al.* Translocator protein (18 kDa): New nomenclature for the peripheral-type benzodiazepine receptor based on its structure and molecular function. *Trends Pharmacol. Sci.* **2006**, *27*, 402–409. [CrossRef] [PubMed]
5. Anholt, R.R.; Pedersen, P.L.; De Souza, E.B.; Snyder, S.H. The peripheral-type benzodiazepine receptor. Localization to the mitochondrial outer membrane. *J. Biol. Chem.* **1986**, *261*, 576–583. [PubMed]
6. McEnery, M.W.; Snowman, A.M.; Trifiletti, R.R.; Snyder, S.H. Isolation of the mitochondrial benzodiazepine receptor: Association with the voltage-dependent anion channel and the adenine nucleotide carrier. *Proc. Natl. Acad. Sci. USA* **1992**, *89*, 3170–3174. [CrossRef] [PubMed]
7. Veenman, L.; Gavish, M. The peripheral-type benzodiazepine receptor and the cardiovascular system. Implications for drug development. *Pharmacol. Ther.* **2006**, *110*, 503–524. [CrossRef] [PubMed]
8. Verma, A.; Nye, J.S.; Snyder, S.H. Porphyrins are endogenous ligands for the mitochondrial (peripheral-type) benzodiazepine receptor. *Proc. Natl. Acad. Sci. USA* **1987**, *84*, 2256–2260. [CrossRef] [PubMed]
9. Li, F.; Liu, J.; Zheng, Y.; Garavito, R.M.; Ferguson-Miller, S. Protein structure. Crystal structures of translocator protein (TSPO) and mutant mimic of a human polymorphism. *Science* **2015**, *347*, 555–558. [CrossRef] [PubMed]
10. Veenman, L.; Papadopoulos, V.; Gavish, M. Channel-like functions of the 18-kDa translocator protein (TSPO): Regulation of apoptosis and steroidogenesis as part of the host-defense response. *Curr. Pharm. Des.* **2007**, *13*, 2385–2405. [CrossRef] [PubMed]

11. Veenman, L.; Gavish, M.; Kugler, W. Apoptosis induction by erucylphosphohomocholine via the 18 kDa mitochondrial translocator protein: Implications for cancer treatment. *Anticancer Agents Med. Chem.* **2014**, *14*, 559–577. [CrossRef] [PubMed]

12. Papadopoulos, V.; Lecanu, L. Translocator protein (18 kDa) TSPO: An emerging therapeutic target in neurotrauma. *Exp. Neurol.* **2009**, *219*, 53–57. [CrossRef] [PubMed]

13. Zeno, S.; Veenman, L.; Katz, Y.; Bode, J.; Gavish, M.; Zaaroor, M. The 18 kDa mitochondrial translocator protein (TSPO) prevents accumulation of protoporphyrin IX. Involvement of reactive oxygen species (ROS). *Curr. Mol. Med.* **2012**, *12*, 494–501. [CrossRef] [PubMed]

14. Zeno, S.; Zaaroor, M.; Leschiner, S.; Veenman, L.; Gavish, M. CoCl$_2$ induces apoptosis via the 18 kDa translocator protein in U118MG human glioblastoma cells. *Biochemistry* **2009**, *48*, 4652–4661. [CrossRef] [PubMed]

15. Fares, F.; Bar-Ami, S.; Brandes, J.M.; Gavish, M. Gonadotropin- and estrogen-induced increase of peripheral-type benzodiazepine binding sites in the hypophyseal-genital axis of rats. *Eur. J. Pharmacol.* **1987**, *133*, 97–102. [CrossRef]

16. Gavish, M.; Laor, N.; Bidder, M.; Fisher, D.; Fonia, O.; Muller, U.; Reiss, A.; Wolmer, L.; Karp, L.; Weizman, R. Altered platelet peripheral-type benzodiazepine receptor in posttraumatic stress disorder. *Neuropsychopharmacology* **1996**, *14*, 181–186. [CrossRef]

17. Veenman, L.; Gavish, M. Peripheral-type benzodiazepine receptors: Their implication in brain disease. *Drug Dev. Res.* **2000**, *50*, 355–370. [CrossRef]

18. Veenman, L.; Gavish, M. The role of 18 kDa mitochondrial translocator protein (TSPO) in programmed cell death, and effects of steroids on TSPO expression. *Curr. Mol. Med.* **2012**, *12*, 398–412. [CrossRef] [PubMed]

19. Lin, R.; Angelin, A.; da Settimo, F.; Martini, C.; Taliani, S.; Zhu, S.; Wallace, D.C. Genetic analysis of dTSPO, an outer mitochondrial membrane protein, reveals its functions in apoptosis, longevity, and Ab42-induced neurodegeneration. *Aging Cell* **2014**, *13*, 507–518. [CrossRef] [PubMed]

20. Milenkovic, V.M.; Rupprecht, R.; Wetzel, C.H. The Translocator protein 18 kDa (TSPO) and its role in mitochondrial biology and psychiatric disorders. *Mini Rev. Med. Chem.* **2015**, *15*, 366–372. [CrossRef] [PubMed]

21. Veenman, L.; Vainshtein, A.; Gavish, M. TSPO as a target for treatments of diseases, including neuropathological disorders. *Cell Death Dis.* **2015**, *6*, e1911. [CrossRef] [PubMed]

22. Vainshtein, A.; Veenman, L.; Shterenberg, A.; Singh, S.; Masarwa, A.; Dutta, B.; Island, B.; Tsoglin, E.; Levin, E.; Leschiner, S.; *et al.* Quinazoline-based tricyclic compounds that regulate programmed cell death, induce neuronal differentiation, and are curative in animal models for excitotoxicity and hereditary brain disease. *Cell Death Discov.* **2015**, *1*, 15027. [CrossRef]

23. Veenman, L.; Levin, E.; Weisinger, G.; Leschiner, S.; Spanier, I.; Snyder, S.H.; Weizman, A.; Gavish, M. Peripheral-type benzodiazepine receptor density and *in vitro* tumorigenicity of glioma cell lines. *Biochem. Pharmacol.* **2004**, *68*, 689–698. [CrossRef] [PubMed]

24. Veenman, L.; Shandalov, Y.; Gavish, M. VDAC activation by the 18 kDa translocator protein (TSPO), implications for apoptosis. *J. Bioenerg. Biomembr.* **2008**, *40*, 199–205. [CrossRef] [PubMed]

25. Levin, E.; Premkumar, A.; Veenman, L.; Kugler, W.; Leschiner, S.; Spanier, I.; Weisinger, G.; Lakomek, M.; Weizman, A.; Snyder, S.H.; *et al.* The peripheral-type benzodiazepine receptor and tumorigenicity: Isoquinoline binding protein (IBP) antisense knockdown in the C6 glioma cell line. *Biochemistry* **2005**, *44*, 9924–9935. [CrossRef] [PubMed]

26. Kugler, W.; Veenman, L.; Shandalov, Y.; Leschiner, S.; Spanier, I.; Lakomek, M.; Gavish, M. Ligands of the mitochondrial 18 kDa translocator protein attenuate apoptosis of human glioblastoma cells exposed to erucylphosphohomocholine. *Cell Oncol.* **2008**, *30*, 435–450. [PubMed]

27. Caballero, B.; Veenman, L.; Gavish, M. Role of mitochondrial translocator protein (18 kDa) on mitochondrial-related cell death processes. *Recent Pat. Endocr. Metab. Immune Drug Discov.* **2013**, *7*, 86–101. [CrossRef] [PubMed]

28. Caballero, B.; Veenman, L.; Bode, J.; Leschiner, S.; Gavish, M. Concentration-dependent bimodal effect of specific 18 kDa translocator protein (TSPO) ligands on cell death processes induced by ammonium chloride: Potential implications for neuropathological effects due to hyperammonemia. *CNS Neurol. Disord. Drug Targets* **2014**, *13*, 574–592. [CrossRef] [PubMed]

29. Mukhin, A.G.; Papadopoulos, V.; Costa, E.; Krueger, K.E. Mitochondrial benzodiazepine receptors regulate steroid biosynthesis. *Proc. Natl. Acad. Sci. USA* **1989**, *86*, 9813–9816. [CrossRef] [PubMed]

30. Nothdurfter, C.; Baghai, T.C.; Schüle, C.; Rupprecht, R. Translocator protein (18 kDa) (TSPO) as a therapeutic target for anxiety and neurologic disorders. *Eur. Arch. Psychiatry Clin. Neurosci.* **2012**, *262*, 107–112. [CrossRef] [PubMed]

31. Papadopoulos, V.; Aghazadeh, Y.; Fan, J.; Campioli, E.; Zirkin, B.; Midzak, A. Translocator protein-mediated pharmacology of cholesterol transport and steroidogenesis. *Mol. Cell. Endocrinol.* **2015**, *408*, 90–98. [CrossRef] [PubMed]

32. Varga, B.; Markó, K.; Hádinger, N.; Jelitai, M.; Demeter, K.; Tihanyi, K.; Vas, A.; Madarász, E. Translocator protein (TSPO 18 kDa) is expressed by neural stem and neuronal precursor cells. *Neurosci. Lett.* **2009**, *462*, 257–262. [CrossRef] [PubMed]

33. Manku, G.; Wang, Y.; Thuillier, R.; Rhodes, C.; Culty, M. Developmental expression of the translocator protein 18 kDa (TSPO) in testicular germ cells. *Curr. Mol. Med.* **2012**, *12*, 467–475. [CrossRef] [PubMed]

34. Veenman, L.; Bode, J.; Gaitner, M.; Caballero, B.; Pe'er, Y.; Zeno, S.; Kietz, S.; Kugler, W.; Lakomek, M.; Gavish, M. Effects of 18-kDa translocator protein knockdown on gene expression of glutamate receptors, transporters, and metabolism, and on cell viability affected by glutamate. *Pharmacogenet. Genom.* **2012**, *22*, 606–619. [CrossRef] [PubMed]

35. Yasin, N.; Veenman, L.; Gavish, M. Regulation of nuclear gene expression by PK 11195, a ligand specific for the 18 kDa mitochondrial translocator protein (TSPO). In Proceedings of the Annual Meeting of the Israel Society for Neuroscience, Eilat, Israel, 20–22 December 2015; Abstract #98.

36. Yasin, N.; Veenman, L.; Gavish, M. NCBI. Available online: http://www.ncbi.nlm.nih.gov/geo/query/acc. cgi?acc=GSE77998 (accessed on 2 June 2016).

37. Verma, A.; Snyder, S.H. Characterization of porphyrin interactions with peripheral type benzodiazepine receptors. *Mol. Pharmacol.* **1988**, *34*, 800–805. [PubMed]

38. Mantione, C.R.; Goldman, M.E.; Martin, B.; Bolger, G.T.; Lueddens, H.W.; Paul, S.M.; Skolnick, P. Purification and characterization of an endogenous protein modulator of radioligand binding to "peripheral-type" benzodiazepine receptors and dihydropyridine Ca^{2+}-channel antagonist binding sites. *Biochem. Pharmacol.* **1988**, *37*, 339–347. [CrossRef]

39. Alho, H.; Fremeau, R.T., Jr.; Tiedge, H.; Wilcox, J.; Bovolin, P.; Brosius, J.; Roberts, J.L.; Costa, E. Diazepam binding inhibitor gene expression: Location in brain and peripheral tissues of rat. *Proc. Natl. Acad. Sci. USA* **1988**, *85*, 7018–7022. [CrossRef] [PubMed]

40. Ball, J.A.; Ghatei, M.A.; Sekiya, K.; Krausz, T.; Bloom, S.R. Diazepam binding inhibitor-like immunoreactivity (51–70): Distribution in human brain, spinal cord and peripheral tissues. *Brain Res.* **1989**, *479*, 300–305. [CrossRef]

41. Slobodyansky, E.; Guidotti, A.; Wambebe, C.; Berkovich, A.; Costa, E. Isolation and characterization of a rat brain triakontatetraneuropeptide, a posttranslational product of diazepam binding inhibitor: Specific action at the Ro 5-4864 recognition site. *J. Neurochem.* **1989**, *53*, 1276–1284. [CrossRef] [PubMed]

42. Bovolin, P.; Schlichting, J.; Miyata, M.; Ferrarese, C.; Guidotti, A.; Alho, H. Distribution and characterization of diazepam binding inhibitor (DBI) in peripheral tissues of rat. *Regul. Pept.* **1990**, *29*, 267–281. [CrossRef]

43. Perrone, M.; Moon, B.S.; Park, H.S.; Laquintana, V.; Jung, J.H.; Cutrignelli, A.; Lopedota, A.; Franco, M.; Kim, S.E.; Lee, B.C.; *et al.* A novel PET imaging probe for the detection and monitoring of translocator protein 18 kDa expression in pathological disorders. *Sci. Rep.* **2016**, *6*, 20422. [CrossRef] [PubMed]

44. Le Fur, G.; Vaucher, N.; Perrier, M.L.; Flamier, A.; Benavides, J.; Renault, C.; Dubroeucq, M.C.; Guérémy, C.; Uzan, A. Differentiation between two ligands for peripheral benzodiazepine binding sites, [^{3}H]RO5-4864 and [^{3}H]PK 11195, by thermodynamic studies. *Life Sci.* **1983**, *33*, 449–457. [CrossRef]

45. Romeo, E.; Cavallaro, S.; Korneyev, A.; Kozikowski, A.P.; Ma, D.; Polo, A.; Costa, E.; Guidotti, A. Stimulation of brain steroidogenesis by 2-aryl-indole-3-acetamidederivatives acting at the mitochondrial diazepam-binding inhibitor receptor complex. *J. Pharmacol. Exp. Ther.* **1993**, *267*, 462–471. [PubMed]

46. Denora, N.; Margiotta, N.; Laquintana, V.; Lopedota, A.; Cutrignelli, A.; Losacco, M.; Franco, M.; Natile, G. Synthesis, characterization, and *in vitro* evaluation of a new TSPO-selective bifunctional chelate ligand. *ACS Med. Chem. Lett.* **2014**, *5*, 685–689. [CrossRef] [PubMed]

47. Schlicke, H.; Richter, A.; Rothbart, M.; Brzezowski, P.; Hedtke, B.; Grimm, B. Function of tetrapyrroles, regulation of tetrapyrrole metabolism and methods for analyses of tetrapyrroles. *Procedia Chem.* **2015**, *14*, 171–175. [CrossRef]

48. Fujiwara, T.; Harigae, H. Biology of heme in mammalian erythroid cells and related disorders. *BioMed Res. Int.* **2015**, *2015*, 278536. [CrossRef] [PubMed]

49. Czarnecki, O.; Grimm, B. Post-translational control of tetrapyrrole biosynthesis in plants, algae, and cyanobacteria. *J. Exp. Bot.* **2012**, *63*, 1675–1687. [CrossRef] [PubMed]

50. Basile, A.S.; Klein, D.C.; Skolnick, P. Characterization of benzodiazepine receptors in the bovine pineal gland: Evidence for the presence of an atypical binding site. *Brain Res.* **1986**, *387*, 127–135. [CrossRef]

51. Awad, M.; Gavish, M. Binding of [^{3}H]Ro 5-4864 and [^{3}H]PK 11195 to cerebral cortex and peripheral tissues of various species: Species differences and heterogeneity in peripheral benzodiazepine binding sites. *J. Neurochem.* **1987**, *49*, 1407–1414. [CrossRef] [PubMed]

52. Katz, Y.; Weizman, A.; Gavish, M. Ligands specific to peripheral benzodiazepine receptors for treatment of porphyrias. *Lancet* **1989**, *1*, 932–933. [CrossRef]

53. Lourenço, C.; Lee, C.; Anderson, K.E. Disorders of haem biosynthesis. In *Inborn Metabolic Diseases: Diagnosis and Treatment*, 5th ed.; Saudubray, J.M., van den Berghe, G., Walter, J.H., Eds.; Springer: New York, NY, USA, 2012; pp. 521–532.

54. Cantoni, L.; Rizzardini, M.; Skorupska, M.; Cagnotto, A.; Codegoni, A.; Pecora, N.; Frigo, L.; Ferrarese, C.; Mennini, T. Hepatic protoporphyria is associated with a decrease in ligand binding for the mitochondrial benzodiazepine receptors in the liver. *Biochem. Pharmacol.* **1992**, *44*, 1159–1164. [CrossRef]

55. Pastorino, J.G.; Simbula, G.; Gilfor, E.; Hoek, J.B.; Farber, J.L. Protoporphyrin IX, an endogenous ligand of the peripheral benzodiazepine receptor, potentiates induction of the mitochondrial permeability transition and the killing of cultured hepatocytes by rotenone. *J. Biol. Chem.* **1994**, *269*, 31041–31046. [PubMed]

56. Gemelli, C.; Dongmo, B.M.; Ferrarini, F.; Grande, A.; Corsi, L. Cytotoxic effect of hemin in colonic epithelial cell line: Involvement of 18 kDa translocator protein (TSPO). *Life Sci.* **2014**, *107*, 14–20. [CrossRef] [PubMed]

57. Gerrish, K.E.; Putnam, C.W.; Laird, H.E., 2nd. Prolactin-stimulated mitogenesis in the Nb2 rat lymphoma cell: Lack of protoporphyrin IX effects. *Life Sci.* **1990**, *47*, 1647–1653. [CrossRef]

58. Fonia, O.; Weizman, R.; Coleman, R.; Kaganovskaya, E.; Gavish, M. PK 11195 aggravates 3,5-diethoxycarbonyl-1,4-dihydrocollidine-induced hepatic porphyria in rats. *Hepatology* **1996**, *24*, 697–701. [CrossRef] [PubMed]

59. Tsankova, V.; Magistrelli, A.; Cantoni, L.; Tacconi, M.T. Peripheral benzodiazepine receptor ligands in rat liver mitochondria: Effect on cholesterol translocation. *Eur. J. Pharmacol.* **1995**, *294*, 601–607. [CrossRef]

60. Awad, M.; Gavish, M. Species differences and heterogeneity of solubilized peripheral-type benzodiazepine binding sites. *Biochem. Pharmacol.* **1989**, *38*, 3843–3849. [CrossRef]

61. Tsankova, V.; Visentin, M.; Cantoni, L.; Carelli, M.; Tacconi, M.T. Peripheral benzodiazepine receptor ligands in rat liver mitochondria: Effect on 27-hydroxylation of cholesterol. *Eur. J. Pharmacol.* **1996**, *299*, 197–203. [CrossRef]

62. Kanegawa, N.; Collste, K.; Forsberg, A.; Schain, M.; Arakawa, R.; Jucaite, A.; Lekander, M.; Olgart Höglund, C.; Kosek, E.; Lampa, J.; *et al. In vivo* evidence of a functional association between immune cells in blood and brain in healthy human subjects. *Brain Behav. Immun.* **2016**, *54*, 149–157. [CrossRef] [PubMed]

63. Owen, D.R.; Gunn, R.N.; Rabiner, E.A.; Bennacef, I.; Fujita, M.; Kreisl, W.C.; Innis, R.B.; Pike, V.W.; Reynolds, R.; Matthews, P.M.; *et al.* Mixed-affinity binding in humans with 18-kDa translocator protein ligands. *J. Nucl. Med.* **2011**, *52*, 24–32. [CrossRef] [PubMed]

64. Owen, D.R.; Lewis, A.J.; Reynolds, R.; Rupprecht, R.; Eser, D.; Wilkins, M.R.; Bennacef, I.; Nutt, D.J.; Parker, C.A. Variation in binding affinity of the novel anxiolytic XBD173 for the 18 kDa translocator protein in human brain. *Synapse* **2011**, *65*, 257–259. [CrossRef] [PubMed]

65. Owen, D.R.; Yeo, A.J.; Gunn, R.N.; Song, K.; Wadsworth, G.; Lewis, A.; Rhodes, C.; Pulford, D.J.; Bennacef, I.; Parker, C.A.; *et al.* An 18-kDa translocator protein (TSPO) polymorphism explains differences in binding affinity of the PET radioligand PBR28. *J. Cereb. Blood Flow Metab.* **2012**, *32*, 1–5. [CrossRef] [PubMed]

66. Jaremko, M.; Jaremko, Ł.; Giller, K.; Becker, S.; Zweckstetter, M. Structural integrity of the A147T polymorph of mammalian TSPO. *ChemBioChem* **2015**, *16*, 1483–1489. [CrossRef] [PubMed]

67. Jaremko, M.; Jaremko, Ł.; Giller, K.; Becker, S.; Zweckstetter, M. Backbone and side-chain resonance assignment of the A147T polymorph of mouse TSPO in complex with a high-affinity radioligand. *Biomol. NMR Assign.* **2016**, *10*, 79–83. [CrossRef] [PubMed]

68. Ratcliffe, S.L.; Matthews, E.K. Modification of the photodynamic action of δ aminolaevulinic acid (ALA) on rat pancreatoma cells by mitochondrial benzodiazepine receptor ligands. *Br. J. Cancer* **1995**, *71*, 300–305. [CrossRef] [PubMed]

69. Guo, Y.; Kalathur, R.C.; Liu, Q.; Kloss, B.; Bruni, R.; Ginter, C.; Kloppmann, E.; Rost, B.; Hendrickson, W.A. Protein structure. Structure and activity of tryptophan-rich TSPO proteins. *Science* **2015**, *347*, 551–555. [CrossRef] [PubMed]

70. Zhao, A.H.; Tu, L.N.; Mukai, C.; Sirivelu, M.P.; Pillai, V.V.; Morohaku, K.; Cohen, R.; Selvaraj, V. Mitochondrial translocator protein (TSPO) function is not essential for heme biosynthesis. *J. Biol. Chem.* **2016**, *291*, 1591–1603. [CrossRef] [PubMed]

71. Wendler, G.; Lindemann, P.; Lacapère, J.J.; Papadopoulos, V. Protoporphyrin IX binding and transport by recombinant mouse PBR. *Biochem. Biophys. Res. Commun.* **2003**, *11*, 847–852. [CrossRef]

72. Bisland, S.K.; Goebel, E.A.; Hassanali, N.S.; Johnson, C.; Wilson, B.C. Increased expression of mitochondrial benzodiazepine receptors following low-level light treatment facilitates enhanced protoporphyrin IX production in glioma-derived cells *in vitro. Lasers Surg. Med.* **2007**, *39*, 678–684. [CrossRef] [PubMed]

73. Ozaki, H.; Zoghbi, S.S.; Hong, J.; Verma, A.; Pike, V.W.; Innis, R.B.; Fujita, M. *In vivo* binding of protoporphyrin IX to rat translocator protein imaged with positron emission tomography. *Synapse* **2010**, *64*, 649–653. [CrossRef] [PubMed]

74. Bender, A.S.; Norenberg, M.D. Effect of benzodiazepines and neurosteroids on ammonia-induced swelling in cultured astrocytes. *J. Neurosci. Res.* **1998**, *54*, 673–680. [CrossRef]

75. Jayakumar, A.R.; Panickar, K.S.; Norenberg, M.D. Effects on free radical generation by ligands of the peripheral benzodiazepine receptor in cultured neural cells. *J. Neurochem.* **2002**, *83*, 1226–1234. [CrossRef] [PubMed]

76. Repalli, J. Translocator protein (TSPO) role in aging and Alzheimer's disease. *Curr. Aging Sci.* **2014**, *7*, 168–175. [CrossRef] [PubMed]

77. Gut, P.; Zweckstetter, M.; Banati, R.B. Lost in translocation: The functions of the 18-kD translocator protein. *Trends Endocrinol. Metab.* **2015**, *26*, 349–356. [CrossRef] [PubMed]

78. Veenman, L.; Leschiner, S.; Spanier, I.; Weisinger, G.; Weizman, A.; Gavish, M. PK 11195 attenuates kainic acid-induced seizures and alterations in peripheral-type benzodiazepine receptor (PBR) protein components in the rat brain. *J. Neurochem.* **2002**, *80*, 917–927. [CrossRef] [PubMed]

79. Pimentel, E.; Luz, M.; Vidal, L.M.; Cruces, M.P.; Janczur, M.K. Action of protoporphyrin-IX (PP-IX) in the lifespan of *Drosophila melanogaster* deficient in endogenous antioxidants, Sod and Cat. *Open J. Anim. Sci.* **2013**, *3*, 1–7. [CrossRef]

80. Vidal, L.M.E.; Pimentel, E.P.; Cruces, M.P.; Sánchez, J.C.M. Genetic damage induced by CrO$_3$ can be reduced by low doses of Protoporphyrin-IX in somatic cells of *Drosophila melanogaster. Toxicol. Rep.* **2014**, *1*, 894–899. [CrossRef]

81. Curtis, C.; Landis, G.N.; Folk, D.; Wehr, N.B.; Hoe, N.; Waskar, M.; Abdueva, D.; Skvortsov, D.; Ford, D.; Luu, A.; *et al.* Transcriptional profiling of MnSOD-mediated lifespan extension in *Drosophila* reveals a species-general network of aging and metabolic genes. *Genome Biol.* **2007**, *8*, R262. [CrossRef] [PubMed]

82. Wright, G.; Reichenbecher, V. The effects of superoxide and the peripheral benzodiazepine receptor ligands on the mitochondrial processing of manganese-dependent superoxide dismutase. *Exp. Cell Res.* **1999**, *246*, 443–450. [CrossRef] [PubMed]

83. Cho, J.H.; Park, J.H.; Chung, C.G.; Shim, H.J.; Jeon, K.H.; Yu, S.W.; Lee, S.B. Parkin-mediated responses against infection and wound involve TSPO-VDAC complex in *Drosophila. Biochem. Biophys. Res. Commun.* **2015**, *463*, 1–6. [CrossRef] [PubMed]

84. Haase, A.; Bicker, G. Nitric oxide and cyclic nucleotides are regulators of neuronal migration in an insect embryo. *Development* **2003**, *130*, 3977–3987. [CrossRef] [PubMed]

85. Caiaffa, C.D.; Stiebler, R.; Oliveira, M.F.; Lara, F.A.; Paiva-Silva, G.O.; Oliveira, P.L. Sn-protoporphyrin inhibits both heme degradation and hemozoin formation in *Rhodnius prolixus* midgut. *Insect Biochem. Mol. Biol.* **2010**, *40*, 855–860. [CrossRef] [PubMed]

86. Rosenberg, N.; Rosenberg, O.; Weizman, A.; Leschiner, S.; Sakoury, Y.; Fares, F.; Soudry, M.; Weisinger, G.; Veenman, L.; Gavish, M. *In vitro* mitochondrial effects of PK 11195, a synthetic translocator protein 18 kDa (TSPO) ligand, in human osteoblast-like cells. *J. Bioenerg. Biomembr.* **2011**, *43*, 739–746. [CrossRef] [PubMed]

87. Rosenberg, N.; Rosenberg, O.; Weizman, A.; Veenman, L.; Gavish, M. *In vitro* catabolic effect of protoporphyrin IX in human osteoblast-like cells: Possible role of the 18 kDa mitochondrial translocator protein. *J. Bioenerg. Biomembr.* **2013**, *45*, 333–341. [CrossRef] [PubMed]

88. Rosenberg, N.; Rosenberg, O.; Weizman, A.; Veenman, L.; Gavish, M. *In vitro* effect of FGIN-1–27, a ligand to 18 kDa mitochondrial translocator protein, in human osteoblast-like cells. *J. Bioenerg. Biomembr.* **2014**, *46*, 197–204. [CrossRef] [PubMed]

89. Rosenberg, N.; Rosenberg, O.; Weizman, A.; Veenman, L.; Gavish, M. *In vitro* effects of the specific mitochondrial TSPO ligand Ro5 4864 in cultured human osteoblasts. *J. Bioenerg. Biomembr.* **2016**, submitted.

90. Gavish, M.; Veenman, J.A.; Shterenberg, A.; Marek, I. Heterocyclic Derivatives, Pharmaceutical Compositions and Methods of Use Thereof. U.S. Patent 8,541,428, 24 September 2013.

91. Gavish, M.; Marek, I.; Avital, A.; Shterenberg, A.; Vainshtein, A.; Veenman, L. Quinazoline Derivatives, Pharmaceutical Compositions and Methods of Use Thereof. WO 2015162615 A1, 29 October 2015.

92. Ferrarese, C.; Appollonio, I.; Frigo, M.; Perego, M.; Pierpaoli, C.; Trabucchi, M.; Frattola, L. Characterization of peripheral benzodiazepine receptors in human blood mononuclear cells. *Neuropharmacology* **1990**, *29*, 375–378. [CrossRef]

93. Berkovich, A.; Ferrarese, C.; Cavaletti, G.; Alho, H.; Marzorati, C.; Bianchi, G.; Guidotti, A.; Costa, E. Topology of two DBI receptors in human lymphocytes. *Life Sci.* **1993**, *52*, 1275–1277. [CrossRef]

94. Kinnally, K.W.; Zorov, D.B.; Antonenko, Y.N.; Snyder, S.H.; McEnery, M.W.; Tedeschi, H. Mitochondrial benzodiazepine receptor linked to inner membrane ion channels by nanomolar actions of ligands. *Proc. Natl. Acad. Sci. USA* **1993**, *90*, 1374–1378. [CrossRef] [PubMed]

95. Kinnally, K.W.; Muro, C.; Campo, M.L. MCC and PSC, the putative protein import channels of mitochondria. *J. Bioenerg. Biomembr.* **2000**, *32*, 47–54. [CrossRef] [PubMed]

96. Taketani, S.; Kohno, H.; Furukawa, T.; Tokunaga, R. Involvement of peripheral-type benzodiazepine receptors in the intracellular transport of heme and porphyrins. *J. Biochem.* **1995**, *117*, 875–880. [PubMed]

97. Nakajima, O.; Hashimoto, Y.; Iwasaki, S. Possible involvement of peripheral-type benzodiazepine receptors in erythroid differentiation of human leukemia cell line, K562. *Biol. Pharm. Bull.* **1995**, *18*, 903–906. [CrossRef] [PubMed]

98. Odber, J.; Cutler, M.; Dover, S.; Moore, M.R. Haem precursor effects on [^{3}H]-PK 11195 binding to platelets. *Neuroreport* **1994**, *5*, 1093–1096. [CrossRef] [PubMed]

99. Li, F.; Xia, Y.; Meiler, J.; Ferguson-Miller, S. Characterization and modeling of the oligomeric state and ligand binding behavior of purified translocator protein 18 kDa from *Rhodobacter sphaeroides*. *Biochemistry* **2013**, *52*, 5884–5889. [CrossRef] [PubMed]

100. Jaremko, L.; Jaremko, M.; Giller, K.; Becker, S.; Zweckstetter, M. Structure of the mitochondrial translocator protein in complex with a diagnostic ligand. *Science* **2014**, *343*, 1363–1366. [CrossRef] [PubMed]

101. Hinsen, K.; Vaitinadapoule, A.; Ostuni, M.A.; Etchebest, C.; Lacapere, J.J. Construction and validation of an atomic model for bacterial TSPO from electron microscopy density, evolutionary constraints, and biochemical and biophysical data. *Biochim. Biophys. Acta* **2015**, *1848*, 568–580. [CrossRef] [PubMed]

102. Wang, J. Comment on "Crystal structures of translocator protein (TSPO) and mutant mimic of a human polymorphism". *Science* **2015**, *350*, 519. [CrossRef] [PubMed]

103. Dougherty, T.J.; Sumlin, A.B.; Greco, W.R.; Weishaupt, K.R.; Vaughan, L.A.; Pandey, R.K. The role of the peripheral benzodiazepine receptor in photodynamic activity of certain pyropheophorbide ether photosensitizers: Albumin site II as a surrogate marker for activity. *Photochem. Photobiol.* **2002**, *76*, 91–97. [CrossRef]

104. Chen, Y.; Zheng, X.; Dobhal, M.P.; Gryshuk, A.; Morgan, J.; Dougherty, T.J.; Oseroff, A.; Pandey, R.K. Methyl pyropheophorbide-a analogues: Potential fluorescent probes for the peripheral-type benzodiazepine receptor. Effect of central metal in photosensitizing efficacy. *J. Med. Chem.* **2005**, *48*, 3692–3695. [CrossRef] [PubMed]

105. Li, X.B.; Guo, H.L.; Shi, T.Y.; Yang, L.; Wang, M.; Zhang, K.; Guo, Y.Y.; Wu, Y.M.; Liu, S.B.; Zhao, M.G. Neuroprotective effects of a novel translocator protein (18 kDa) ligand, ZBD-2, against focal cerebral ischemia and NMDA-induced neurotoxicity. *Clin. Exp. Pharmacol. Physiol.* **2015**, *42*, 1068–1074. [CrossRef] [PubMed]

106. Wang, D.S.; Tian, Z.; Guo, Y.Y.; Guo, H.L.; Kang, W.B.; Li, S.; Den, Y.T.; Li, X.B.; Feng, B.; Feng, D.; *et al.* Anxiolytic-like effects of translocator protein (TSPO) ligand ZBD-2 in an animal model of chronic pain. *Mol. Pain* **2015**, *11*. [CrossRef] [PubMed]

107. Jaremko, Ł.; Jaremko, M.; Giller, K.; Becker, S.; Zweckstetter, M. Conformational flexibility in the transmembrane protein TSPO. *Chemistry* **2015**, *21*, 16555–16563. [CrossRef] [PubMed]

108. Armstrong, G.A.; Alberti, M.; Leach, F.; Hearst, J.E. Nucleotide sequence, organization, and nature of the protein products of the carotenoid biosynthesis gene cluster of *Rhodobacter. capsulatus*. *Mol. Gen. Genet.* **1989**, *216*, 254–268. [CrossRef] [PubMed]

109. Bui, E.T.; Bradley, P.J.; Johnson, P.J. A common evolutionary origin for mitochondria and hydrogenosomes. *Proc. Natl. Acad. Sci. USA* **1996**, *93*, 9651–9656. [CrossRef] [PubMed]

110. Baker, M.E.; Fanestil, D.D. Mammalian peripheral-type benzodiazepine receptor is homologous to CrtK protein of *Rhodobacter capsulatus*, a photosynthetic bacterium. *Cell* **1991**, *65*, 721–722. [CrossRef]

111. Yeliseev, A.A.; Kaplan, S. A sensory transducer homologous to the mammalian peripheral-type benzodiazepine receptor regulates photosynthetic membrane complex formation in *Rhodobacter sphaeroides* 2.4.1. *J. Biol. Chem.* **1995**, *270*, 21167–21175. [CrossRef] [PubMed]

112. Chapalain, A.; Chevalier, S.; Orange, N.; Murillo, L.; Papadopoulos, V.; Feuilloley, M.G.J. Bacterial ortholog of mammalian translocator protein (TSPO) with virulence regulating activity. *PLoS ONE* **2009**, *4*, e6096. [CrossRef] [PubMed]

113. Yeliseev, A.A.; Krueger, K.E.; Kaplan, S. A mammalian mitochondrial drug receptor functions as a bacterial "oxygen" sensor. *Proc. Natl. Acad. Sci. USA* **1997**, *94*, 5101–5106. [CrossRef] [PubMed]

114. Verissimo, A.F.; Daldal, F. Cytochrome c biogenesis System I: An intricate process catalyzed by a maturase supercomplex? *Biochim. Biophys. Acta* **2014**, *1837*, 989–998. [CrossRef] [PubMed]

115. Yeliseev, A.A.; Kaplan, S. A novel mechanism for the regulation of photosynthesis gene expression by the TspO outer membrane protein of *Rhodobacter sphaeroides* 2.4.1. *J. Biol. Chem.* **1999**, *274*, 21234–21243. [CrossRef] [PubMed]

116. Zeng, X.; Kaplan, S. TspO as a modulator of the repressor/antirepressor (PpsR/AppA) regulatory system in *Rhodobacter sphaeroides* 2.4.1. *J. Bacteriol.* **2001**, *183*, 6355–6364. [CrossRef] [PubMed]

117. Yeliseev, A.A.; Kaplan, S. TspO of *Rhodobacter sphaeroides*. A structural and functional model for the mammalian peripheral benzodiazepine receptor. *J. Biol. Chem.* **2000**, *275*, 5657–5667. [CrossRef] [PubMed]

118. Scarf, A.M.; Auman, K.M.; Kassiou, M. Is there any correlation between binding and functional effects at the translocator protein (TSPO) (18 kDa)? *Curr. Mol. Med.* **2012**, *12*, 387–397. [CrossRef] [PubMed]

119. Li, H.; Yao, Z.; Degenhardt, B.; Teper, G.; Papadopoulos, V. Cholesterol binding at the cholesterol recognition/interaction amino acid consensus (CRAC) of the peripheral-type benzodiazepine receptor and inhibition of steroidogenesis by an HIV TAT-CRAC peptide. *Proc. Natl. Acad. Sci. USA* **2001**, *98*, 1277–1272. [CrossRef]

120. Lindemann, P.; Koch, A.; Degenhardt, B.; Hause, G.; Grimm, B.; Papadopoulos, V. A novel *Arabidopsis thaliana* protein is a functional peripheral-type benzodiazepine receptor. *Plant Cell Physiol.* **2004**, *45*, 723–733. [CrossRef] [PubMed]

121. Guillaumot, D.; Guillon, S.; Morsomme, P.; Batoko, H. ABA, porphyrins and plant TSPO-related protein. *Plant Signal. Behav.* **2009**, *4*, 1087–1090. [CrossRef] [PubMed]

122. Corsi, L.; Avallone, R.; Geminiani, E.; Cosenza, F.; Venturini, I.; Baraldi, M. Peripheral benzodiazepine receptors in potatoes (*Solanum tuberosum*). *Biochem. Biophys. Res. Commun.* **2004**, *313*, 62–66. [CrossRef] [PubMed]

123. Guillaumot, D.; Guillon, S.; Déplanque, T.; Vanhee, C.; Gumy, C.; Masquelier, D.; Morsomme, P.; Batoko, H. The *Arabidopsis* TSPO-related protein is a stress and abscisic acid-regulated, endoplasmic reticulum-Golgi-localized membrane protein. *Plant J.* **2009**, *60*, 242–256. [CrossRef] [PubMed]

124. Balsemão-Pires, E.; Jaillais, Y.; Olson, B.J.; Andrade, L.R.; Umen, J.G.; Chory, J.; Sachetto-Martins, G. The *Arabidopsis* translocator protein (AtTSPO) is regulated at multiple levels in response to salt stress and perturbations in tetrapyrrole metabolism. *BMC Plant Biol.* **2011**, *11*, 108. [CrossRef]

125. Vanhee, C.; Zapotoczny, G.; Masquelier, D.; Ghislain, M.; Batoko, H. The *Arabidopsis* multistress regulator TSPO is a heme binding membrane protein and a potential scavenger of porphyrins via an autophagy-dependent degradation mechanism. *Plant Cell* **2011**, *23*, 785–805. [CrossRef] [PubMed]

126. Tanaka, R.; Tanaka, A. Tetrapyrrole biosynthesis in higher plants. *Annu. Rev. Plant Biol.* **2007**, *58*, 321–346. [CrossRef] [PubMed]

127. Vanhee, C.; Batoko, H. *Arabidopsis* TSPO and porphyrins metabolism: A transient signaling connection? *Plant Signal. Behav.* **2011**, *6*, 1383–1385. [CrossRef] [PubMed]

128. Vanhee, C.; Batoko, H. Autophagy involvement in responses to abscisic acid by plant cells. *Autophagy* **2011**, *7*, 655–656. [CrossRef] [PubMed]

129. Frank, W.; Baar, K.M.; Qudeimat, E.; Woriedh, M.; Alawady, A.; Ratnadewi, D.; Gremillon, L.; Grimm, B.; Reski, R. A mitochondrial protein homologous to the mammalian peripheral-type benzodiazepine receptor is essential for stress adaptation in plants. *Plant J.* **2007**, *51*, 1004–1018. [CrossRef] [PubMed]

130. Lehtonen, M.T.; Akita, M.; Frank, W.; Reski, R.; Valkonen, J.P. Involvement of a class III peroxidase and the mitochondrial protein TSPO in oxidative burst upon treatment of moss plants with a fungal elicitor. *Mol. Plant Microbe Interact.* **2012**, *25*, 363–371. [CrossRef] [PubMed]

131. Tripathy, B.C.; Sherameti, I.; Oelmuller, R. Siroheme: An essential component for life on earth. *Plant Signal. Behav.* **2010**, *5*, 14–20. [CrossRef] [PubMed]

132. Martin, W.F.; Garg, S.; Zimorski, V. Endosymbiotic theories for eukaryote origin. *Philos. Trans. R. Soc. Lond. B Biol. Sci.* **2015**, *370*, 20140330. [CrossRef] [PubMed]

133. Leister, D. Towards understanding the evolution and functional diversification of DNA containing plant organelles. *F1000Research* **2016**, *5*. [CrossRef] [PubMed]

134. French, K.L.; Hallmann, C.; Hope, J.M.; Schoon, P.L.; Zumberge, J.A.; Hoshino, Y.; Peters, C.A.; George, S.C.; Love, G.D.; Brocks, J.J.; *et al.* Reappraisal of hydrocarbon biomarkers in Archean rocks. *Proc. Natl. Acad. Sci. USA* **2015**, *112*, 5915–5920. [CrossRef] [PubMed]

135. Weisinger, G.; Kelly-Hershkovitz, E.; Veenman, L.; Spanier, I.; Leschiner, S.; Gavish, M. Peripheral benzodiazepine receptor antisense knockout increases tumorigenicity of MA-10 Leydig cells *in vivo* and *in vitro*. *Biochemistry* **2004**, *43*, 12315–12321. [CrossRef] [PubMed]

136. Kobayashi, Y.; Ando, H.; Hanaoka, M.; Tanaka, K. Abscisic acid participates in the control of cell cycle initiation through heme homeostasis in the unicellular red alga *Cyanidioschyzon merolae*. *Plant Cell Physiol.* **2016**, *57*, 953–960. [CrossRef] [PubMed]

137. Busch, A.W.; Montgomery, B. The tryptophan-rich sensory protein (TSPO) is involved in stress-related and light-dependent processes in the cyanobacterium *Fremyella diplosiphon*. *Front. Microbiol.* **2015**, *6*, 1393. [CrossRef] [PubMed]

138. Stowe-Evans, E.L.; Ford, J.; Kehoe, D.M. Genomic DNA microarray analysis: Identification of new genes regulated by light color in the cyanobacterium *Fremyella diplosiphon*. *J. Bacteriol.* **2004**, *186*, 4338–4349. [CrossRef] [PubMed]

139. Pecoits, E.; Smith, M.L.; Catling, D.C.; Philippot, P.; Kappler, A.; Konhauser, K.O. Atmospheric hydrogen peroxide and Eoarchean iron formations. *Geobiology* **2015**, *13*, 1–14. [CrossRef] [PubMed]

140. Leneveu-Jenvrin, C.; Connil, N.; Bouffartigues, E.; Papadopoulos, V.; Feuilloley, M.G.; Chevalier, S. Structure-to-function relationships of bacterial translocator protein (TSPO): A focus on Pseudomonas. *Front. Microbiol.* **2014**, *5*, 631. [CrossRef] [PubMed]

141. Kobayashi, K.; Masuda, T.; Tajima, N.; Wada, H.; Sato, N. Molecular phylogeny and intricate evolutionary history of the three isofunctional enzymes involved in the oxidation of protoporphyrinogen IX. *Genome Biol. Evol.* **2014**, *6*, 2141–2155. [CrossRef] [PubMed]

142. Boldareva-Nuianzina, E.N.; Bláhová, Z.; Sobotka, R.; Koblízek, M. Distribution and origin of oxygen-dependent and oxygen-independent forms of Mg-protoporphyrin monomethylester cyclase among phototrophic proteobacteria. *Appl. Environ. Microbiol.* **2013**, *79*, 2596–2604. [CrossRef] [PubMed]

International Journal of
Molecular Sciences

MDPI

Article

TSPO Ligand-Methotrexate Prodrug Conjugates: Design, Synthesis, and Biological Evaluation

Valentino Laquintana [1,*], Nunzio Denora [1], Annalisa Cutrignelli [1], Mara Perrone [1], Rosa Maria Iacobazzi [1,2], Cosimo Annese [3], Antonio Lopalco [4], Angela Assunta Lopedota [1] and Massimo Franco [1]

[1] Department of Pharmacy–Pharmaceutical Sciences, University of Bari "Aldo Moro", Bari 70125, Italy; nunzio.denora@uniba.it (N.D.); annalisa.cutrignelli@uniba.it (A.C.); mara.perrone@uniba.it (M.P.); rosamaria.iacobazzi@uniba.it (R.M.I.); angelaassunta.lopedota@uniba.it (A.A.L.); massimo.franco@uniba.it (M.F.)

[2] Istituto tumori IRCCS "Giovanni Paolo II", Flacco, St. 65, Bari 70124, Italy

[3] Department of Chemistry, University of Bari "Aldo Moro", Bari 70125, Italy; cosimo.annese@uniba.it

[4] Department of Pharmaceutical Chemistry, University of Kansas, Lawrence, KS 66047, USA; lopalco@ku.edu

* Correspondence: valentino.laquintana@uniba.it; Tel.: +39-080-544-2767

Academic Editor: Genxi Li

Received: 9 May 2016; Accepted: 13 June 2016; Published: 18 June 2016

Abstract: The 18-kDa translocator protein (TSPO) is a potential mitochondrial target for drug delivery to tumors overexpressing TSPO, including brain cancers, and selective TSPO ligands have been successfully used to selectively deliver drugs into the target. Methotrexate (MTX) is an anticancer drug of choice for the treatment of several cancers, but its permeability through the blood brain barrier (BBB) is poor, making it unsuitable for the treatment of brain tumors. Therefore, in this study, MTX was selected to achieve two TSPO ligand-MTX conjugates (TSPO ligand α-MTX and TSPO ligand γ-MTX), potentially useful for the treatment of TSPO-rich cancers, including brain tumors. In this work, we have presented the synthesis, the physicochemical characterizations, as well as the *in vitro* stabilities of the new TSPO ligand-MTX conjugates. The binding affinity for TSPO and the selectivity *versus* central-type benzodiazepine receptor (CBR) was also investigated. The cytotoxicity of prepared conjugates was evaluated on MTX-sensitive human and rat glioma cell lines overexpressing TSPO. The estimated coefficients of lipophilicity and the stability studies of the conjugates confirm that the synthesized molecules are stable enough in buffer solution at pH 7.4, as well in physiological medium, and show an increased lipophilicity compared to the MTX, compatible with a likely ability to cross the blood brain barrier. The latter feature of two TSPO ligand-MTX conjugates was also confirmed by *in vitro* permeability studies conducted on Madin-Darby canine kidney cells transfected with the human MDR1 gene (MDCK-MDR1) monolayers. TSPO ligand-MTX conjugates have shown to possess a high binding affinity for TSPO, with IC_{50} values ranging from 7.2 to 40.3 nM, and exhibited marked toxicity against glioma cells overexpressing TSPO, in comparison with the parent drug MTX.

Keywords: translocator protein; methotrexate; TSPO-ligand; bio-conjugate; glioma

1. Introduction

The subcellular 18-kDa translocator protein (TSPO) [1], is an attractive biomarker for molecular imaging and a potential therapeutic target for drug delivery to tumors overexpressing TSPO [2,3]. TSPO is mostly located in the outer mitochondria membrane of steroid-synthesizing cells in peripheral organs system. In contrast, its presence in the central nervous system is delimited to ependymal cells including the glia. TSPO is involved in several pathophysiological processes, such as steroidogenesis, immunomodulation, apoptosis, brain injury, neurodegeneration, and cancer [4–7]. TSPO is upregulated in neuroinflammation and its overexpression has been proved in several types of cancers, including gliomas, whereas expression in the healthy brain is low [8,9]. Thus, TSPO could serve as potential therapeutic tool to target brain tumors using TSPO-ligands conjugates with anti-cancer drugs. Classical synthetic TSPO ligands include phenylisoquinoline carboxamides (e.g., PK 11195), benzodiazepines (e.g., RO5-4864), phenoxyphenyl acetamide (e.g., DAA1106), pyrazolo[1,5-*a*]pyrimidine acetamides (e.g., DPA-713), indoleacetamides (FGIN-1-27), and imidazo[1,2-*a*]pyridines (e.g., alpidem) [5,10] (Figure 1a). Recently, we have synthesized new 2-phenylimidazo[1,2-*a*]pyridine acetamides, designed from alpidem by introducing hydrophilic substituents on the imidazopyridine nucleus, that bind the mitochondrial protein with high affinity and selectivity (Figure 1b) [10]. The structure-activity analysis has shown that the substitution at the 8-position of the imidazopyridine nucleus with appropriate lipophilic or hydrophilic groups, combined with a chlorine atom added at the *para*-position of the phenyl ring are key factors to increase the affinity and the selectivity toward the TSPO binding sites [10]. Additionally, the incorporation of amino-, hydroxyl-, or carboxylic groups allows for convenient bioconjugation with anticancer drugs.

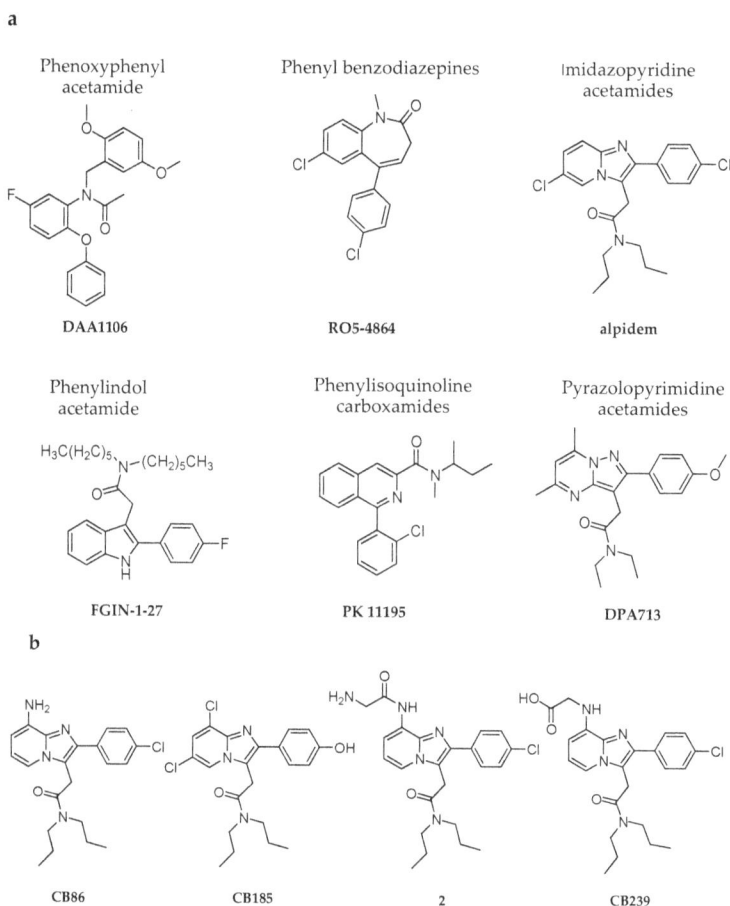

Figure 1. Chemical structures of the typical synthetic TSPO ligands presented in structural classes (**a**); and selective TSPO ligands with hydrophilic substituents (**b**).

Methotrexate (MTX; 2,4-diamino-N^{10}-methylpteroylglutamic acid; **1**) is an anticancer drug of choice for the treatment of several cancers such as acute lymphocytic leukemia, choriocarcinoma, non-Hodgkin's lymphoma, gastric, breast, head, and neck cancers [11]. MTX and its active metabolites (polyglutamates) are competitive inhibitors of the enzyme dihydrofolate reductase (DHFR) that lead to blockage of tetrahydrofolate synthesis and the depletion of nucleotide precursors. MTX is a class IV drug in the FDA's Biopharmaceutical Classification System (BCS, Amidon, cFederal Drug and Food Administration, Rockville, MD, USA) with a low permeability (see Table 1) and poor aqueous solubility [12]. It has a short plasma half-life and poor permeability across blood-brain barrier (BBB), when used in normal dosages of the protocols, making it unsuitable to brain tumors [13]. The use of high-dose MTX alone was proposed as the first-line treatment for primary central nervous system (CNS) lymphoma, but the side effects were more obvious. In fact, the myelosuppression, a common side effect and meningitis that often appears in the intrathecal treatment are dose-dependent and are more severe in patients receiving MTX in high dose. Moreover, CNS toxicities, including acute encephalopathy, also occur when the drug is used in high doses [13]. For these reasons, MTX is a suitable candidate for this study and therefore it was selected as a model drug to achieve TSPO ligand-MTX conjugates, potentially useful for the treatment of TSPO-rich cancers, including brain tumors overexpressing the TSPO. In this paper, we describe the synthesis, the physicochemical characterization, as well as the *in vitro* stabilities of the new TSPO ligand-MTX conjugates. The binding affinity for TSPO and selectivity *versus* the central-type benzodiazepine receptors (CBR) of conjugates were also evaluated. Furthermore, the cytotoxicity of prepared compounds was investigated on human SF126 and SF188, and rat RG2 and C6 glioma cell lines, together with their ability to permeate MDCK-MDR1 cell monolayers.

Table 1. Lipophilicity, permeation of blood-brain barrier (BBB) and stability in phosphate buffer solution and in rat serum solution of TSPO ligand-MTX (methotrexate) conjugates **3** and **4**.

Compound	CLogP [1]	$t_{1/2}$ (h) PBS	$t_{1/2}$ (h) Diluted Rat Serum
2	+3.25 ± 1.10	-	-
3	+5.21 ± 1.37	144 ± 25	4.1 ± 0.9
4	+4.40 ± 1.37	74 ± 12	3.3 ± 0.5
MTX	−0.24 ± 0.72	-	-

[1] Estimated according to ACD/Labs 10 (LogP v.14.04) software (Toronto, ON, Canada); PBS: phosphate buffered saline.

2. Results and Discussion

A drug delivery system, such as a bio-conjugate that carries an anticancer drug to brain tumors overexpressing the mitochondrial protein, should have favorable biopharmaceutical properties that include membrane permeability, high receptor binding affinity and selectivity, cytotoxicity or ability to convert itself into a cytotoxic moiety. The conjugate strategy is widely implemented to optimize biopharmaceutical and pharmacokinetic characteristics of drugs, including the transport across biological barriers and the reduction of adverse side effects [14]. Usually, the design of a conjugate involves the formation of a covalent chemical bond between the drug and a pharmacologically-inactive portion (e.g., a backbone polymer), whilst activation occurs after *in vivo* administration (conjugate prodrugs) or after reaching the target site upon cellular internalization. The further development of the conjugate is the strategy of the bioconjugate, which contributes in the direct linkage of the drug to a pharmacologically-active portion (e.g., a selective ligand or a peptide), or by the intermediacy of a spacer. In this regard, a number of bioconjugates with selective TSPO ligands have been developed for molecular imaging or for the delivery of hydrophilic anticancer drugs into brain tumors across the BBB. Bioconjugation of nanodevices with TSPO ligands (bio-conjugates with low molecular weight, TSPO decorated nanoparticles, and TSPO-targeted dendrimers) have also been described [15–18]. Moreover, in our previous studies we pointed out that some selective TSPO-ligands showed apoptotic effects and that the simultaneous transport of a TSPO-ligand with an anticancer drug may result in synergistic effects, precisely the synergism of the antitumor activity of the anticancer drug and of the TSPO ligand [16]. Thus, the aim of this study was to synthesize two new bio-conjugates of the anticancer drug MTX with the potent and highly selective TSPO ligand **2** (Scheme 1).

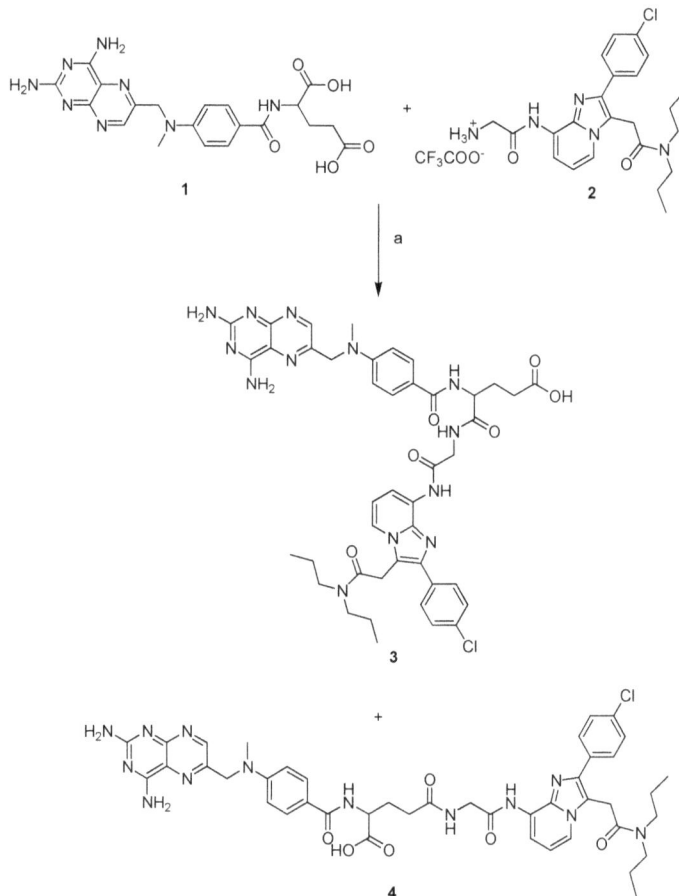

Scheme 1. Synthesis of TSPO-ligand α-MTX (methotrexate) conjugate **3** and TSPO-ligand γ-MTX conjugate **4**. Reagents and conditions: (a) 1,1'-carbonyldiimidazole (CDI), anhydrous *N,N*-dimethylformamide DMF, room temperature.

2.1. Chemistry

Conjugation of MTX (non-hydrated form and with unprotected carboxylic groups) with TSPO ligand **2** as trifluoroacetic acid salt was performed in anhydrous *N,N*-dimethylformamide (DMF) at room temperature, using carbonyldiimidazole (CDI) as the coupling agent. Under these conditions, TSPO-ligand α-MTX conjugate **3** and TSPO-ligand γ-MTX conjugate **4** (Scheme 1), could be obtained in 70:30 ratio (α/γ).

The TSPO-ligand MTX conjugates were completely characterized by spectroscopic techniques and mass spectrometry. The ESI-HRMS spectra showed a peak at $m/z = 876.3459$ (TSPO-ligand MTX conjugate **3**, $[M - H]^-$) or at 876.3430 (TSPO-ligand MTX conjugate **4**, $[M - H]^-$), both in agreement with the expected chemical formula, $C_{43}H_{48}ClN_{13}O_6$. Additionally, the one-dimensional

(1D-) and two-dimensional (2D-) nuclear magnetic resonance (NMR) characterization (^{1}H, correlation spectroscopy (COSY), and nuclear overhauser effect spectroscopy (NOESY)) provided spectra in full agreement with the structures assigned to **3** and **4**. The interpretation of combined 2D spectra can prove extremely useful in discriminating the structure of regioisomers [19]. In the case at hand, the NOESY spectra of the two regioisomers **3** and **4** show major differences in the intensity of cross-peaks occurring between the Gly NH of the TSPO moiety and the protons of Glu side-chain. Figure 2 summarizes these relevant NOESY correlations. In one case, the strong NH/α-CH correlation and the weak NH/β-CH$_2$ are consistent with conjugation of TSPO ligand to the α-COOH of MTX. In the other case, the absence of the NH/α-CH correlation and the strong NH/γ-CH$_2$ correlation are distinctive features of the conjugation to the γ-COOH of MTX.

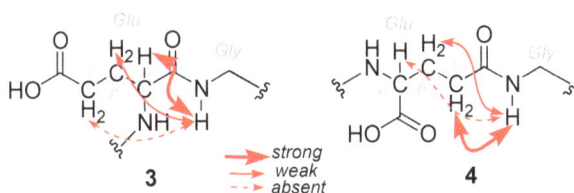

Figure 2. Relevant nuclear overhauser effect spectroscopy (NOESY) correlations for TSPO-ligand α-/γ-MTX conjugates (**3** and **4**, respectively).

2.2. Lipophilicity

The lipophilicity of a molecule, quantitatively expressed as LogP can be useful to predict its permeability through biological barriers. The lipophilicity of TSPO-ligand **2**, MTX and of the TSPO-ligand MTX conjugates **3** and **4** was estimated by calculating their 1-octanol/water partition coefficients, using CLOGP software (Toronto, ON, Canada), based on the fragmental method of Hansch and Leo [20]. Compounds that possess a value of LogP $\geqslant$2.5 are able to cross the BBB. As it can be seen from data in Table 1, both TSPO-ligand MTX conjugates have logP values higher than +2.5 (*i.e.*, +5.21 for **3** and +4.40 for **4**), and appeared to be suitably lipophilic to cross the BBB. Conversely, the parent drug has a negative value of CLogP (-0.24), in line with the experimental evidence (as mentioned above) that MTX is not able to cross the BBB effectively.

2.3. Stability Studies

The chemical stability of the TSPO-ligand MTX conjugates **3** and **4** was evaluated in 50 mM phosphate buffer solution at pH 7.4 and 37 °C; the physiological stability was also determined using a dilute rat serum solution (50% v/v) at 37 °C. The half-lives were calculated by the disappearance of the starting conjugate and are reported in Table 1. The aim of this work is the synthesis of TSPO ligand-MTX conjugates in order to improve the plasma stability and having less side effects than MTX. In addition TSPO ligand-MTX conjugates could enhance the transport across BBB and due a tumor targeting effect for the TSPO moiety, might selectively deliver the antineoplastic agent to brain tumors and prolong its efficacy. The TSPO-ligand MTX conjugates **3** and **4** are stable in buffer solution at pH 7.4 and enough stable in physiological medium with their half-lives exceeding 3 h. The conjugates show a similar stability in serum while TSPO-ligand α-MTX conjugate **3** is the most stable in phosphate buffer with a doubled half-life to the TSPO-ligand γ-MTX conjugate **4**. The liquid chromatography mass spectrometry (LC-MS) (ESI$^+$) analysis of the degradation mixture, obtained from enzymatic kinetics samples after 2 h, shows that the degradation of the conjugates occurs through the hydrolysis of the amide bond leading to free MTX (m/z = 477, [M + Na]$^+$) and TSPO ligand **2** (m/z = 464, [M + Na]$^+$). In addition, CB86 (m/z = 407, [M + Na]$^+$) (Figure 1), most likely originating from the degradation of ligand **2**, as previously shown [18], could be detected. Other products arising from the further degradation of the anticancer drug were found.

2.4. Radioligand Binding Assays

Binding affinities of the TSPO-ligand MTX conjugates **3** and **4** for TSPO and central-type benzodiazepine receptor (CBR) were assessed on membrane preparations from rat cerebral cortex. The IC_{50} values were determined from the displacement curves using the reference compound [^{3}H]-PK 11195, for TSPO, or [^{3}H]flunitrazepam for CBR. The data obtained, expressed as IC_{50} values, as well as the ratios between CBR and TSPO affinity, respectively (Selectivity Index, SI), as a measure of the *in vitro* selectivity, are shown in Table 2 and were compared with unlabeled TSPO selective ligand PK 11195, and with CBR selective ligand flunitrazepam. The results of the binding affinities indicate that TSPO-ligand MTX conjugates **3** and **4** show high affinity for TSPO, with IC_{50} values ranging from 7.2 to 40.3 nM. Moreover, the SI of TSPO-ligand MTX conjugate **3** is similar to the reference compound PK 11195. Radioligand binding assays on human brain tissue showed a genetic polymorphism and the variation of binding affinity of selective ligands for the TSPO, due to presence of differing binders in the general population (*i.e.*, high-affinity binders, HABs, low-affinity binders, LABs and mixed-affinity binders, MABs) [21]. Thus, the binding quantification of TSPO ligands is confounded by the expression of different forms of TSPO and to assess the potentially therapeutic use of TSPO-ligands would be appropriate to genetically test the patients.

2.5. Cytotoxicity Studies

Cytotoxicity assays of the TSPO-ligand MTX conjugates **3** and **4** were conducted against SF126 and SF188 human glioma cells and, RG2 and C6 rat glioma cells and results are also shown in Table 2. These cellular lines were selected for their high expression level of TSPO and, therefore, have been used extensively for *in vitro* experiments on the brain tumor models and for cytotoxicity studies relating to anticancer drugs, as well as TSPO-ligands and its nanodevices. The tested compounds show high cytotoxicity on all cell lines, especially for C6 rat glioma cells. In particular, the TSPO-ligand MTX α-conjugates **3** appears to be the most cytotoxic displaying a lower IC_{50} value compared to MTX on C6 cells.

2.6. Transport Studies

Transport studies of the TSPO-ligand MTX conjugates **3** and **4** were carried out on a MDCKII-MDR1 cell monolayer and results are shown in Table 2. This approach is an established and versatile *in vitro* method to evaluate the drug permeability through the BBB, as well MDCKII-MDR1 was identified as the most promising cell line for qualitative predictions of brain distribution [22]. However He and co-worker have reported the brain microvascular endothelial cells (BMECs) as a more established approach to address the permeability through the BBB [23]. This method was previously described by Audus and has been also used in our recent work [24,25].

Table 2. Affinity for rat cerebrocortical TSPO and central-type benzodiazepine receptors (CBR), cytotoxicity against glioma cells and transport across Madin-Darby canine kidney cells transfected with the human MDR1 gene (MDCKII-MDR1) cells of MTX and TSPO ligand-MTX conjugates **3** and **4**.

Compound	IC$_{50}$ (nM)		SI [3]	IC$_{50}$ (nM) Rat Glioma Cells		IC$_{50}$ (nM) Human Glioma Cells		Transport across MDCKII-MDR1 [4]
	TSPO Cortex	CBR Cerebral		C6	RG2	SF126	SF188	P_{app} (AP) (cm/s)
2 [1]	24.0 ± 4.2	$>10^5$	$>4.1 \times 10^3$	-	-	-	-	-
3	7.2 ± 2.1	$>10^5$	$>1.4 \times 10^4$	1.2 ± 0.7*	4.2 ± 3.2*	14.2 ± 4.2*	9.2 ± 3.4*	$3.65 \pm 0.14 \times 10^{-6}$
4	40.3 ± 3.8	$>10^5$	$>2.5 \times 10^3$	2.2 ± 0.6*	15.2 ± 2.9**	25.0 ± 2.6**	18.2 ± 2.0**	$3.32 \pm 0.28 \times 10^{-6}$
MTX	-	-	-	2.6 ± 0.8	3.8 ± 1.5	1.2 ± 0.3	1.2 ± 0.2	$9.10 \pm 0.44 \times 10^{-8}$
PK11195 [2]	1.5 ± 1.5	$>10^5$	$>6.7 \times 10^4$	-	-	-	-	-
Flunitrazepam [2]	-	5.1 ± 0.5	-	-	-	-	-	-

Values are means ± SEM of three experiments performed in duplicate; [1] According to [18]; [2] According to [5]; [3] SI: selectivity index; [4] Values are means ± SEM of three experiments performed in triplicate. The statistical differences between TSPO ligand-MTX conjugates **3** and **4**, with respect to MTX was calculated by two-way ANOVA followed by Tukey's multiple comparison test (* $p < 0.05$, ** $p < 0.01$, *ns.* MTX).

The apparent permeability coefficient (P_{app}) of each compound was calculated on the basis of the amounts permeated through MDCKII-MDR1 monolayer by the equation:

$$P_{app} = \frac{V_A \, dC}{A \, [D]_{in} \, dt}$$

where V_A (dC/dt) is the linear appearance rate of mass in the receiver solution; A is the filter/cell surface area and $[D]_{in}$ is the initial tested compound concentration in the apical (AP) chamber. The P_{app} values were evaluated in apical to basolateral direction. Moreover, the flux of fluorescein isothiocyanate-dextran (FD4, Sigma, Milano, Italy) (200 µg/mL) and diazepam (75 µM) as paracellular and transcellular markers, respectively, was measured as an internal control to verify the cell monolayer's integrity and tight junction integrity during the assay. On an experimental day, the integrity of the cell monolayer was assessed by measuring the transendothelial electrical resistance (TEER) value. The TEER of the BBB was assessed close to 200 Ω/cm^2, precluding the access to a wide number of common anticancer drugs to brain tissue, including MTX. The results of the transport studies demonstrate that both TSPO-ligand MTX conjugates **3** and **4** were characterized by higher P_{app} values than MTX and paracellular marker FD4 ($3.54 \pm (0.39 \times 10^{-10})$), with a P_{app} values being about. 30 times greater than the parent drug. Moreover, the permeability of MTX-conjugates was found to be lower than that of the transcellular marker diazepam ($6.40 \pm (0.31 \times 10^{-5})$). Finally, these results demonstrate that although the MTX-conjugates possessed an adequate calculated lipophilic character, which result in a better permeation capacity, compared to the MTX, the P_{app} values were predictive of a modest ability to overcome the BBB by the transcellular pathway. However, due to the development of brain tumors, the structure of tumor vasculature changes and the BBB is gradually replaced by the more permeable blood-brain tumor barrier (BBTB) [26].

3. Experimental Section

3.1. Materials and Methods

Anhydrous *N*,*N*-dimethylformamide (DMF), 1,1′-carbonyldiimidazole (CDI), *N*-(3-dimeth ylaminopropyl)-*N*′-ethylcarbodiimide hydrochloride (EDAC), 2,4-diamino-N^{10}-methylpteroylglutamic acid (methotrexate, MTX), dichloromethane (CH$_2$Cl$_2$), hexadeuterodimethyl sulfoxide (DMSO-d_6), L-glutamine, penicillin (100 U/mL), and streptomycin (100 µg/mL) were purchased from Sigma (Milan, Italy). Reagent grade chemicals were used without further purification. TSPO ligand, *N*,*N*-dipropyl-[2-(8-(2-aminoacetamido)-2-(4-chlorophenyl)imidazo[1,2-a]pyridin-3-yl)]acetamide (**2**) was synthesized as trifluoroacetate salt, in accord to the procedures described previously [18]. Melting points were determined in open capillary tubes on a Büchi apparatus and are uncorrected. Elemental analyses were performed on a Eurovector elemental analyzer EA 3000. Electrospray mass spectrometry (ESI-MS) were carried out with an Agilent 1100 LC-MSD trap system instrument (Agilent, Santa Clara, CA, USA). IR spectra were obtained using a FT-IR spectrophotometer (Perkin-Elemer, Waltham, MA, USA) using KBr pellets. ^{1}H-NMR, COSY, and NOESY NMR spectra were recorded with an Agilent (VNMRS500) 500 MHz spectrometer (Agilent). In the experimental section the values of the coupling constant, *J*, are given in hertz (Hz) and ^{1}H chemical shifts are reported as δ values (*i.e.*, in parts per million, ppm), relative to the residual isotopic impurity of DMSO-d_6 solvent (2.50 ppm). The following codes are used for the description of signal multiplicity: singlet (s), doublet (d), triplet (t), doublet of doublets (dd), multiplet (m), br (broad signal), Ar (aromatic H). Silica gel 60 (70–230 mesh, (Merck, Darmstadt, Germania)) was used for column chromatography. Reactions were performed under an argon atmosphere.

3.2. High-Performance Liquid Chromatography (HPLC) Analyses

HPLC analyses were performed with an Agilent 1260 Infinity Quaternary LC System equipped with an Agilent variable wavelength UV detector, a Rheodyne injector (Rheodyne, Model 7725i,

Agilent) equipped with a 20 μL loop and a OpenLAB CDS ChemStation software (Agilent). A reversed phase Symmetry C18 column (25 cm × 3.9 mm; 5 μm particles) as the stationary phase and a mixture of methanol and deionized water 80:20 (*v*/*v*) as the mobile phase with the flow rate of 1.0 mL/min were utilized and the column effluent was monitored continuously at 254 nm. The compounds were quantified by measuring the areas of the peaks, and using, as references, suitable standard solutions, chromatographed under the same conditions. The data were processed using Microsoft Excel 2010 or GraphPad Software (La Jolla, CA, USA).

3.3. Stability Studies in Phosphate Buffer Solution

TSPO ligand-MTX conjugates **3** and **4** chemical stabilities were studied at pH 7.4 in 0.05 M phosphate buffer at 37 °C. The kinetic studies were carried out by adding 100 μL of a stock solution of the conjugate (1.0 mg/mL in DMSO) to 1.9 mL of the buffer solution preheated at 37 °C. Thus, the final concentration of the compounds in the tested solutions was 1 M. The resulting solutions were vortexed and maintained in a shaker water bath at the constant temperature of 37 °C ($\pm$0.2). At appropriate time intervals, aliquots of 100 μL of each sample were removed, filtered through cellulose acetate membranes (0.22 μm, Advantec MFS, Pleasanton, CA, USA), and then 20 μL of filtrates were immediately analyzed by HPLC. The studies were done in triplicate and the pseudo-first-order rate constants were obtained from the slopes of the linear plot of the logarithms of the residual concentrations of conjugates against time.

3.4. Stability Studies in Rat Serum Solution

The TSPO ligand-MTX conjugates **3** and **4** stabilities in physiological medium were studied at 37 °C in 0.05 M phosphate buffer and 0.14 M NaCl at pH 7.4, containing 50% *v*/*v* of rat serum. The *in vitro* stabilities were assayed by adding 100 μL of the stock solution of conjugates in DMSO (1 mg/mL) to 1.1 mL of preheated serum solution, and the mixture was maintained in water bath at 37 °C ($\pm$0.2). Thus, the final concentration of the compounds in the tested solutions was 1 M. At different time points over a period of 0–120 min, aliquots of 100 μL of seach sample were withdrawn and added to 500 μL of cold acetonitrile for deproteinize the serum. Samples were centrifuged at 3500 rpm for 10 min, and the supernatant was removed, filtered through cellulose acetate membranes (0.22 μm, Advantec MFS) and then 20 μL of filtrates were immediately analyzed by HPLC. The studies were done in triplicate and the pseudo-first-order rate constants were obtained from the slopes of the linear plot of the logarithms of the residual concentrations of conjugates against time.

3.5. Synthesis of TSPO-Ligand-MTX Conjugates Prodrugs

To a solution of MTX (0.25 g, 0.55 mmol) in anhydrous DMF (2.5 mL) cooled at 0 °C in ice bath was added CDI (0.10 g, 0.68 mmol), and the reaction mixture stirred for 30 min. After this time, a solution of compound **2** (0.306 g, 0.55 mmol) in anhydrous DMF (2.5 mL) was added dropwise, and stirring prolonged for 12 h at room temperature in the dark. Solvent was evaporated under reduced pressure, and the residue dissolved in CH_2Cl_2 (20 mL). The resulting solution was washed with 5% $NaHCO_3$, dried over Na_2SO_4, and rotovapored to dryness. Purification of the residue by silica-gel column chromatography using 15% $MeOH/CH_2Cl_2$ as the eluting solvent, afforded regioisomers **3** (0.34 g, ~70% yield) and **4** (0.1 g, ~25% yield) as orange powders.

5-(2-(2-(4-Chlorophenyl)-3-(2-(dipropylamino)-2-oxoethyl)imidazo[1,2-*a*]pyridin-8-ylamino)-2-oxoethylamino)-4-(4-(((2,4-diaminopteridin-6-yl)methyl)(methyl)amino)benzamido)-5-oxopentanoic acid (TSPO-ligand α-MTX conjugate **3**). m.p. 210 °C (dec.). Anal. calcd. for $C_{43}H_{48}ClN_{13}O_6$: C, 58.80; H, 5.51; N, 20.73%. Found: C, 58.94; H, 5.48; N, 20.64%. ^{1}H-NMR (DMSO-d_6, 500 MHz): δ 0.80 (t, *J* = 7.5 Hz, 3H, CH₃), 0.85 (t, *J* = 7.5 Hz, 3H, CH₃), 1.48 (m, 2H, CH₂), 1.58 (m, 2H, CH₂), 1.98 (m, 1H, Glu β-CHH), 2.13 (m, 1H, Glu β-CHH), 2.33 (m, 2H, Glu γ-CH₂), 3.21 (s, 3H, CH₃N), 3.23 (m, 2H, CH₂N), 3.30 (m, 2H, CH₂N), 4.03 (dd, J_1 = 17.0, J_2 = 6.0 Hz, 1H, Gly α-CHH), 4.10 (dd, J_1 = 17.0,

J_2 = 6.0 Hz, 1H, Gly α-CHH), 4.22 (s, 2H, CH$_2$CO), 4.50 (m, 1H, Glu α-CH), 4.78 (s, 2H, CH$_2$N), 6.62 (br. s, 2H, NH$_2$), 6.81 (d, *J* = 9.1 Hz, 2H, Ar CH), 6.90 (m, 1H, Ar CH), 7.04 (br. s, 2H, NH$_2$), 7.51 (d, *J* = 8.6 Hz, 2H, Ar CH), 7.70 (d, *J* = 8.6 Hz, 2H, Ar CH), 7.75 (d, *J* = 9.1 Hz, Ar C*H*), 7.96 (m, 2H, Ar CH), 8.16 (d, *J* = 7.5 Hz, Glu NH), 8.38 (t, *J* = 6.0 Hz, Gly NH), 8.56 (s, 1H, Ar CH), 9.82 (s, 1H, NH). HRMS (ESI) *m/z* [M − H]$^-$: calcd. for C$_{43}$H$_{47}$ClN$_{13}$O$_6{}^-$ 876.3466. Found 876.3459. IR (KBr): 1698 cm^{-1}.

5-(2-(2-(4-Chlorophenyl)-3-(2-(dipropylamino)-2-oxoethyl)imidazo[1,2-*a*]pyridin-8-ylamino)-2-oxoethylamino)-2-(4-(((2,4-diaminopteridin-6-yl)methyl)(methyl)amino)benzamido)-5-oxopentanoic acid. (TSPO-ligand γ-MTX conjugate **4**). m.p. 190 °C (dec.). Anal. calcd. for C$_{43}$H$_{48}$ClN$_{13}$O$_6$: C, 58.80; H, 5.51; N, 20.73%. Found: C, 58.94; H, 5.48; N, 20.64%. ^{1}H-NMR (DMSO-*d$_6$*, 500 MHz): δ 0.80 (t, *J* = 7.5 Hz, 3H, CH$_3$), 0.85 (t, *J* = 7.5 Hz, 3H, CH$_3$), 1.48 (m, 2H, CH$_2$), 1.58 (m, 2H, CH$_2$), 2.00 (m, 1H, Glu β-CHH), 2.12 (m, 1H, Glu β-CHH), 2.34 (m, 2H, Glu γ-CH$_2$), 3.21 (s, 3H, CH$_3$N), 3.23 (m, 2H, CH$_2$N), 3.30 (m, 2H, CH$_2$N), 4.04 (m, 2H, Gly α-CH$_2$), 4.22 (s, 2H, CH$_2$CO), 4.31 (m, 1H, Glu α-CH), 4.77 (s, 2H, CH$_2$N), 6.60 (br. s, 2H, NH$_2$), 6.81 (d, *J* = 9.0 Hz, 2H, Ar CH), 6.90 (m, 1H, Ar CH), 7.42 (br. s, 2H, NH$_2$), 7.52 (d, *J* = 8.6 Hz, 2H, Ar CH), 7.68 (d, *J* = 8.6 Hz, 2H, Ar CH), 7.72 (d, *J* = 9.0 Hz, Ar CH), 7.95 (m, 2H, Ar CH), 8.18 (d, *J* = 7.0 Hz, Glu NH), 8.46 (t, *J* = 6.0 Hz, Gly NH), 8.56 (s, 1H, Ar CH), 9.80 (s, 1H, NH). HRMS (ESI) *m/z* [M − H]$^-$: calcd. for C$_{43}$H$_{47}$ClN$_{13}$O$_6{}^-$ 876.3466. Found 876.3430. IR (KBr): 1698 cm^{-1}.

3.6. Biology

3.6.1. Radioligand Binding Assays

Binding assays ([^{3}H]Flunitrazepam and [^{3}H]PK 11195 bindings) were assessed to membrane preparations from rat cerebral cortex as described previously [15]. The IC$_{50}$ values (GraphPad Prism software, (La Jolla, CA, USA)) were determined from the displacement curves using the reference compound [^{3}H]-PK11195, for TSPO or [^{3}H]flunitrazepam, for CBR, respectively. The selectivity index (SI) was calculated from: IC$_{50}$(CBR)/IC$_{50}$(TSPO).

3.6.2. Cell Cultures and Cell Viability Analysis by MTT Assay

Rat RG2 and C6 glioma cells were grown in DMEM (Dulbecco's modified Eagle's medium, Sigma Aldrich, Milano, Italy) nutrient supplemented with 10% heat inactivated FBS (Fetal Bovine Serum, Sigma Aldrich), 2 mM, L-glutamine, penicillin (100 U/mL), and streptomycin (100 µg/mL) at 37 °C in a humidified 5% CO$_2$ atmosphere. Cells were fed every day and seeded in 96-well plates at a density of ~5000 cells/well for assays. MTX and TSPO ligand-MTX conjugates were dissolved in DMSO at initial concentrations of 1000 µg/mL and then diluted with medium. Cells were exposed to different concentrations of compounds (0.1–1000 nM) for 24 h. The percentage of organic solvent (DMSO), used to dissolve the tested compounds d did not exceed 1% (*v/v*) in the samples.

Cell viability was evaluated by MTT (3-(4,5-dimethylthiazol-2-yl)-2,5-diphenyltetrazolium bromide) assay as previously described [15]. In brief, 10 µL of MTT (5.0 mg/mL) was added to the cells and cultured at 37 °C, in a humidified 5% CO$_2$ atmosphere for 2 h. After this period, medium was removed and the cells were replaced with 150 µL of a DMSO/ethanol (1:1) solution per well. The absorbance of the individual well was measured by microplate reader (Wallac Victor3, 1420 Multilabel Counter, Perkin-Elmer (Waltham, MA, USA). All experiments were repeated three times and each compound's concentration was tested in triplicate.

3.6.3. Bi-Directional Permeability Study

The bi-directional transport study was conducted as previously reported [15]. A Transwell® 12 plate consisting of differentiated MDCKII-MDR1cells plated on polyester microporous filters (12 mm diameter, 0.4 µm pore size, 0.5 mL apical volume, 1.5 mL basolateral volume) at a density of 100,000 cells/cm^2, was used for the permeability assay. The integrity of the cell monolayer was assessed by measuring the TEER value in each well, using an epithelial volt-ohm meter (EVOM). Cell monolayers with TEER 120–140 Ohm/cm^2 were used. In each experiment the cells were equilibrated for 30 min at 37 °C in transport medium which include 0.4 mM K_2HPO_4, 25 mM $NaHCO_3$, 3 mM KCl, 122 mM NaCl, 10 mM glucose, and the pH was 7.4, with osmolarity of 300 mOsm as determined by a freeze point-based osmometer. For apical-to-basal permeability (AP-BL), test and control compounds solutions were prepared in transport medium at concentration of 75 μM (or 200 μg/mL for FD4), and added to the apical side of the cell monolayer (0.5 mL). Fresh assay medium was placed in the receiver compartment. The transport experiments were carried out in incubator at 37 °C, 5% CO_2, and 95% humidity. After incubation time of 2 h, samples were removed from the apical side of the monolayer and then analyzed. MTX and TSPO ligand-MTX conjugates **3** and **4** were analyzed by HPLC as described above in the stability studies, while diazepam was analyzed with a PerkinElmer double-beam UV-visible spectromphotometer Lamba Bio 20 (Milan, Italy), equipped with 10 mm path-length-matched quartz cells. The FD4 samples were analyzed with a Victor3 fluorometer (Wallac Victor3, 1420 Multilabel Counter, Perkin-Elmer) at excitation and emission wavelengths of 485 and 535 nm, respectively. All experiments were repeated three times and each compound was tested in triplicate.

3.7. Statistical Analysis

The statistical analysis was accomplished using one-way analysis of variance (ANOVA) followed by the Tukey post *hoc* tests (GraphPad Prism version 5.04 for Windows, GraphPad Software, San Diego, CA, USA). Differences were considered statistically significant at $p < 0.05$.

4. Conclusions

TSPO ligand-MTX conjugates have shown to possess a high binding affinity and selectivity for TSPO, and exhibited marked toxicity against glioma cells, in comparison with the parent drug MTX. These results also highlight the ability of the TSPO-ligand to transport the hydrophilic drug through biological membranes and determine its accumulation in target cells overexpressing the TSPO. The present work describes the proof-of-concept demonstration of the strength of the bio-conjugate strategy that simultaneously carries inside of cancer cells two agents with distinct modes of action, in the treatment of brain tumors. For this reason, the TSPO ligand-MTX conjugates could be potential tools to increase the effectiveness of the drug in the treatment of brain tumors overexpressing the mitochondrial target TSPO.

Acknowledgments: We thank Antonio Palermo and Giovanni Dipinto for their skillful technical assistance in the field of NMR and LC-HRMS studies, respectively. The University of Bari (Bari, Italy), the Inter-University Consortium for Research on the Chemistry of Metal Ions in Biological Systems (C.I.R.C.M.S.B.) are gratefully acknowledged.

Author Contributions: Nunzio Denora, Valentino Laquintana and Massimo Franco conceived and designed the experiments; Mara Perrone and Antonio Lopalco performed the synthesis and analyzed the experiment data; Rosa Maria Iacobazzi and Mara Perrone performed the biological studies and analyzed the experiment data; Cosimo Annese designed the NMR studies and analyzed the experiment data. Annalisa Cutrignelli and Angela Assunta Lopedota performed the stability studies, analyzed the experiment data and contributed analysis tools; Valentino Laquintana wrote the paper.

Conflicts of Interest: The authors declare no conflict of interest.

Abbreviations

BBB	blood-brain barrier
BBTB	blood-brain tumor barrier
CBR	Central-type Benzodiazepine Receptor
CDI	1,1′-carbonyldiimidazole
CNS	Central Nervous System
COSY	COrrelation SpectroscopY

EDAC	*N*-(3-Dimethylaminopropyl)-*N'*-ethylcarbodiimide hydrochloride
Gly	Glycine
Glu	Glutamic acid
MTX	Methotrexate
NOESY	Nuclear Overhauser Effect SpectroscopY
PBR	Peripheral-type Benzodiazepine Receptor
TEER	Transendothelial electrical resistance
TSPO	Translocator protein (18 kDa)

References

1. Papadopoulos, V.; Baraldi, M.; Guilarte, T.R.; Knudsen, T.B.; Lacapère, J.J.; Limdermann, P.; Norenberg, M.D.; Nutt, D.; Weizman, A.; Zhang, M.R.; *et al.* Translocator protein (18 kDa): New nomenclature for the peripheral-type benzodiazepine receptor based on its structure and molecular function. *Trends Pharmacol. Sci.* **2006**, *27*, 402–409. [CrossRef] [PubMed]

2. Rupprecht, R.; Papadopoulos, V.; Rammes, G.; Baghai, T.C.; Fan, J.; Akula, N.; Groyer, G.; Adams, D.; Schumacher, M. Translocator protein (18 kDa) (TSPO) as a therapeutic target for neurological and psychiatric disorders. *Nat. Rev. Drug Discov.* **2010**, *9*, 971–988. [CrossRef] [PubMed]

3. Perrone, M.; Moon, B.S.; Park, H.S.; Laquintana, V.; Jung, J.H.; Cutrignelli, A.; Lopedota, A.; Franco, M.; Kim, S.E.; Lee, B.C.; *et al.* A novel PET imaging probe for the detection and monitoring of translocator protein 18 kDa expression in pathological disorders. *Sci. Rep.* **2016**, *6*, 20422. [CrossRef] [PubMed]

4. Midzak, A.; Zirkin, B.; Papadopoulos, V. Translocator protein: Pharmacology and steroidogenesis. *Biochem. Soc. Trans.* **2015**, *43*, 572–578. [CrossRef] [PubMed]

5. Midzak, A.; Denora, N.; Laquintana, V.; Cutrignelli, A.; Lopedota, A.; Franco, M.; Altomare, C.D.; Papadopoulos, V. 2-Phenylimidazo[1,2-*a*]pyridine-containing ligands of the 18-kDa translocator protein (TSPO) behave as agonists and antagonists of steroidogenesis in a mouse leydig tumor cell line. *Eur. J. Pharm. Sci.* **2015**, *76*, 231–237. [CrossRef] [PubMed]

6. Austin, C.J.; Kahlert, J.; Kassiou, M.; Rendina, L.M. The translocator protein (TSPO): A novel target for cancer chemotherapy. *Int. J. Biochem. Cell Biol.* **2013**, *45*, 1212–1216. [CrossRef] [PubMed]

7. Werry, E.L.; Barron, M.L.; Kassiou, M. TSPO as a target for glioblastoma therapeutics. *Biochem. Soc. Trans.* **2015**, *43*, 531–536. [CrossRef] [PubMed]

8. Veenman, L.; Levin, E.; Weisinger, G.; Leschiner, S.; Spanier, I.; Snyder, S.H.; Weizman, A.; Gavish, M. Peripheral-type benzodiazepine receptor density and *in vitro* tumorigenicity of glioma cell lines. *Biochem. Pharmacol.* **2004**, *68*, 689–698. [CrossRef] [PubMed]

9. Rechichi, M.; Salvetti, A.; Chelli, B.; Costa, B.; da Pozzo, E.; Spinetti, F.; Lena, A.; Evangelista, M.; Rainaldi, G.; Martini, C.; *et al.* TSPO over-expression increases motility, transmigration and proliferation properties of C6 rat glioma cells. *Biochim. Biophys. Acta* **2008**, *2*, 118–125. [CrossRef] [PubMed]

10. Denora, N.; Laquintana, V.; Pisu, M.G.; Dore, R.; Murru, L.; Latrofa, A.; Trapani, G.; Sanna, E. 2-Phenyl-imidazo[1,2-*a*]pyridine compounds containing hydrophilic groups as potent and selective ligands for peripheral benzodiazepine receptors: Synthesis, binding affinity and electrophysiological studies. *J. Med. Chem.* **2008**, *51*, 6876–6888. [CrossRef] [PubMed]

11. Khan, Z.A.; Tripathi, R.; Mishra, B. Methotrexate: A detailed review on drug delivery and clinical aspects. *Expert Opin. Drug Deliv.* **2012**, *9*, 151–169. [CrossRef] [PubMed]

12. Kasim, N.A.; Whitehouse, M.; Ramachandran, C.; Bermejo, M.; Lennernäs, H.; Hussain, A.S.; Junginger, H.E.; Stavchansky, S.A.; Midha, K.K.; Shah, V.P.; *et al.* Molecular properties of WHO essential drugs and provisional biopharmaceutical classification. *Mol. Pharm.* **2004**, *1*, 85–96. [CrossRef] [PubMed]

13. Jacob, L.A.; Sreevatsa, A.; Chinnagiriyappa, L.K.; Dasappa, L.; Suresh, T.M.; Babu, G. Methotrexate-induced chemical meningitis in patients with acute lymphoblastic leukemia/lymphoma. *Ann. Indian Acad Neurol.* **2015**, *8*, 206–209.

14. Trapani, G.; Denora, N.; Trapani, A.; Laquintana, V. Recent advances in ligand targeted therapy. *J. Drug Target.* **2012**, *1*, 1–22. [CrossRef] [PubMed]

15. Denora, N.; Laquintana, V.; Trapani, A.; Lopedota, A.; Latrofa, A.; Gallo, J.M.; Trapani, G. Translocator protein (TSPO) ligand-Ara-C (cytarabine) conjugates as a strategy to deliver antineoplastic drugs and to enhance drug clinical potential. *Mol. Pharm.* **2010**, *7*, 2255–2269. [CrossRef] [PubMed]

16. Laquintana, V.; Denora, N.; Lopalco, A.; Lopedota, A.; Cutrignelli, A.; Lasorsa, F.M.; Agostino, G.; Franco, M. Translocator protein ligand-PLGA conjugated nanoparticles for 5-fluorouracil delivery to glioma cancer cells. *Mol. Pharm.* **2014**, *11*, 859–871. [CrossRef] [PubMed]

17. Denora, N.; Laquintana, V.; Lopalco, A.; Iacobazzi, R.M.; Lopedota, A.; Cutrignelli, A.; Iacobellis, G.; Annese, C.; Cascione, M.; Leporatti, S.; *et al. In vitro* targeting and imaging the translocator protein TSPO 18-kDa through G(4)-PAMAM-FITC labeled dendrimer. *J. Control. Release* **2013**, *172*, 1111–1125. [CrossRef] [PubMed]

18. Laquintana, V.; Denora, N.; Musacchio, T.; Lasorsa, M.; Latrofa, A.; Trapani, G. Peripheral benzodiazepine receptor ligand-PLGA polymer conjugates potentially useful as delivery systems of apoptotic agents. *J. Control. Release* **2009**, *137*, 185–195. [CrossRef] [PubMed]

19. Annese, C.; Fanizza, I.; Calvano, C.D.; D'Accolti, L.; Fusco, C.; Curci, R.; Williard, P.G. Selective synthesis of hydroxy analogues of valinomycin using methyl(trifluoromethyl)dioxirane. *Org. Lett.* **2011**, *13*, 5096–5099. [CrossRef] [PubMed]

20. Norinder, U.; Haeberlein, M. Computational approaches to the prediction of the blood-brain distribution. *Adv. Drug Deliv. Rev.* **2002**, *54*, 291–313. [CrossRef]

21. Kreisl, W.C.; Jenko, J.K.; Hines, C.S.; Lyoo, H.C.; Corona, W.; Morse, C.L.; Zoghbi, S.S.; Hyde, T.; Kleinman, J.E.; Pike, V.W.; *et al.* A genetic polymorphism for translocator protein 18 kDa affects both *in vitro* and *in vivo* radioligand binding in human brain to this putative biomarker of neuroinflammation. *J. Cereb. Blood Flow Metab.* **2013**, *33*, 53–58. [CrossRef] [PubMed]

22. Garberg, P.; Ball, M.; Borg, N.; Cecchelli, R.; Fenart, L.; Hurst, R.D.; Lindmark, T.; Mabondzo, A.; Nilsson, J.E.; Raub, T.J.; *et al. In vitro* models for the blood-brain barrier. *Toxicol. Vitr.* **2005**, *19*, 299–334. [CrossRef] [PubMed]

23. He, Y.; Yao, Y.; Tsirka, S.E.; Cao, Y. Cell-culture models of the blood-brain barrier. *Stroke* **2014**, *45*, 2514–2526. [CrossRef] [PubMed]

24. Audus, K.L.; Borchardt, R.T. Bovine brain microvessel endothelial cell monolayers as a model system for the blood–brain barrier. *Ann. N. Y. Acad. Sci.* **1987**, *507*, 9–18. [CrossRef] [PubMed]

25. Denora, N.; Cassano, T.; Laquintana, V.; Lopalco, A.; Trapani, A.; Cimmino, C.S.; Laconca, L.; Giuffrida, A.; Trapani, G. Novel codrugs with GABAergic activity for dopamine delivery in the brain. *Int. J. Pharm.* **2012**, *437*, 221–231. [CrossRef] [PubMed]

26. Laquintana, V.; Trapani, A.; Denora, N.; Wang, F.; Gallo, J.M.; Trapani, G. New strategies to deliver anticancer drugs to brain tumors. *Expert Opin. Drug Deliv.* **2009**, *6*, 1017–1032. [CrossRef] [PubMed]

International Journal of
Molecular Sciences

MDPI

Article

Synthesis, Characterization, and Cytotoxicity of the First Oxaliplatin Pt(IV) Derivative Having a TSPO Ligand in the Axial Position

Salvatore Savino [1], Nunzio Denora [2], Rosa Maria Iacobazzi [2,3], Letizia Porcelli [3], Amalia Azzariti [3], Giovanni Natile [1] and Nicola Margiotta [1,*]

[1] Dipartimento di Chimica, Università degli Studi di Bari Aldo Moro, via E. Orabona 4, 70125 Bari, Italy; salvatoresavino.s@libero.it (S.S.); giovanni.natile@uniba.it (G.N.)

[2] Dipartimento di Farmacia-Scienze del Farmaco, Università degli Studi di Bari Aldo Moro, via E. Orabona 4, 70125 Bari, Italy; nunzio.denora@uniba.it (N.D.); rosamaria.iacobazzi@libero.it (R.M.I.)

[3] Istituto Tumori IRCCS Giovanni Paolo II, viale O. Flacco 65, 70124 Bari, Italy; porcelli.letizia@gmail.com (L.P.); a.azzariti@oncologico.bari.it (A.A.)

* Correspondence: nicola.margiotta@uniba.it; Tel.: +39-080-544-2759

Academic Editor: Nick Hadjiliadis
Received: 15 April 2016; Accepted: 20 June 2016; Published: 25 June 2016

Abstract: The first Pt(IV) derivative of oxaliplatin carrying a ligand for TSPO (the 18-kDa mitochondrial translocator protein) has been developed. The expression of the translocator protein in the brain and liver of healthy humans is usually low, oppositely to steroid-synthesizing and rapidly proliferating tissues, where TSPO is much more abundant. The novel Pt(IV) complex, *cis,trans,cis*-[Pt(ethanedioato)Cl{2-(2-(4-(6,8-dichloro-3-(2-(dipropylamino)-2-oxoethyl)imidazo[1,2-*a*]pyridin-2-yl)phenoxy)acetate)-ethanolato}(1*R*,2*R*-DACH)] (DACH = diaminocyclohexane), has been fully characterized by spectroscopic and spectrometric techniques and tested in vitro against human MCF7 breast carcinoma, U87 glioblastoma, and LoVo colon adenocarcinoma cell lines. In addition, affinity for TSPO (IC_{50} = 18.64 nM), cellular uptake (ca. 2 times greater than that of oxaliplatin in LoVo cancer cells, after 24 h treatment), and perturbation of cell cycle progression were investigated. Although the new compound was less active than oxaliplatin and did not exploit a synergistic proapoptotic effect due to the presence of the TSPO ligand, it appears to be promising in a receptor-mediated drug targeting context towards TSPO-overexpressing tumors, in particular colorectal cancer (IC_{50} = 2.31 µM after 72 h treatment).

Keywords: oxaliplatin; antitumor drugs; platinum(IV); translocator protein 18 kDa; TSPO; colorectal cancer

1. Introduction

The discovery of the antitumor activity of cisplatin, *cis*-diamminedichloridoplatinum(II) (Figure 1) [1,2], was a corner stone that triggered the interest in the development of platinum(II)/(IV) complexes in oncology. However, only a few of these complexes—carboplatin and oxaliplatin (Figure 1)—have been approved worldwide for the use in the clinics [3–5].

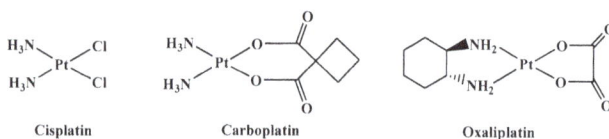

Figure 1. Pt(II) complexes worldwide clinically used.

The second generation platinum drug carboplatin, *cis*-diammine(1,1-cyclobutanedicarboxylato) platinum(II), contains a more stable leaving group (1,1-cyclobutanedicarboxylato) with respect to the chlorides present in cisplatin. This modification was introduced to lower the toxicity without affecting the spectrum of antitumor activity. Effectively, carboplatin is less neuro- and nephro-toxic and can be administered at higher doses than cisplatin [6].

The third generation platinum drug oxaliplatin—*trans*-1*R*,2*R*-diaminocyclohexane(oxalato) platinum(II)—contains the diaminocyclohexane (1*R*,2*R*-DACH) chelating ligand and received worldwide approval for the treatment of colorectal cancer [7]. Although oxaliplatin produces the same type of lesions on DNA as cisplatin, its spectrum of activity is different from that of the first-generation drug and also the occurrence of resistance to oxaliplatin is different from that of cisplatin and carboplatin. The mechanism of action responsible for the different activity of oxaliplatin, with respect to cisplatin, has not yet been completely understood. However, it has been demonstrated that the non-leaving diamine has a fundamental role in the recognition of {(DACH)Pt}-DNA adducts by DNA-repair proteins. These adducts, in turn, could contribute to the absence of cross-resistance with the other two Pt-drugs cisplatin and carboplatin [8]. Neurotoxicity is the major dose-limiting feature associated with the use of oxaliplatin [9].

Due to their scarce water solubility, platinum(II) complexes are administered by intravenous infusion in the clinics, with low compliance in treated patients [10]. This drawback can be overcome with platinum(IV)-based drugs which are known to be much more resistant (with respect to Pt(II) complexes) to substitution from biomolecules and are hence suitable for oral administration [11,12]. Pt(IV) compounds are generally considered prodrugs since they must be activated by intracellular reduction to their Pt(II) counterparts, therefore the structure, substituents and reduction potential of Pt(IV) complexes are strictly correlated to their antitumor activity. Hambley and colleagues showed that in a series of ethylenediamine-based Pt(IV) complexes the cathodic potential for the reduction of Pt(IV) to Pt(II) depends upon the nature of the axial ligands and decreases in the order Cl > OCOR > OH [13].

Axial ligands also represent a possible way to link biovectors with the ability to direct the complex toward a tumor. Overall activity would not be altered since these ligands are released upon reduction with generation of the active Pt(II) metabolite [14].

The peripheral-type benzodiazepine receptor (PBR) has been recently re-named 18-kDa mitochondrial translocator protein (TSPO) and has been used as target for several potential drugs with therapeutic and imaging uses [15–18].

The expression of TSPO in the brain and liver of healthy humans is usually low. On the contrary, in steroid-synthesizing and rapidly proliferating tissues, TSPO is more abundant. In several pathologies such as brain, breast, colon, prostate, and ovarian cancers and in astrocytomas and hepatocellular and endometrial carcinomas, an overexpression of the TSPO has been found [19,20]. Neurodegenerative diseases such as Alzheimer, Parkinson, Huntington, and multiple sclerosis, that are generally associated with inflammatory processes, also show high levels of TSPO expression in activated microglial cells [21].

It appears clear why TSPO gained much attraction for its use as an intracellular target for imaging of pathologic tissues overexpressing this protein and also for selectively targeting the functions associated with mitochondrial activity [22–25]. Several classes of ligands having high affinity for TSPO have been developed for diagnostic or therapeutic uses and, in some cases, for both (theranostic agents) [21]. Some of us have exploited the 2-phenyl-imidazo[1,2-*a*]pyridine-*N*,*N*-dipropylacetamide scaffold specific of the anxyolitic drug alpidem [22–24] and, as an example, the PET imaging ^{18}F-labeled agent 2-(2-(4-(2-[^{18}F]fluoroethoxy)phenyl)-6,8-dichloroimidazo[1,2-*a*]pyridine-3-yl)-*N*,*N*-dipropylacetamide was designed as a biomarker for the diagnosis of neuroinflammation, neurodegeneration, and tumor progression [26].

Pursuing our interest for tissue-specific targeting of metal complexes, in recent papers we have synthesized platinum(II) compounds with ligands specific for TSPO, such as 2-[6,8-dichloro-2-(1,3-thiazol-2-yl)*H*-imidazo-[1,2-*a*]pyridin-3-yl]-*N*,*N*-dipropylacetamide (TZ6). The resulting cisplatin-like compound, *cis*-[PtCl$_2$(TZ6)] (Figure 2), has been shown to possess affinity and selectivity for the TSPO comparable to those of TZ6. In solvents with low dielectric constants, we also observed the formation of a dimeric aggregate formed through non-covalent intermolecular interactions of the planar aromatic cycles of the ligands. This finding further supports the potential intercalating ability of *cis*-[PtCl$_2$(TZ6)] toward DNA [27]. The poor aqueous solubility of *cis*-[PtCl$_2$(TZ6)], which is typical of platinum(II) complexes with bidentate aromatic ligands [28], compelled us to pursue the synthesis of Pt compounds with TSPO ligands endowed with enhanced water solubility. To this end, in a subsequent work we synthesized two new Pt compounds structurally analogous to picoplatin and differing for the anionic ligands, *cis*-[PtI$_2$(NH$_3$){[2-(4-chlorophenyl)-8-aminoimidazo[1,2-*a*]pyridin-3-yl]-*N*,*N*-di-*n*-propylacetamide}] and *cis*-[PtCl$_2$(NH$_3$) {[2-(4-chlorophenyl)-8-aminoimidazo[1,2-*a*]pyridin-3-yl]-*N*,*N*-di-*n*-propylacetamide}] (Figure 2) [29]. Both complexes contained the imidazopyridinic ligand [2-(4-chlorophenyl)-8-aminoimidazo-[1,2-*a*]pyridin-3-yl]-*N*,*N*-di-*n*-propylacetamide and were endowed with high affinity and selectivity for TSPO [30].

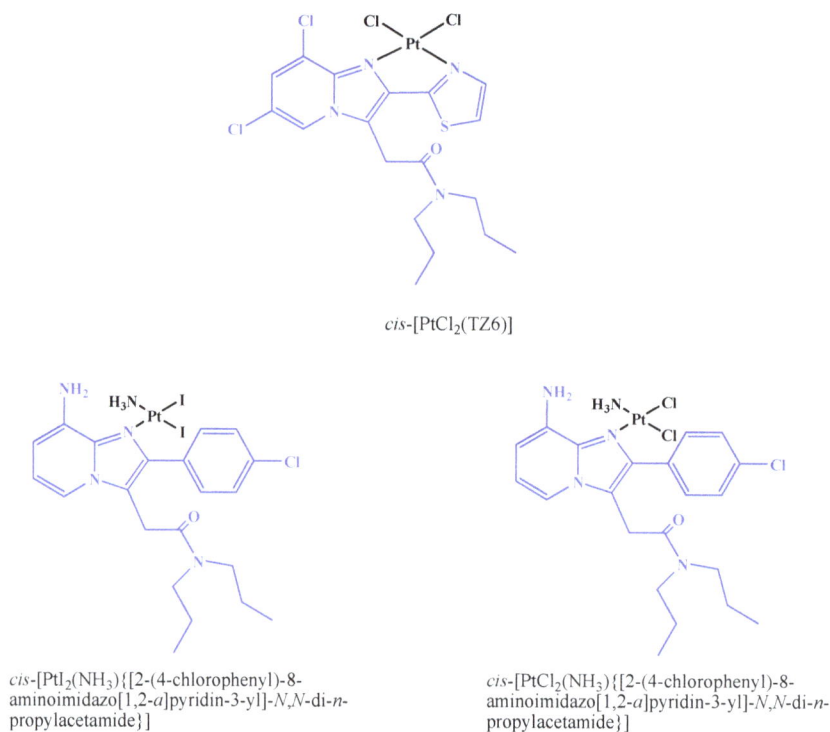

cis-[PtCl$_2$(TZ6)]

cis-[PtI$_2$(NH$_3$){[2-(4-chlorophenyl)-8-aminoimidazo[1,2-*a*]pyridin-3-yl]-*N*,*N*-di-*n*-propylacetamide}]

cis-[PtCl$_2$(NH$_3$){[2-(4-chlorophenyl)-8-aminoimidazo[1,2-*a*]pyridin-3-yl]-*N*,*N*-di-*n*-propylacetamide}]

Figure 2. Pt(II) complexes with TSPO ligands (highlighted with a blue color).

In addition, the two compounds were massively accumulated in glioma cells (10- to 100-fold enhanced accumulation) and were capable of inducing apoptosis similar to cisplatin [30].

In this work we have developed the first Pt(IV) complex containing a TSPO ligand in axial position (the two axial positions are those perpendicular to the square-planar coordination plane of the platinum(II) precursor).

The design of a Pt(IV) complex with a TSPO ligand in the axial position was motivated by the reasons mentioned earlier in this paragraph: the targeting of the tumor site and the potential synergistic proapoptotic effect caused by the TSPO ligand released by intracellular reduction of the Pt(IV) conjugate. We selected the oxaliplatin derivative *cis,trans,cis*-[Pt(ethanedioato)Cl(2-hydroxyethanolato)

(1*R*,2*R*-DACH)] (DACH = diaminocyclohexane) as Pt(IV) precursor (compound **1** in Figure 3) [31]. The choice of this monofunctional Pt(IV) derivative of oxaliplatin was guided by several considerations: (i) it contains a single reactive group (the terminal hydroxyl moiety of 2-hydroxyethanolato), thus preventing the formation of different condensation products (mono- and di-substituted) in the conjugation with biovectors; (ii) the presence of an axial chloride makes the reduction potential less negative (i.e., "easier" reduction from a thermodynamic point of view) and increases the reduction rate [12,32,33]; (iii) the 2-hydroxyethanolato residue should act as a spacer increasing the distance between the conjugated TSPO ligand and the Pt center, allowing an easier interaction with the receptor.

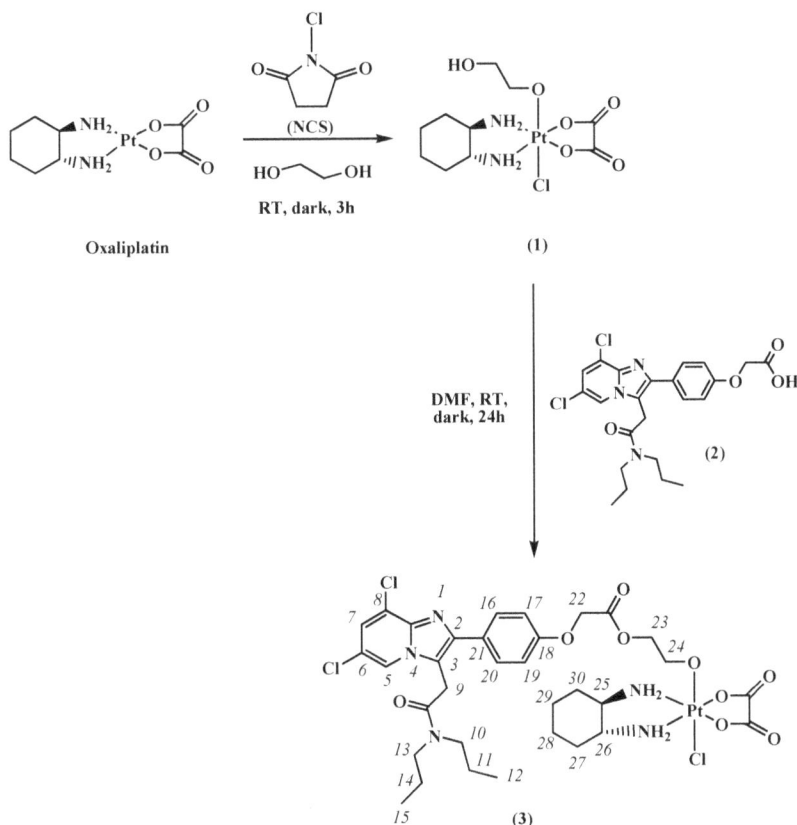

Figure 3. Synthesis of the novel Pt(IV) derivative of oxaliplatin with TSPO ligand in axial position (**3**) and numbering of protons. RT = room temperature.

Hence, the Pt(IV) precursor was tethered to a potent TSPO ligand characterized by a 2-phenyl-imidazo[1,2-*a*]pyridine acetamide structure and containing a terminal carboxylic group useful for its further conjugation, 2-(4-(6,8-dichloro-3-(2-(dipropylamino)-2-oxoethyl)imidazo [1,2-*a*]pyridin-2-yl)phenoxy)acetic acid (compound **2** in Figure 3) [34]. In this way we obtained a novel Pt(IV) derivative of oxaliplatin carrying a ligand for TSPO, *cis,trans,cis*-[Pt(ethanedioato)Cl {2-(2-(4-(6,8-dichloro-3-(2-(dipropylamino)-2-oxoethyl)imidazo[1,2-*a*]pyridin-2-yl)phenoxy)acetate)-ethanolato}(1*R*,2*R*-DACH)] (compound **3** in Figure 3).

The novel Pt(IV) complex has been fully characterized by spectroscopic and spectrometric techniques and tested in vitro against human MCF7 breast carcinoma, human U87 glioblastoma, and human LoVo colon adenocarcinoma cell lines. In addition, the affinity of **3** for TSPO was assessed by receptor binding assays, measuring its ability to displace [³H]-1-(2-chlorophenyl)-*N*-methyl-*N*-(1-methylpropyl)-3-isoquinolinecarboxamide ([³H]-PK 11195). Cellular uptake experiments were performed in order to correlate the cytotoxicity of **3** with its cellular uptake ability. Finally, cell cycle analysis was performed to evaluate the capability of **3** to perturb the LoVo cells cycle progression.

2. Results and Discussion

2.1. Synthesis and Characterization of 3

Although platinum(IV) complexes were identified by Rosenberg as having anticancer activity at the same time of the discovery of that of cisplatin [1], their clinical efficacy has been tested only recently. The properties of Pt(IV) complexes are substantially different from those of Pt(II) species since six-coordinate octahedral platinum(IV) complexes have a saturated coordination sphere that renders them less prone to substitution reactions than platinum(II) complexes. In addition, the possibility to coordinate two additional ligands in the axial positions enables tuning of the chemical properties and conjugation of cancer-targeting ligands. The preliminar intracellular reduction to Pt(II), with simultaneous release of the two axial ligands, is necessary for Pt(IV) complexes to exert their antitumor activity. Thus, Pt(IV) complexes are generally defined as prodrugs of platinum-based drugs. In addition, the intracellular reduction of Pt(IV) complexes releases the axial ligands that can be themselves biologically active. The Pt(IV) precursor is indeed a dual-threating pharmaceutical agent, which combines two biologically active components into a single molecule [12,14]. In specifically designed Pt(IV) complexes, the Pt-core and the released axial ligands may have different intracellular targets; moreover, the axial ligands may exhibit a targeting property towards substrates that are specifically overexpressed in tumor cells. This is the rationale that prompted us to prepare a Pt(IV) complex with a TSPO ligand in the axial position.

Most of the Pt(IV) complexes present in the literature are prepared via oxidation of predesigned Pt(II) compounds with chlorine [35–37] or hydrogen peroxide [38], leading to Pt(IV) complexes with symmetric axial ligands. However, more recently, unsymmetric Pt(IV) complexes have also been synthesized [39,40].

For the preparation of the unsymmetric Pt(IV) derivative of oxaliplatin with just one TSPO ligand in axial position, we first treated oxaliplatin (Figure 3) with *N*-chlorosuccinimide (NCS)—a source of "positive chlorine"—in ethane-1,2-diol, obtaining the asymmetric compound **1**, as previously reported [31,41]. This synthetic strategy is based on the observation that the oxidative addition to Pt(II) complexes is generally assisted by the solvent coordinating opposite to the attacking "positive chlorine" [37,42,43]. In the present case, by using ethane-1,2-diol as solvent, this latter coordinates *trans* to chlorine [41]. The Pt(IV) complex **1** was then conjugated with **2** [34] forming the ester complex *cis,trans,cis*-[Pt(ethanedioato)Cl{2-(2-(4-(6,8-dichloro-3-(2-(dipropylamino)-2-oxoethyl)imidazo[1,2-*a*]pyridin-2-yl)phenoxy)acetate)-ethanolato}(1*R*,2*R*-DACH)] (**3**).

With reference to the coupling reaction, the best result was obtained using EDC and HOBt as coupling reagents in the presence of TEA in DMF; a similar synthetic approach was used for the conjugation of the glycolic monomer of PEG with PLGA [44].

Compound **3** was characterized by elemental analysis, ESI-MS, and NMR spectroscopy. In the ESI-MS spectrum a peak at $m/z = 976.33$, corresponding to [**3** + Na]⁺, was evident (data not shown). The experimental isotopic pattern of this peak was in good agreement with the theoretical one.

The NMR characterization of the compound started from the assignment of methyl protons 12 and 15 (triplets falling at 0.82 and 0.87 ppm) of the dipropylacetamidic chains (see Figure 3 for numbering of protons), that show TOCSY cross-peaks (Figure 4) with methylenic protons 11 and 14 (overlapping multiplets at 1.51 and 1.61 ppm). These signals show TOCSY cross-peaks with two signals, overlapping with the signal of the solvent, assigned to methylenes 10 and 13 falling at 3.34 and 3.25 ppm. The two singlets falling at 4.25 and 4.83 ppm were assigned to the methylene groups 9 and 22, respectively.

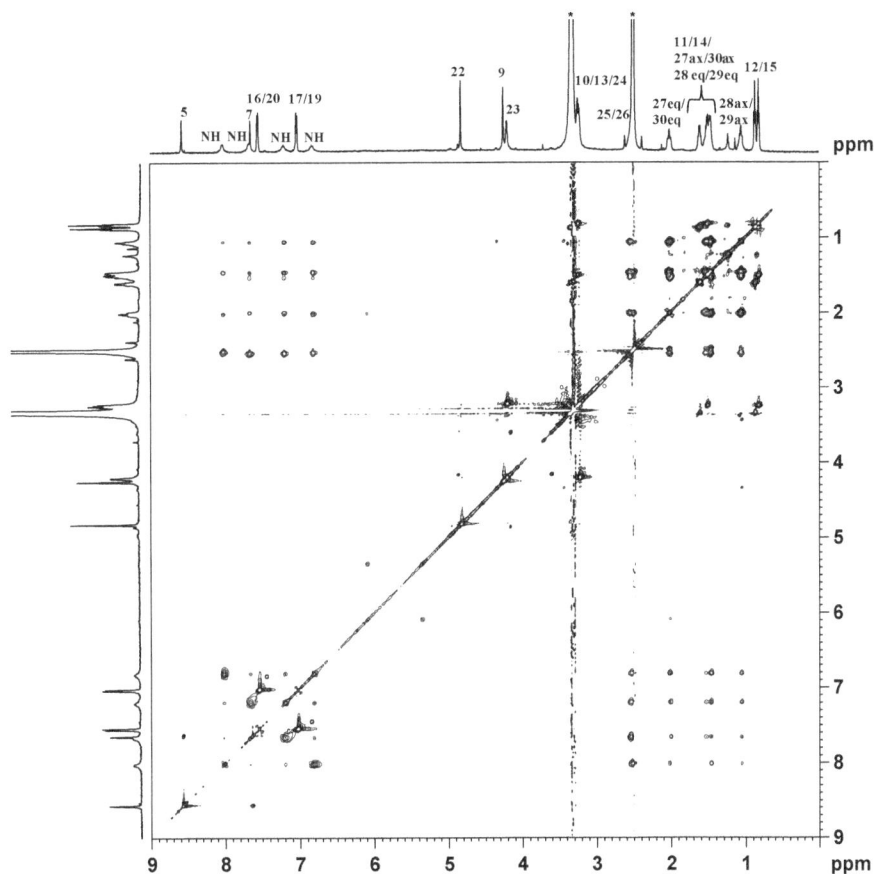

Figure 4. ^{1}H (**top** and **left**) and TOCSY 2D (**center**) NMR spectra (600 MHz, ^{1}H) of **3** in DMSO-d_{6}. The asterisks indicate residual solvent peaks.

This assignment was confirmed by a NOESY 2D NMR experiment that shows a NOESY cross-peak between the methylene group 9 and the imidazopyridine proton 5 located at 8.57 ppm (cross-peak **A** in Figure 5), and an additional NOESY cross-peak of the methylene 22 with the protons 17 and 19 of the phenoxy ring located at 7.03 ppm (cross-peak **B** in Figure 5).

Proton 7 of the imidazopyridine ring was assigned to the singlet at $\delta = 7.65$ ppm while the doublet located at 7.55 ppm, integrating for two protons, was assigned to protons 16 and 20 of the phenoxy ring (Figure 4, TOCSY cross-peak with protons 17 and 19). With reference to the hydroxyethanolato spacer, the multiplet at 4.20 ppm was assigned to the methylenic protons 23. This signal shows a TOCSY cross-peak (Figure 4) with the multiplet located at 3.23 ppm (overlapping with the signal of water) assigned to methylene 24. The 1R,2R-cychlohexanediamine is in a chair conformation and, owing to the axial asymmetry of the Pt complex, different signals are expected for both the axial and the equatorial protons. Therefore, the multiplet at 1.06 ppm was assigned to the axial protons 28 and 29, which showed TOCSY cross-peaks with the multiplet at 1.43–1.50 ppm (assigned to the axial protons 27 and 30 and to the equatorial protons 28 and 29 of 1R,2R-DACH), the multiplet at 2.01 ppm (attributed to equatorial protons 27 and 30), and two signals overlapping with the solvent (assigned to the methynic protons 25 and 26). All the chemical shifts found for the methylenic and methynic protons of DACH are in good agreement with those reported for similar asymmetric Pt(IV) octahedral complexes having a chlorido and glycolic moiety in the axial positions [41].

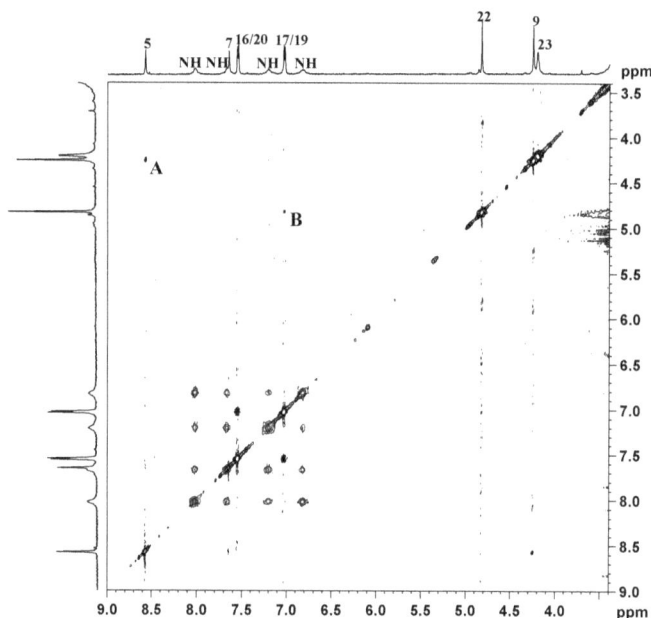

Figure 5. Portion of ^{1}H (**top** and **left**) and NOESY 2D (**center**) NMR spectra (600 MHz, ^{1}H) of **3** in DMSO-d_6.

As far as the amine groups are concerned, it is possible to observe (Figure 4, top) the presence of four different signals for the NH$_2$ protons of coordinated DACH located at 6.83, 7.21, 7.67 and 8.02 ppm (only two signals are observed in the case of complexes with equal axial ligands).

The [^{1}H-^{195}Pt]-HSQC 2D NMR spectrum recorded in DMSO-d_6 is reported in Figure 6. The spectrum shows four cross peaks falling at 6.83/887.2, 7.21/887.2, 7.67/887.2, and 8.02/887.2 ppm (^{1}H/^{195}Pt) due to the coupling of ^{195}Pt with the four magnetically non-equivalent NH$_2$ protons of DACH. In addition, a cross-peak located at 3.23/887.2 ppm is assigned to the coupling between ^{195}Pt and the CH$_2$ of the glycolic linker. The ^{195}Pt chemical shift is in good agreement with those reported for similar asymmetric Pt(IV) octahedral complexes with a PtClN$_2$O$_3$ core [41]. The assignment of ^{13}C signals has been accomplished by a [^{1}H-^{13}C]-HSQC 2D NMR spectrum (Figure 7 and Table 1). The chemical shifts found for 1*R*,2*R*-DACH and the glycolic linker are in good agreement with those reported for similar Pt(IV) complexes [41]. In particular, the methylene groups of DACH fall at 23.5 and 30.1 ppm, two values very similar to those reported for the product of esterification between BOC-L-alanine and *cis,trans,cis*-[Pt(cyclobutane-1,1′-dicarboxylate)Cl(2-hydroxyethanolato)(1*R*,2*R*-DACH)] (23.5, 30.1 and 30.7 ppm) [41]. The cross peak at 2.54/61.5 ppm (^{1}H/^{13}C) was assigned to the methynic group of DACH while the cross-peaks at 3.23/65.5 and 4.20/64.3 ppm (^{1}H/^{13}C) were assigned, respectively, to C24 and C23 of the glycolic moiety (66.0 and 64.5 ppm in the ester compound between BOC-L-alanine and *cis,trans,cis*-[Pt(cyclobutane-1,1′-dicarboxylate)Cl(2-hydroxyethanolato)(1*R*,2*R*-DACH)] mentioned above) [41]. The cross-peaks located at 122.3 and 123.5 ppm (^{13}C) were assigned to C5 and C7 of the imidazopyridine ring, respectively. The cross peaks falling at 7.03/114.5 and 7.55/129.0 ppm (^{1}H/^{13}C) were assigned to carbon atoms 16/20 and 17/19 of the phenyl ring, respectively. The ^{13}C chemical shift of methylene groups 9 and 22 fall at 28.7 and 64.4 ppm, respectively. The dipropylacetamidic chain gives two cross peaks at 0.82/10.9 and 0.87/10.9 ppm for the two methyl groups, two signals at

20.3 and 21.5 ppm belonging to methylene groups 11 and 14, and two signals at 46.9 and 48.9 ppm belonging to methylene groups 10 and 13.

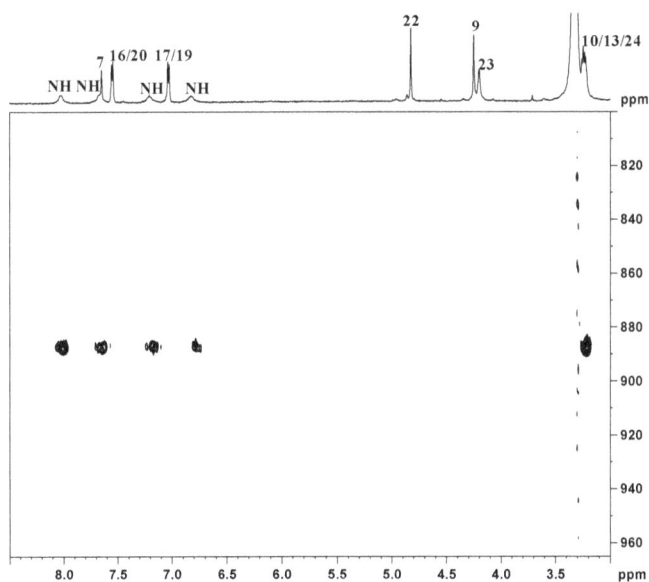

Figure 6. [^{1}H-^{195}Pt]-HSQC 2D NMR (300 MHz, ^{1}H) of **3** in DMSO-d_6.

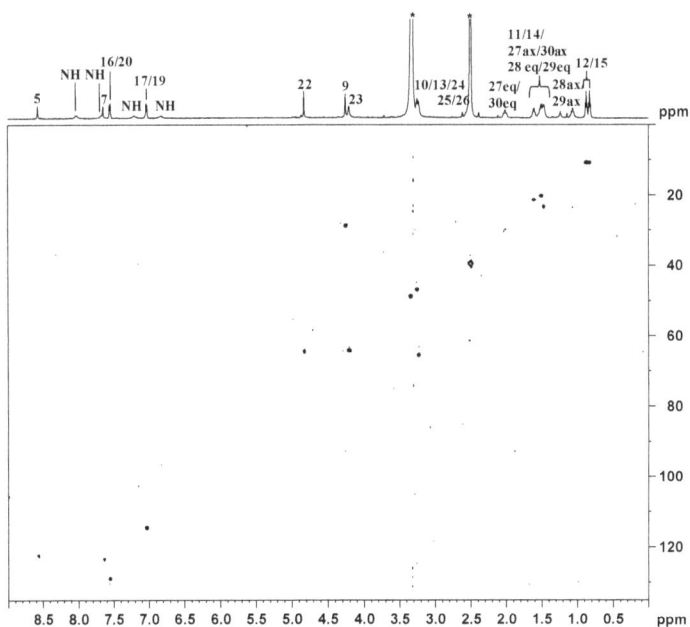

Figure 7. ^{1}H (**top**) and [^{1}H-^{13}C] HSQC 2D (**bottom**) NMR (600 MHz, ^{1}H) spectra of **3** in DMSO-d_6. The asterisks indicate residual solvent peaks.

Table 1. Summary of ^{13}C chemical shifts of **3** in DMSO-d_6.

C	δ ^{13}C (ppm)
5	122.3
7	123.5
9	28.7
10 or 13	46.9
10 or 13	48.9
11 or 14	20.3
11 or 14	21.5
12/15	10.9
16/20	129
17/19	114.5
22	64.4
23	64.3
24	65.5
CH$_2$ DACH	23.5
CH$_2$ DACH	30.1
CH DACH	61.5

2.2. Stability of 3

The stability of **3** was investigated by ^{1}H-NMR spectroscopy in DMSO-d_6/D$_2$O (5:95, v/v) at pH 7.4 (2 mM phosphate buffer) and 37 °C. A portion of the spectra, recorded at different times, is reported in Figure 8. The spectrum acquired soon after dissolution shows only the peaks of **3** (H5, 8.11 ppm; H7, 7.45 ppm; H16/20, 7.39 ppm; marked with ✖ in Figure 8a). After 8 h of incubation we observed a decrease in intensity of the peaks of **3** and the concomitant increase of a new set of peaks, marked with ● in Figure 8b (H5, 8.09 ppm; H7, 7.44 ppm; H16/20 7.35 ppm), that was assigned to compound **2** by comparison with a spectrum of the TSPO ligand recorded in similar conditions. This new set of signals indicates that hydrolysis of the ester bond linking **2** to the ethanolato linker occurs in physiological-like conditions. The spectra recorded at 24, and 56 h showed a further decrease of the signals of **3**, which disappeared completely after 108 h. The half-life of the complex, as calculated from the NMR experiment, was determined to be approximately 24 h.

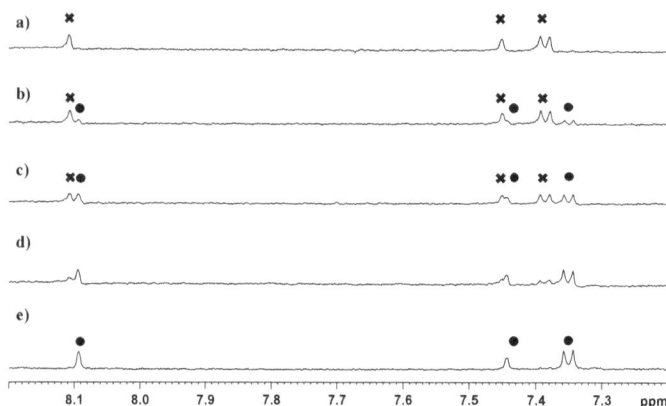

Figure 8. Portion of the ^{1}H NMR (600 MHz, ^{1}H) spectra of compound **3** dissolved in DMSO-d$_6$/D$_2$O (5:95 v/v) and incubated at pH 7.4 (2 mM phosphate buffer) and 37 °C. Spectra were recorded at zero time (**a**) and after 8 (**b**); 24 (**c**); 56 (**d**); and 108 h (**e**). ✖ indicates peaks relevant to the portion of TSPO ligand in complex **3**. ● indicates peaks of free TSPO ligand (**2**). The amine protons were deuterated due to the fast H/D exchange.

An HPLC analysis of the sample was performed after 108 h incubation at 37 °C. Two peaks were visible in the chromatogram (not shown) having retention times of 8.7 and 21.81 min, which are comparable to those found for the Pt(IV) precursor **1** and the free TSPO ligand **2**, respectively. Therefore, the HPLC investigation confirmed that **3** undergoes a slow hydrolysis of the ester bond in physiological-like conditions. No reduction of the Pt(IV) complex occurred under the conditions used in this experiment.

2.3. Biological Assays

The affinity towards TSPO of complex **3** and of the free ligand **2** was evaluated by testing their ability to prevent binding to the rat cerebral cortex of the selective TSPO ligand [^{3}H]-PK 11195. The results, expressed as inhibitory concentration (IC$_{50}$), are listed in Table 2. Compound **3** showed high affinity for the TSPO receptor (IC$_{50}$ of 18.64 nM), although it was lower than that found for the free ligand **2** (IC$_{50}$ of 2.12 nM). The free ligand **2**, in turn, was more affine than the reference ligand PK 11195 (IC$_{50}$ of 4.27 nM). These results indicate that the choice of a terminal carboxylic residue on the free ligand **2** for conjugation to a Pt(IV) complex was correct and does not significantly alter the affinity of **2** for the TSPO.

Table 2. Affinities of compounds **2** and **3** for TSPO from rat cerebral cortex.

Compound	IC$_{50}$ (nM) [a]
	TSPO [b]
2	2.12 ± 0.10
3	18.64 ± 0.84
PK 11195	4.27 ± 0.22

[a] Data are means of three separate experiments performed in duplicate which differed by less than 10%;
[b] PK 11195, a selective ligand for TSPO 18-kDa, was used for comparison.

The cytotoxic activity of complex **3** and of its precursors was assessed by the MTT in vitro assay performed in human MCF7 breast carcinoma, human U87 glioblastoma, and human LoVo colon adenocarcinoma cell lines. The cytotoxicity of the compounds, expressed as IC$_{50}$ values after 72 h incubation (except in two cases where it is expressed as the % of cell viability at the maximum concentration tested of 100 μM) are reported in Table 3.

Table 3. Concentration inducing 50% cell survival inhibition (IC$_{50}$) (except in two cases where is given the % of cell viability at the maximum concentration of 100 μM) after 72 h treatment.

Cell Lines	IC$_{50}$ (μM) or % Cell Viability at 100 μM			
	1	2	3	Oxaliplatin
MCF7	8.2 ± 0.4	53% ± 1%	14.1 ± 0.1	5. 4 ± 0.4
U87	9.1 ± 0.4	70.2 ± 0.3 (μM)	16.1 ± 0.3	3.1 ± 0.2
LoVo	2.5 ± 0.5	65% ± 4%	2.3 ± 0.1	0.46 ± 0.01

Compound **2** showed only a marginal cytotoxic effect at 100 μM concentration (the highest concentration used in the investigation) in the case of MCF7 and LoVo cells and a weak activity against U87 cell line (IC$_{50}$ = 70.2 μM). Moreover, we could assess that compound **2** causes cell death through induction of apoptosis as evidenced by the formation of a sub-G0/G1 cell population (cell cycle analysis of LoVo cell line, see following discussion). These results are in line with previous results indicating that TSPO ligands, constructed on the imidazopyridinic scaffold, are endowed with good proapoptotic activity [30] although, in some cases, high concentrations are needed to achieve cytotoxic and proapoptotic effects, as also demonstrated for PK 11195 [45]. In particular, Kugler et al. [45] evidenced a peculiar property of TSPO ligands: being inert when no challenge is present but counteracting programmed cell death when lethal agents are present. They are therefore recognized to be potentially useful for the treatment of brain trauma and neurodegenerative brain diseases [45]. In the present case it is difficult to say if the low activity of compound **3** is due to the TSPO ligand **2** preventing cell death induction by oxaliplatin or to the intrinsic lower activity in cellular experiments of Pt(IV) species as compared to their Pt(II) counterparts. In this regard, further experiments will be carried out by giving the ligand **2** and the oxaliplatin derivative **1** simultaneously but as individual components.

In all three cell lines compound **3** had activity comparable to that of the Pt(IV) precursor complex **1** (IC$_{50}$ ratios of 1.7, 1.8, and 0.92 for MCF7, U87, and LoVo cell lines, respectively). In the above mentioned cell lines both Pt(IV) compounds were less active than oxaliplatin. This latter result is not surprising since Pt(IV) compounds usually show higher IC$_{50}$ values in in vitro investigation with respect to the related Pt(II) compounds since they require a preliminary intracellular reduction process before exerting their cytotoxic effect [46].

As already reported [8], oxaliplatin is a chemotherapeutic drug mainly used for the treatment of colorectal cancer, therefore, we extended our investigation only to LoVo cells. First we measured the intracellular accumulation of platinum by ICP-MS. LoVo cells were exposed to IC$_{50}$ concentrations of compounds **3**, **1**, and oxaliplatin for a short (4 h) and a long (24 h) period. Interestingly, the results (Table 4) indicate that for 24 h incubation the uptake of compound **3** is greater than that of **1** and that the latter is greater than that of oxaliplatin (with a ratio of 2 between the uptake of **3** and that of oxaliplatin). Interestingly, for the shorter time (4 h) the Pt uptake follows the order **3** > oxaliplatin > **1** indicating that factors other than passive diffusion might play a role at short time exposure. Most likely, organic cation transporters (OCT) may facilitate the uptake of oxaliplatin [47] at short time exposure and then, after saturation, the passive diffusion of the most lipophilic compounds becomes prominent.

Table 4. Uptake by LoVo colon cancer cells of oxaliplatin and compounds **1** and **3**.

Treatment Time	Uptake by LoVo Cells (ppb of Pt)		
	Oxaliplatin	1	3
after 4 h treatment	0.24 ± 0.04	0.18 ± 0.01	0.33 ± 0.02
after 24 h treatment	0.38 ± 0.01	0.59 ± 0.03	0.77 ± 0.02

There does not appear to be a direct correlation between Pt uptake and cytotoxic effect (72 h) but, in our opinion, this is an intrinsic limitation of cellular testing where Pt(IV) compounds, which require reductive activation, generally result in less cytotoxicity than Pt(II) species and the directing role of TSPO ligands might remain hidden. In this case a more reliable answer can come from in vivo experiments which are being planned.

The capability of the new compounds **1** and **3** to perturb the cell cycle progression of LoVo cells at their IC$_{50}$ concentrations was investigated by FCM analysis and compared to that of oxaliplatin and of the TSPO ligand **2** (Figure 9). After 24 h treatment with the test compounds, cells exhibited only a slight modification of the cell cycle distribution. As expected, oxaliplatin delayed the S-phase and started to accumulate cells in G2/M-phase and G0/G1 phase [48]. Compounds **3** and **1** induced no notable changes compared to oxaliplatin. The TSPO ligand **2** induced a sub-G0/G1 cell population after 24 h, consistent with the formation of apoptotic cells as previously demonstrated for this class of compounds [30].

Compound **3** did not induce apoptosis as the free TSPO ligand **2** did, however, this is because the concentration of compound **2** released after intracellular hydrolysis of compound **3** is not sufficient to exert apoptosis (note that compound **2** was administered at a much higher dose due to its lower cytotoxicity).

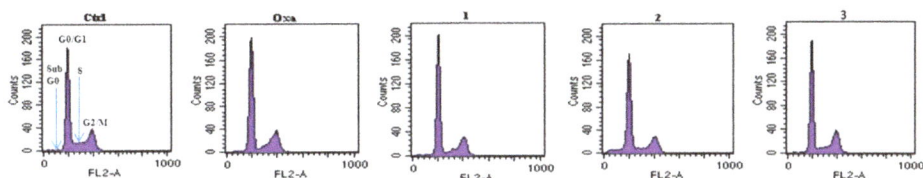

Figure 9. Flow cytometric analysis of LoVo cells stained with propidium iodide after 24 h of treatment with test compounds.

3. Experimental Section

3.1. Materials and Methods

Commercial reagent grade chemicals and solvents were used as received without further purification. ^{1}H-NMR, COSY, TOCSY, NOESY, and [^{1}H-^{13}C]-HSQC 2D NMR spectra were recorded on a Bruker Avance III 600 MHz instrument (Bruker Italia S.r.l., Milano, Italy). Spectra of [^{1}H-^{195}Pt] HSQC 2D NMR were recorded on a Bruker Avance DPX 300 MHz instrument (Bruker Italia S.r.l., Milano, Italy). Chemical shifts of ^{1}H and ^{13}C were referenced using the internal residual peak of the solvent (DMSO-d_6: 2.50 ppm for ^{1}H and 39.51 ppm for ^{13}C). ^{195}Pt NMR spectra were referenced relative to K_2PtCl_4 (external standard placed at -1620 ppm with respect to $Na_2[PtCl_6]$) [49].

Electrospray Mass Spectrometry (ESI-MS) experiments were performed with a dual electrospray interface and a quadrupole time-of-flight mass spectrometer (Agilent 6530 Series Accurate-Mass Quadrupole Time-of-Flight (Q-TOF) LC/MS, Agilent Technologies Italia S.p.A., Cernusco sul Naviglio, Italy).

Elemental analyses were carried out with an Eurovector EA 3000 CHN instrument (Eurovector S.p.A., Milano, Italy).

Oxaliplatin was synthesized as previously reported [50].

Cis,trans,cis-[Pt(ethanedioato)Cl(2-hydroxyethanolato)(1*R*,2*R*-DACH)] (**1**) [31] and 2-(4-(6,8-dichloro-3-(2-(dipropylamino)-2-oxoethyl)imidazo[1,2-*a*]pyridin-2-yl)phenoxy)acetic acid (**2**) [34] were prepared according to already reported procedures.

3.2. Synthesis of cis,trans,cis-[Pt(ethanedioato)Cl{2-(2-(4-(6,8-dichloro-3-(2-(dipropylamino)-2-oxoethyl) imidazo[1,2-a]pyridin-2-yl)phenoxy)acetate)-ethanolato}(1R,2R-DACH)] (3)

A solution of **2** (37.6 mg, 0.079 mmol) and 1-Ethyl-3-(3-dimethylaminopropyl)carbodiimide (EDC, 18 µL, 0.102 mmol) in 2 mL of dry DMF was stirred at room temperature. After 5 min, 1-hydroxybenzotriazole hydrate (HOBt·H_2O, 15.6 mg, 0.102 mmol) was added and the mixture was stirred at room temperature for 10 min. Then, a solution containing **1** (38.8 mg, 0.079 mmol) and triethylamine (TEA, 14 µL, 0.102 mmol), dissolved in 3 mL of dry DMF, was added dropwise and the reaction mixture was left stirring in the dark at room temperature. After 24 h, the solvent was removed by evaporation under reduced pressure. Compound **3** was purified by direct-phase chromatography using silica gel as stationary phase and a solution of 90:10 chloroform/methanol as eluent (Rf = 0.4). Yield: 21 mg (28%). *Anal.: calculated for* $C_{33}H_{42}Cl_3N_5O_9Pt \cdot 2H_2O$ (**3**·$2H_2O$): C, 40.03; H, 4.68; N, 7.07%. *Found*: C, 39.84; H, 4.60; N, 6.87%. ESI-MS: *calculated for* $C_{33}H_{42}Cl_3N_5O_9PtNa$ $[1 + Na]^+$: 976.15. *Found*: *m*/*z* 976.33. ^{1}H NMR (DMSO-d_6): δ = 0.82 (t, 3H, *J* = 7.2 Hz, CH_3, H_{12} or H_{15}), 0.87 (t, 3H, *J* = 7.2 Hz, CH_3, H_{12} or H_{15}), 1.06 (m, 2H, $CH_{2(ax)}$, H_{28ax}/H_{29ax}), 1.43–1.50 (m, 4H, $CH_{2(ax)}$, $CH_{2(eq)}$, H_{27ax}/H_{30ax}, H_{28eq}/H_{29eq}), 1.51 (m, 4H, CH_2, H_{11} or H_{14}), 1.61 (m, 4 H, CH_2, H_{11} or H_{14}), 2.01 (m, 2H, $CH_{2(eq)}$, H_{27eq}/H_{30eq}), 2.54 (overlapping with the signal of DMSO-d_6, 2H, CH, H_{25}/H_{26}), 3.23 (m, 2H, CH_2, H_{24}), 3.25 (m, 2H, CH_2, H_{10} or H_{13}), 3.34 (overlapping with the signal of the residual water, 2H, CH_2, H_{10} or H_{13}), 4.20 (m, 2H, CH_2, H_{23}), 4.25 (s, 2H, CH_2, H_9), 4.83 (s, 2H, CH_2, H_{22}), 6.83 (b, 1H, NH_2), 7.03 (d, *J* = 8.2 Hz, 2H, CH, H_{17}/H_{19}), 7.21 (b, 1H, NH_2), 7.55 (d, *J* = 8.2 Hz, 2H, CH, H_{16}/H_{20}), 7.65

(s, 1H, CH, H_7), 7.67 (b, 1H, NH_2), 8.02 (b, 1H, NH_2), 8.57 (s, 1H, CH, H_5). ^{13}C NMR (DMSO-d_6): δ = 10.9 (C_{12}/C_{15}), 20.3 (C_{11} or C_{14}), 21.5 (C_{11} or C_{14}), 23.5 (CH_2 DACH), 28.7 (C_9), 30.1 (CH_2 DACH), 46.9 (C_{10} or C_{13}), 48.9 (C_{10} or C_{13}), 61.5 (CH DACH), 64.3 (C_{23}), 64.4 (C_{22}), 65.5 (C_{24}), 114.5 (C_{17}/C_{19}), 122.3 (C_5), 123.5 (C_7), 129 (C_{16}/C_{20}) ppm. See Figure 3 for numbering of protons.

3.3. Stability of Compound 3

The stability of compound **3** in a physiological-like solution was investigated by NMR and HPLC. Because of the poor water solubility of **3**, the complex was previously dissolved in DMSO-d_6 and then diluted with deuterated aqueous phosphate buffered saline solution (2 mM, pH 7.4) obtaining a final concentration of 52 µM in DMSO-d_6/D_2O (5:95, *v/v*). The sample was incubated at 37 °C in the dark and ^{1}H-NMR spectra recorded at different time intervals.

The NMR solution was further analyzed by HPLC analysis. Stationary phase: Waters Symmetry RP-C18 column, 5 µm, 4.6 × 250 mm, 100 Å. Mobile phase: phase A = water and phase B = Acetonitrile; isocratic elution 5% phase B for 5 min, linear gradient from 5% to 50% phase B in 10 min, isocratic elution 50% phase B for 7 min, linear gradient from 50% to 5% phase B in 2 min, isocratic elution 5% phase B for 6 min. Flow rate = 0.7 mL·min^{-1}. UV-visible detector set at 220 nm.

3.4. Biological Assays

3.4.1. Cell Lines

Human MCF7 breast carcinoma, human U87 glioblastoma, human LoVo colon adenocarcinoma cell lines and C6 rat glioma cells were used. As recommended for these cell lines, the base medium used (EuroClone, Pero (MI), Italy) was RPMI for MCF7 cells, DMEM for U87 and C6 cells, and F-12/HAM for LoVo cells, enriched with fetal bovine serum (10%), penicillin (100 U/mL)/streptomycin (100 µg/mL) (1%) and glutamine 200 mM (1%) (EuroClone, Pero (MI), Italy). Cells were incubated at 37 °C under an atmosphere containing 5% CO_2.

3.4.2. Membrane Preparation

Membranes from C6 rat glioma cells were prepared as described by Piccinonna et al. [51] with minor modifications. In brief, cells scraped in PBS and harvested, were then homogenized with a Brinkman Polytron (Kinematica AG, Münstertäler, Germany). After centrifugation at 37,000× *g* for 30 min at 4 °C the resulting pellet was resuspended in ice-cold 10 mM PBS and stored at −80 °C until use.

3.4.3. Receptor Binding Assays

The ability of complex **3** and the precursor TSPO ligand **2** to bind with high affinity to TSPO was assessed by in vitro receptor binding assays performed as described in our previous works [49,52]. In brief, a suspension of C6 rat glioma cells membranes (100 µg) in PBS was treated with 0.7 nM [^{3}H]-PK11195 (a selective ligand for TSPO) and compound **2** or **3** at different concentrations. After an incubation time of 90 min at 25 °C, the samples were rapidly filtered through Whatman GF/C glass microfiber filters and the filters washed with 3 × 1 mL of ice-cold PBS. For the determination of the nonspecific binding the compound PK 11195 (10 µM) was used. A specific binding equal to 90% was determined under these experimental conditions.

3.4.4. Cell Proliferation Assay

Determination of cell growth inhibition was performed on MCF7, U87 and LoVo cell lines, using the 3-[4,5-dimethylthiazol-2-yl]-2,5-diphenyltetrazoliumbromide (MTT) assay. The determination of the % of cell viability after treatment with the test compounds, was performed as described by Denora et al. [34]. Briefly, cells were dispensed on 96 microtiter plates at a density of 5000 cells/well and, after overnight incubation, were exposed for 72 h to solutions of the compounds with concentration in the range 0.01–100 µM. After addition of MTT (10 µL of 0.5% *w/v* in PBS) and incubation for an additional 3 h at 37 °C, the cells were lysed with 100 μL of DMSO/EtOH 1:1 (*v/v*) solution, dispensed in each well. A PerkinElmer 2030 multilabel reader Victor TM X3 (manufactured for WALLAC Oy, Turku, Finland by PerkinElmer Singapore was used for the absorbance determination at 570 nm.

3.4.5. Cell Cycle Analysis

Cells were dispensed in 60 mm tissue culture dishes at a density of 500,000 cells/dish and after 24 h were treated with compounds **1**, **3**, or oxaliplatin at their IC_{50} concentrations or with compound **2** at 100 μM for an additional 24 h. Then, the experimental procedure described in our previous work was performed [53].

3.4.6. Cellular Uptake

A Varian 820-MS ICP mass spectrometer (Varian Italia, Cernusco sul Naviglio (MI), Italy) was used to perform the inductively coupled plasma mass spectrometry (ICP-MS) analyses for the determination of the platinum cellular uptake in LoVo colon cancer cells. Cells were seeded in 60 mm tissue culture dishes at a density of 500,000 cells/dish and incubated at 37 °C in a humidified atmosphere with 5% CO_2. After 1 day, the culture medium was replaced with 3 mL of medium containing complex **3**, the precursor **1**, or oxaliplatin at their IC_{50} concentrations, and incubated for 4 and 24 h. After the incubation period, the cell monolayer was washed twice with ice-cold PBS and then digested with 2 mL of an HNO_3(67%)/H_2O_2(30%), 1:1 (*v/v*), solution for 4 h at 60 °C in a stove. The platinum content was determined by ICP-MS.

4. Conclusions

The first Pt(IV) complex containing a TSPO ligand in the axial position has been synthesized, fully characterized, and its cytotoxic activity tested in vitro. The TSPO ligand in the axial position of the conjugate **3** maintains the high affinity for the TSPO receptor in C6 rat glioma cells at nanomolar level. Therefore, the new compound appears to be very promising in a receptor-mediated drug targeting context towards TSPO-overexpressing tumors, in particular colorectal cancer.

Indeed, the best cytotoxic effect of **3** was observed in LoVo colon cancer cells. However, the cytotoxicity of **3** was 5-fold lower than that of oxaliplatin, which is a good result considering the different oxidation state of the metal in **3** and in oxaliplatin.

Remarkably, compound **3**, which in MCF7 and U87 cell lines was less active than **1**, proved to be more active than **1** in LoVo cells. Since TSPO is overexpressed in colorectal cancer [19], an additional increase of potency of **3** with respect to **1** could, in principle, derive from the presence of a TSPO ligand in axial position.

We expect a clearer answer about the potential of these combination compounds from in vivo testing, where the required reductive activation of the Pt(IV) precursor, and the hydrolysis necessary for the release of the TSPO ligand might not represent a handicap as it does for in vitro cellular testing.

Acknowledgments: We acknowledge the University of Bari (Italy), the Italian Ministero dell'Università e della Ricerca, the Inter-University Consortium for Research on the Chemistry of Metal Ions in Biological Systems (C.I.R.C.M.S.B.), and the European Union (COST CM1105: Functional metal complexes that bind to biomolecules) for support.

Author Contributions: Nicola Margiotta and Nunzio Denora conceived and designed the experiments; Salvatore Savino, Rosa Maria Iacobazzi, Letizia Porcelli and Amalia Azzariti performed the experiments; Salvatore Savino, Nicola Margiotta, Giovanni Natile, Nunzio Denora, Rosa Maria Iacobazzi, Letizia Porcelli and Amalia Azzariti analyzed the data; Salvatore Savino, Nicola Margiotta, Rosa Maria Iacobazzi and Giovanni Natile wrote the paper.

Conflicts of Interest: The authors declare no conflict of interest.

Abbreviations

COSY	correlation spectroscopy
DACH	diaminocyclohexane
DMF	dimethylformamide
DMSO	dimethylsulfoxide
EDC	1-Ethyl-3-(3-dimethylaminopropyl)carbodiimide
ESI-MS	electrospray mass spectrometry
HOBt	1-hydroxybenzotriazole hydrate
HSQC	heteronuclear single quantum coherence spectroscopy
ICP-MS	inductively coupled plasma mass spectrometry
NOESY	nuclear overhauser enhanced spectroscopy
OCT	organic cation transporters
PEG	polyethylene glycol
PLGA	poly(lactic-co-glycolic acid)
PET	positron emission tomography
TEA	triethylamine
TOCSY	total correlation spectroscopy
TSPO	18-kDa mitochondrial translocator protein

References

1. Rosenberg, B.; van Camp, L.; Trosko, J.E.; Mansour, V.H. Platinum compounds: A new class of potent antitumour agents. *Nature* **1969**, *222*, 385–386. [CrossRef] [PubMed]
2. Gordon, M.; Hollander, S. Review of platinum anticancer compounds. *J. Med.* **1993**, *24*, 209–265. [PubMed]
3. Weiss, R.B.; Christian, M.C. New cisplatin analogues in development a review. *Drugs* **1993**, *46*, 360–377. [CrossRef] [PubMed]
4. Lebwohl, D.; Canetta, R. Clinical development of platinum complexes in cancer therapy: An historical perspective and an update. *Eur. J. Cancer* **1998**, *34*, 1522–1534. [CrossRef]
5. Wong, E.; Giandomenico, C.M. Current status of platinum-based antitumor drugs. *Chem. Rev.* **1999**, *99*, 2451–2566. [CrossRef] [PubMed]
6. Goddard, P.; Valenti, M.; Kelland, L.R. The role of glutathione (GSH) in determining sensitivity to platinum drugsin vivo in platinum-sensitive and -resistant murine leukaemia and plasmacytoma and human ovarian carcinoma xenografts. *Anticancer Res.* **1994**, *14*, 1065–1070. [PubMed]
7. Faivre, S.; Chan, D.; Salinas, R.; Woynarowska, B.; Woynarowski, J.M. DNA strand breaks and apoptosis induced by oxaliplatin in cancer cells. *Biochem. Pharmacol.* **2003**, *66*, 225–237. [CrossRef]
8. Di Francesco, A.M.; Ruggiero, A.; Riccardi, R. Cellular and molecular aspects of drugs of the future: Oxaliplatin. *CMLS* **2002**, *59*, 1914–1927. [CrossRef] [PubMed]
9. Extra, J.M.; Espie, M.; Calvo, F.; Ferme, C.; Mignot, L.; Marty, M. Phase I study of oxaliplatin in patients with advanced cancer. *Cancer Chemother. Pharmacol.* **1990**, *25*, 299–303. [CrossRef] [PubMed]
10. Ho, Y.-P.; Au-Yeung, S.C.F.; To, K.K.W. Platinum-based anticancer agents: Innovative design strategies and biological perspectives. *Med. Res. Rev.* **2003**, *23*, 633–655. [CrossRef] [PubMed]
11. Drougge, L.I.; Elding, L. Mechanisms for acceleration of halide anation reactions of platinum(IV) complexes. REOA versus ligand assistance and platinum(II) catalysis without central ion exchange. *Inorg. Chim. Acta* **1986**, *121*, 175–183. [CrossRef]
12. Hall, M.D.; Hambley, T.W. Platinum(IV) antitumour compounds: Their bioinorganic chemistry. *Coord. Chem. Rev.* **2002**, *232*, 49–67. [CrossRef]
13. Ellis, L.A.; Er, H.M.; Hambley, T.W. The influence of the axial ligands of a series of platinum(IV) anti-cancer complexes on their reduction to platinum(II) and reaction with DNA. *Aust. J. Chem.* **1995**, *48*, 793–806. [CrossRef]
14. Wang, X.; Guo, Z. Targeting and delivery of platinum-based anticancer drugs. *Chem. Soc. Rev.* **2013**, *42*, 202–224. [CrossRef] [PubMed]

15. Rupprecht, R.; Papadopoulos, V.; Rammes, G.; Baghai, T.C.; Fan, J.; Akula, N.; Groyer, G.; Adams, D.; Schumacher, M. Translocator protein (18 kDa) (TSPO) as a therapeutic target for neurological and psychiatric disorders. *Nat. Rev. Drug Discov.* **2010**, *9*, 971–988. [CrossRef] [PubMed]

16. Trapani, G.; Denora, N.; Trapani, A.; Laquintana, V. Recent advances in ligand targeted therapy. *J. Drug Target.* **2012**, *20*, 1–22. [CrossRef] [PubMed]

17. Denora, N.; Cassano, T.; Laquintana, V.; Lopalco, A.; Trapani, A.; Cimmino, C.S.; Laconca, L.; Giuffrida, A.; Trapani, G. Novel codrugs with GABAergic activity for dopamine delivery in the brain. *Int. J. Pharm.* **2012**, *437*, 221–231. [CrossRef] [PubMed]

18. Galiegue, S.; Tinel, N.; Casellas, P. The peripheral benzodiazepine receptor: A promising therapeutic drug target. *Curr. Med. Chem.* **2003**, *10*, 1563–1572. [CrossRef] [PubMed]

19. Maaser, K.; Grabowski, P.; Sutter, A.P.; Höpfner, M.; Foss, H.D.; Stein, H.; Berger, G.; Gavish, M.; Zeitz, M.; Scherübl, H. Overexpression of the peripheral benzodiazepine receptor is a relevant prognostic factor in stage III colorectal cancer. *Clin. Cancer Res.* **2002**, *8*, 3205–3209. [PubMed]

20. Veenman, L.; Levin, E.; Weisinger, G.; Leschiner, S.; Spanier, I.; Snyder, S.H.; Weizman, A.; Gavish, M. Peripheral-type benzodiazepine receptor density and in vitro tumorigenicity of glioma cell lines. *Biochem. Pharmacol.* **2004**, *68*, 689–698. [CrossRef] [PubMed]

21. Scarf, A.M.; Ittner, L.M.; Kassiou, M. The translocator protein (18 kDa): Central nervous system disease and drug design. *J. Med. Chem.* **2009**, *52*, 581–592. [CrossRef] [PubMed]

22. Denora, N.; Laquintana, V.; Trapani, A.; Lopedota, A.; Latrofa, A.; Gallo, J.M.; Trapani, G. Translocator protein (TSPO) ligand–Ara-C (Cytarabine) conjugates as a strategy to deliver antineoplastic drugs and to enhance drug clinical potential. *Mol. Pharm.* **2010**, *7*, 2255–2269. [CrossRef] [PubMed]

23. Laquintana, V.; Denora, N.; Lopedota, A.; Suzuki, H.; Sawada, M.; Serra, M.; Biggio, G.; Latrofa, A.; Trapani, G.; Liso, G. N-benzyl-2-(6,8-dichloro-2-(4-chlorophenyl)imidazo [1,2-a]pyridin-3-yl)-N-(6-(7-nitrobenzo[c] [1,2,5]oxadiazol 4 ylamino)hexyl)acetamide as a new fluorescent probe for peripheral benzodiazepine receptor and microglial cell visualization. *Bioconj. Chem.* **2007**, *18*, 1397–1407. [CrossRef] [PubMed]

24. Laquintana, V.; Denora, N.; Musacchio, T.; Lasorsa, M.; Latrofa, A.; Trapani, G. Peripheral benzodiazepine receptor ligand–PLGA polymer conjugates potentially useful as delivery systems of apoptotic agents. *J. Control. Release* **2009**, *137*, 185–195. [CrossRef] [PubMed]

25. Bai, M.; Bornhop, D.J. Recent avances in receptor-targeted fluorescent probes for in vivo cancer imaging. *Curr. Med. Chem.* **2012**, *19*, 4742–4758. [CrossRef] [PubMed]

26. Perrone, M.; Moon, B.S.; Park, H.S.; Laquintana, V.; Jung, J.H.; Cutrignelli, A.; Lopedota, A.; Franco, M.; Kim, S.E.; Lee, B.C.; et al. A Novel PET Imaging Probe for the Detection and Monitoring of Translocator Protein 18 kDa Expression in Pathological Disorders. *Sci. Rep.* **2016**, *6*, 20422. [CrossRef] [PubMed]

27. Margiotta, N.; Ostuni, R.; Ranaldo, R.; Denora, N.; Laquintana, V.; Trapani, G.; Liso, G.; Natile, G. Synthesis and Characterization of a Platinum(II) Complex Tethered to a Ligand of the Peripheral Benzodiazepine Receptor. *J. Med. Chem.* **2007**, *50*, 1019–1027. [CrossRef] [PubMed]

28. Romaniewska, A.; Jasztold-Howorko, R.; Regiec, A.; Lis, T.; Kuduk-Jaworska, J. Synthesis, structure and characterization of new olivacine derivatives and their platinum(II) complexes. *Eur. J. Inorg. Chem.* **2003**, *2003*, 4043–4054. [CrossRef]

29. Bentzion, D.; Lipatov, O.; Polyakov, I.; MacKintosh, R.; Eckardt, J.; Breitz, H. A phase II study of picoplatin (pico) as second-line therapy for patients (pts) with small cell lung cancer (SCLC) who have resistant or refractory disease or have relapsed within 180 days of completing first-line, platinum (plat)-containing chemotherapy. In Proceedings of the ASCO Annual Meeting (Part I), Chicago, IL, USA, 1–5 June 2007; p. 7722.

30. Margiotta, N.; Denora, N.; Ostuni, R.; Laquintana, V.; Anderson, A.; Johnson, S.W.; Trapani, G.; Natile, G. Platinum(II) complexes with bioactive carrier ligands having high affinity for the translocator protein. *J. Med. Chem.* **2010**, *53*, 5144–5154. [CrossRef] [PubMed]

31. Mailliet, P.; Bourrie, B.; Normand, A. Derives de Platine(IV)—Couples a un Agent de Ciblage Antitumoral. French Patent FR 2 954 321 A1, 15 July 2010.

32. Choi, S.; Filotto, C.; Bisanzo, M.; Delaney, S.; Lagasee, D.; Whitworth, J.L.; Jusko, A.; Li, C.R.; Wood, N.A.; Willingham, J.; et al. reduction and anticancer activity of platinum(IV) complexes. *Inorg. Chem.* **1998**, *37*, 2500–2504. [CrossRef]

33. Chen, L.; Lee, P.F.; Ranford, J.D.; Vittal, J.J.; Wong, S.Y. Reduction of the anti-cancer analogue *cis, trans, cis*-[PtCl$_2$(OCOCH$_3$)NH$_3$)$_2$] by L-methionine and L-cysteine and its crystal structure. *J. Chem. Soc. Dalton Trans.* **1999**, 1209–1212. [CrossRef]

34. Denora, N.; Laquintana, V.; Lopalco, A.; Iacobazzi, R.M.; Lopedota, A.; Cutrignelli, A.; Iacobellis, G.; Annese, C.; Cascione, M.; Leporatti, S.; et al. In vitro targeting and imaging the translocator protein TSPO 18-kDa through G(4)-PAMAM-FITC labeled dendrimer. *J. Control. Release* **2013**, *172*, 1111–1125. [CrossRef] [PubMed]

35. Rendina, L.M.; Puddephatt, R. Oxidative Addition Reactions of Organoplatinum(II) Complexes with Nitrogen-Donor Ligands. *J. Chem. Rev.* **1997**, *97*, 1735–1754. [CrossRef]

36. Gossage, R.A.; Ryabov, A.D.; Spek, A.L.; Stufkens, D.J.; van Beek, J.A.M.; van Eldik, R.; van Koten, G. Models for the Initial Stages of Oxidative Addition. Synthesis, Characterization, and Mechanistic Investigation of η^1-I$_2$ Organometallic "Pincer" Complexes of Platinum. X-ray Crystal Structures of [PtI(C$_6$H$_3${CH$_2$NMe$_2$}$_2$–2,6)(η^1-I$_2$)] and exo-meso-[Pt(η^1-I$_3$)(η^1-I$_2$)(C$_6$H$_3${CH$_2$N(t-Bu)Me}$_2$–2,6)]. *J. Am. Chem. Soc.* **1999**, *121*, 2488–2497.

37. Margiotta, N.; Ranaldo, R.; Intini, F.P.; Natile, G. Cationic intermediates in oxidative addition reactions of Cl$_2$ to [PtCl$_2$(cis-1,4-DACH)]. *Dalton Trans.* **2011**, *40*, 12877–12885. [CrossRef] [PubMed]

38. Giandomenico, C.M.; Abrams, M.J.; Murrer, B.A.; Vollano, J.F.; Rheinheimer, M.I.; Wyer, S.B.; Bossard, G.E.; Higgins, J.D. Carboxylation of Kinetically Inert Platinum(IV) Hydroxy Complexes. An Entr.acte.ee into Orally Active Platinum(IV) Antitumor Agents. *Inorg. Chem.* **1995**, *34*, 1015–1021. [CrossRef] [PubMed]

39. Pichler, V.; Valiahdi, S.M.; Jakupec, M.A.; Arion, V.B.; Galanski, M.; Keppler, B.K. Mono-carboxylated diaminedichloridoplatinum(IV) complexes—Selective synthesis, characterization, and cytotoxicity. *Dalton Trans.* **2011**, *40*, 8187–8192. [CrossRef] [PubMed]

40. Zhang, J.Z.; Bonnitcha, P.; Wexselblatt, E.; Klein, A.V.; Najajreh, Y.; Gibson, D.; Hambley, T.W. Facile preparation of mono-, di- and mixed-carboxylato platinum(IV) complexes for versatile anticancer prodrug design. *Chem. Eur. J.* **2013**, *19*, 1672–1676. [CrossRef] [PubMed]

41. Ravera, M.; Gabano, E.; Pelosi, G.; Fregonese, F.; Tinello, S.; Osella, D. A new entry to asymmetric platinum(IV) complexes via oxidative chlorination. *Inorg. Chem.* **2014**, *53*, 9326–9335. [CrossRef] [PubMed]

42. Dunham, S.O.; Larsen, R.D.; Abbott, E.H. Nuclear magnetic resonance investigation of the hydrogen peroxide oxidation of platinum(II) complexes. Crystal and molecular structures of sodium *trans*-dihydroxobis (malonato)platinate(IV) hexahydrate and sodium *trans*-dihydroxobis(oxalato)platinate(IV) hexahydrate. *Inorg. Chem.* **1993**, *32*, 2049–2055.

43. Petruzzella, E.; Margiotta, N.; Ravera, M.; Natile, G. NMR Investigation of the Spontaneous Thermal- and/or Photoinduced Reduction of trans Dihydroxido Pt(IV) Derivatives. *Inorg. Chem.* **2013**, *52*, 2393–2403. [CrossRef] [PubMed]

44. Cheng, J.; Teply, B.A.; Sherifi, I.; Sung, J.; Luther, G.; Gu, F.X.; Levy-Nissenbaum, E.; Radovic-Moreno, A.F.; Langer, R.; Farokhzad, O.C. Formulation of functionalized PLGA–PEG nanoparticles forin vivo targeted drug delivery. *Biomaterials* **2007**, *28*, 869–876. [CrossRef] [PubMed]

45. Kugler, W.; Veenman, L.; Shandalov, Y.; Leschiner, S.; Spanier, I.; Lakomek, M.; Gavish, M. Ligands of the mitochondrial 18 kDa translocator protein attenuate apoptosis of human glioblastoma cells exposed to erucylphosphohomocholine. *Cell. Oncol.* **2008**, *30*, 435–450. [PubMed]

46. Cubo, L.; Hambley, T.W.; Miguel, P.J.S.; Carnero, A.; Navarro-Ranninger, C.; Quiroga, A.G. The preparation and characterization of trans-platinum(IV) complexes with unusually high cytotoxicity. *Dalton Trans.* **2011**, *40*, 344–347. [CrossRef] [PubMed]

47. Zhang, S.; Lovejoy, K.S.; Shima, J.E.; Lagpacan, L.L.; Shu, Y.; Lapuk, A.; Chen, Y.; Komori, T.; Gray, J.W.; Chen, X.; et al. Organic Cation Transporters Are Determinants of Oxaliplatin Cytotoxicity. *Cancer Res.* **2006**, *66*, 8847–8857. [CrossRef] [PubMed]

48. William-Faltaos, S.; Rouillard, D.; Lechat, P.; Bastian, G. Cell cycle arrest by oxaliplatin on cancer cells. *Fundam. Clin. Pharmacol.* **2007**, *21*, 165–172. [CrossRef] [PubMed]

49. Pregosin, P.S. Platinum-195 nuclear magnetic resonance. *Coord. Chem. Rev.* **1982**, *44*, 247–291. [CrossRef]

50. Kidani, Y.; Inagaki, K. Cis-platinum(ii) complex of trans-l-1,2-diaminocyclohexane. Patent US 4169846 A, 2 October 1979.

51. Piccinonna, S.; Denora, N.; Margiotta, N.; Laquintana, V.; Trapani, G.; Natile, G. Synthesis, characterization, and binding to the translocator protein (18 kDa, TSPO) of a new rhenium complex as a model of radiopharmaceutical agents. *Z. Anorg. Allg. Chem.* **2013**, *639*, 1606–1612. [CrossRef]
52. Denora, N.; Laquintana, V.; Trapani, A.; Suzuki, H.; Sawada, M.; Trapani, G. New fluorescent probes targeting the mitochondrial-located translocator protein 18 kDa (TSPO) as activated microglia imaging agents. *Pharm. Res.* **2011**, *28*, 2820–2832. [CrossRef] [PubMed]
53. Porcelli, L.; Guida, G.; Tommasi, S.; Guida, M.; Azzariti, A. Metastatic melanoma cells with BRAF G469A mutation: Nab-paclitaxel better than vemurafenib? *Cancer Chemother. Pharmacol.* **2015**, *76*, 433–438. [CrossRef] [PubMed]

International Journal of
Molecular Sciences

MDPI

Article

TSPO PIGA Ligands Promote Neurosteroidogenesis and Human Astrocyte Well-Being

Eleonora Da Pozzo, Chiara Giacomelli, Barbara Costa, Chiara Cavallini, Sabrina Taliani, Elisabetta Barresi, Federico Da Settimo and Claudia Martini *

Department of Pharmacy, University of Pisa, Via Bonanno Pisano 6, 56126 Pisa, Italy; eleonora.dapozzo@unipi.it (E.D.P.); chiara.giacomelli@for.unipi.it (C.G.); barbara.costa@unipi.it (B.C.); chiara.cavallini@farm.unipi.it (C.C.); sabrina.taliani@unipi.it (S.T.); elisabetta.barresi@for.unipi.it (E.B.); federico.dasettimo@unipi.it (F.D.S.)
* Correspondence: claudia.martini@unipi.it; Tel.: +39-050-221-9509

Academic Editor: Giovanni Natile
Received: 11 May 2016; Accepted: 23 June 2016; Published: 29 June 2016

Abstract: The steroidogenic 18 kDa translocator protein (TSPO) is an emerging, attractive therapeutic tool for several pathological conditions of the nervous system. Here, 13 high affinity TSPO ligands belonging to our previously described N,N-dialkyl-2-phenylindol-3-ylglyoxylamide (PIGA) class were evaluated for their potential ability to affect the cellular Oxidative Metabolism Activity/Proliferation index, which is used as a measure of astrocyte well-being. The most active PIGA ligands were also assessed for steroidogenic activity in terms of pregnenolone production, and the values were related to the metabolic index in rat and human models. The results showed a positive correlation between the increase in the Oxidative Metabolism Activity/Proliferation index and the pharmacologically induced stimulation of steroidogenesis. The specific involvement of steroid molecules in mediating the metabolic effects of the PIGA ligands was demonstrated using aminoglutethimide, a specific inhibitor of the first step of steroid biosynthesis. The most promising steroidogenic PIGA ligands were the 2-naphthyl derivatives that showed a long residence time to the target, in agreement with our previous data. In conclusion, TSPO ligand-induced neurosteroidogenesis was involved in astrocyte well-being.

Keywords: translocator protein; neurosteroidogenesis; PIGA ligands; cellular proliferation; oxidative metabolism; astrocytes

---

1. Introduction

Neuroactive steroids, which are mainly synthesized by glial cells, exert peculiar actions to influence the development and function of the nervous system through both genomic and non-genomic mechanisms [1,2]. The classic genomic action involves steroid binding to intracellular receptors and the regulation of protein translation [3]. Neuroactive steroids can also show rapid effects, occurring within seconds to minutes, via the activation of membrane neurotransmitter receptors. It has been shown that neuroactive steroids determine the allosteric modulations on ligand-gated channels, including type-A γ-aminobutyric acid (GABA), N-methyl-D-aspartate (NMDA), and nicotinic receptors [1,4–11]. These different interactions lead to the multiple actions of neuroactive steroids, affecting both glia and neurons in a concerted manner [12]. For instance, oestrogens act as transcriptional regulators to modulate the synthesis of various proteins and growth factors in astrocytes [13–25]. Interestingly, oestrogens increase glutamate transporter expression in astrocytes via the nuclear factor κ-light-chain-enhancer of activated B cells (NF-κB) and the cAMP response element binding protein (CREB) pathways [26]. Among the actions exerted by steroids, the increase in the expression of the mitochondria-encoded subunits of the respiratory chain influences the mitochondrial respiratory function, and this activity may be of particular interest for enhancing the functional efficiency of astrocytes [27,28].

Notably, bidirectional glia-neuron communication was suggested by several scientific reports showing that glial cells and neurons can respond to the same signals and that they can mutually modulate the cellular response (for a review see [12]). For this reason, glial cell well-being is of particular importance for the efficiency of the whole brain.

Translocator protein 18 kDa (TSPO), which is primarily located in the outer mitochondrial membrane, is highly expressed in steroid-synthesizing tissues, including glial cells. Although few reports questioned the role of TSPO in steroidogenesis [29–31], most studies, including the most recent ones, propose that TSPO is an important protein for steroid synthesis [32–34]. TSPO binds cholesterol with high affinity and, in a combined action with the steroidogenic acute regulatory protein (StAR), allows cholesterol to translocate into mitochondria, which represents the rate-limiting step of steroidogenesis [35–37]. The steroid biosynthetic pathway is triggered by the cleavage of the cholesterol aliphatic side chain, which is catalyzed by the cytochrome P450 side chain cleavage (P450scc) enzyme, producing pregnenolone. Then, pregnenolone is converted to other neurosteroids by enzymes located in the endoplasmic reticulum, such as hydroxysteroid dehydrogenases [38,39].

A number of TSPO-targeted molecules have been reported as neuroprotective, anti-inflammatory, and regenerating agents in different in vitro and in vivo models, suggesting their possible development as effective therapeutic tools (for a review see [40]). For instance, in gliosis, TSPO ligands were able to decrease reactive gliosis and prevent neuronal loss [41,42]. The stimulation of neurosteroidogenesis has been hypothesized as the basis for the positive actions of the TSPO ligand [43], and, for these reasons, TSPO ligands are currently under investigation as therapeutic tools to preserve a functional brain environment and the glia-neuron bidirectional interactions [44]. Very recently, we have found that the *N,N*-dialkyl-2-phenylindol-3-ylglyoxylamide class (PIGAs) of TSPO ligands reduces oxidative stress and the activity of pro-inflammatory enzymes in rat glial cells through the de novo neurosteroid synthesis [45].

Based on the previously described pro-survival activity of TSPO ligands in neurons and glia, in the present work, the effects of the TSPO ligands on astrocyte well-being were assessed by focusing on the involvement of steroidogenesis. Therefore, the residence time of some investigated ligands were also assessed because we have recently shown that the time over which a ligand interacts with TSPO directly affects its steroidogenic efficacy [46]. The human glioblastoma–astrocytoma cell line U87MG and normal human astrocytes were used as cellular models. U87MG cells express the astrocyte cell marker glial fibrillary acidic protein (GFAP) and are widely used as an in vitro astrocyte model [47–50]. Recent data have shown comparable responses of U87MG cells and primary human astrocytes after inflammatory insult, highlighting the potential use of U87MG cells in drug discovery stages, as it is not feasible to screen compounds in primary human cells [50]. However, data from healthy human astrocytes were crucial for validating TSPO activity under normal conditions. Thus, 13 high affinity, selective TSPO ligands belonging to our previously described PIGA class [51,52] were selected and evaluated for their ability to increase the Oxidative Metabolism Activity/Proliferation index in human astrocyte models. The most promising compounds were then assessed for their steroidogenic activity and residence time. Finally, the relation between oxidative metabolism, proliferation activity, and the induction of neurosteroidogenesis was investigated.

2. Results

2.1. N,N-Dialkyl-2-phenylindol-3-ylglyoxylamide (PIGA) Ligands Increase the Oxidative Metabolism Activity/Proliferation Index in a Human Astrocyte Model

TSPO expression has previously been established in U87MG cells [53]. To assess the potential effects of the PIGA ligands on the activation of oxidative metabolism, U87MG cells were cultured under serum-reduced growth conditions; serum starvation is a well-known method to arrest the cells in a basal metabolic state (G0/G1 phase) [54,55]. The metabolic activity of the astrocyte models was estimated using the (3-(4,5-dimethylthiazol-2-yl)-5-(3-carboxymethoxyphenol)-2-(4-sulfophenyl)-2*H*-tetrazolium, inner salt) (MTS) assay [56]. This tetrazolium dye can be reduced by the metabolic reducing agents NADH and NADPH to a water-soluble formazan salt; the amount of produced formazan has been considered a marker of the Oxidative Metabolic Activity index [57]. The redox reactions can occur in both the mitochondria and cytosol; in particular, it has been shown that tetrazolium reduction mainly reflects cytosolic redox activity in astroglia and is dependent on glyceraldehyde-3-phosphate dehydrogenase activity [56]. Furthermore, as the reduction of the tetrazolium compound can only be achieved in viable cells, the tetrazolium assay has also been widely used for the quantitative assessment of cellular proliferation for over three decades [56,57].

The effects of PIGAs and the TSPO reference standard ligand PK11195 (ranging from nanomolar to micromolar concentrations) on the Oxidative Metabolism Activity/Proliferation (OMAP) index in U87MG cells were evaluated after 48 h of incubation. The derivatives PIGA1128, PIGA1130, PIGA1136, PIGA1137, PIGA1138, PIGA1165, PIGA1174, PIGA1175, and PIGA1212 significantly increased the OMAP index, with the maximal mean value (163%) observed for 1 µM PIGA1138 ($p < 0.001$ vs. the control) (Figure 1).

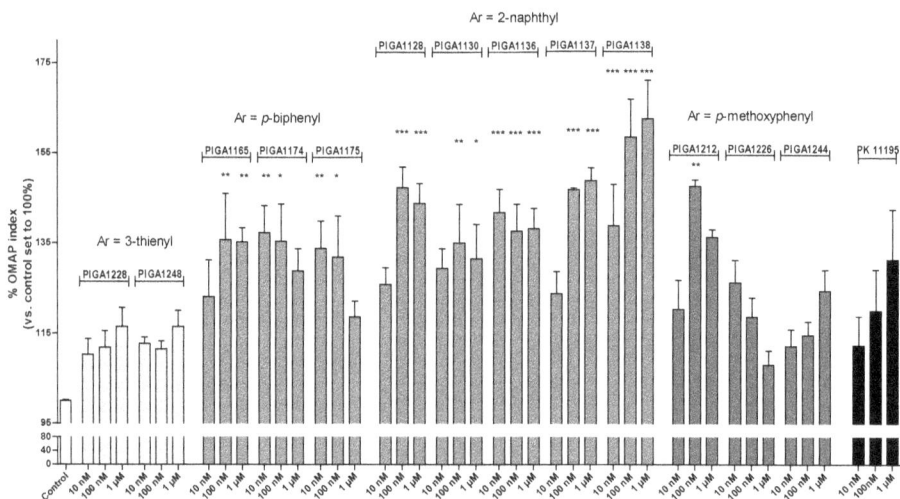

Figure 1. Effects of the PIGA ligands on the Oxidative Metabolism Activity/Proliferation index in U87MG human glioma cells. U87MG cells were treated with different concentrations of the compounds (10 nM–1 µM) in serum-reduced media (1% fetal bovine serum (FBS)), and the [3-(4,5-dimethylthiazol-2-yl)-5-(3-carboxymethoxyphenol)-2-(4-sulfophenyl)-2*H*-tetrazolium, inner salt] (MTS) assay was performed after 48 h of treatment. The data are expressed as percentages of the proliferative/oxidative metabolism activity index compared to the control (0.1% dimethyl sulfoxide (DMSO), a concentration that not interfered with the assay), which was set to 100%, and represent the means ± standard error of the mean (SEM) of three different experiments performed in duplicate. The statistical analysis was performed using one-way analysis of variance (ANOVA) and Bonferroni's post-test; * $p < 0.05$, ** $p < 0.01$, *** $p < 0.001$ vs. the control.

The most promising derivatives were those featuring a 2-naphthyl substituent as an aryl group (PIGA1128, PIGA1136, PIGA1137, and PIGA1338; see Table 1 for the chemical structures). In particular, PIGA1136 and PIGA1138 yielded statistically significant results at all tested concentrations. PIGA1174 and PIGA1175 were significantly effective at 10 and 100 nM, the two lowest concentrations tested. Conversely, PIGA1128, PIGA1130, PIGA1137, and PIGA1165 were active at the higher doses of 100 nM and 1 µM. Statistically significant results were not observed for PIGA1226, PIGA1228, PIGA1244,

PIGA1248, and the reference compound PK11195. These results suggested that the TSPO ligands positively affected the well-being of an astrocytic cell line when the cells were maintained in a constrained low metabolic state.

Table 1. TSPO binding affinity of the compounds. The concentration of the tested compounds that inhibited [^{3}H]PK11195 binding to rat kidney mitochondrial membranes (IC_{50}) by 50% was determined using six concentrations of the displacers, each performed in triplicate. The K_i values are the means $\pm$ SEM of three determinations.

Compound	R_1	R_2	Ar	K_i (nM) [a]
PIGA1228	$(CH_2)_3CH_3$	$(CH_2)_3CH_3$	3-Thienyl	2.83 ± 0.30
PIGA1248	$(CH_2)_5CH_3$	$(CH_2)_5CH_3$	3-Thienyl	0.89 ± 0.10
PIGA1175	$(CH_2)_2CH_3$	$(CH_2)_2CH_3$	*p*-Biphenyl	0.53 ± 0.05
PIGA1165	$(CH_2)_3CH_3$	$(CH_2)_3CH_3$	*p*-Biphenyl	5.50 ± 1.00
PIGA1174	$(CH_2)_5CH_3$	$(CH_2)_5CH_3$	*p*-Biphenyl	1.84 ± 0.20
PIGA1128	$(CH_2)_2CH_3$	$(CH_2)_2CH_3$	2-Naphthyl	0.30 ± 0.04
PIGA1136	$(CH_2)_3CH_3$	$(CH_2)_3CH_3$	2-Naphthyl	0.53 ± 0.06
PIGA1130	$(CH_2)_5CH_3$	$(CH_2)_5CH_3$	2-Naphthyl	0.52 ± 0.06
PIGA1137	CH_3	$(CH_2)_3CH_3$	2-Naphthyl	0.56 ± 0.06
PIGA1138	CH_3	$(CH_2)_4CH_3$	2-Naphthyl	0.37 ± 0.04
PIGA1226	$(CH_2)_3CH_3$	$(CH_2)_3CH_3$	*p*-Methoxyphenyl	20.3 ± 2.21
PIGA1244	$(CH_2)_5CH_3$	$(CH_2)_5CH_3$	*p*-Methoxyphenyl	4.04 ± 0.44
PIGA1212 [a]	$(CH_2)_2CH_3$	$(CH_2)_2CH_3$	*p*-Methylphenyl	5.50 ± 0.38
PK 11195 [b]				9.3 ± 0.50

[a] Data taken from ref. [52]; [b] Data taken from ref. [51].

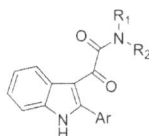

2.2. PIGA Ligands Effectively Stimulate Steroidogenesis in Vitro

To assess the ability of PIGA ligands to stimulate steroidogenesis in vitro, the synthesis of the first steroid metabolite pregnenolone was evaluated in the presence of inhibitors of pregnenolone metabolism. As a first step, the assessment was performed in a rat C6 cell line, the glial cell model that is conventionally used to measure mitochondrial receptor-regulated steroidogenesis [58]. The amount of pregnenolone released from the C6 cells was measured after a 2 h incubation with a fixed concentration of the most promising PIGAs in terms of metabolic activation (PIGA1128, PIGA1130, PIGA1136, PIGA1137, and PIGA1138).

The obtained results showed that all PIGA derivatives significantly increased pregnenolone synthesis in the C6 cells compared to the control (cells treated with DMSO and set to 100%) (Figure 2A and Table 2). The highest pregnenolone level was observed after the C6 cells were treated with PIGA1137 and PIGA1138 (increase in pregnenolone synthesis of 208% and 215%, respectively, $p < 0.001$) (Figure 2A and Table 2).

To explore whether the promising steroidogenic effects of the PIGA ligands were maintained in a human astrocytic model, the pregnenolone assessment was performed in U87MG cells. As shown in Figure 2B and Table 2, the PIGA ligands also significantly induced steroidogenesis in the U87MG cells, and the results were comparable to those obtained in the C6 cells. The best performing derivatives were PIGA1137 and PIGA1138, showing an increase in pregnenolone synthesis of 288% and 299%, respectively ($p < 0.001$; Figure 2B and Table 2). Notably, these two PIGA ligands also presented the best metabolic activation profile in the U87MG cells. The standard PK11195 similarly increased pregnenolone production in the C6 and U87MG cells (139% and 144%, respectively, $p < 0.01$) (Figure 2).

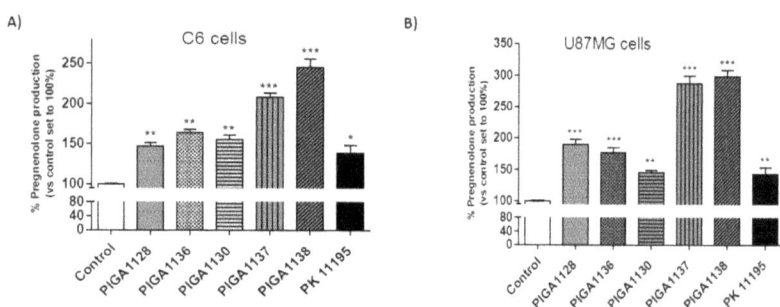

Figure 2. Effects of the PIGA ligands on pregnenolone production. C6 glioma cells (**A**) and U87MG cells (**B**) were incubated with different PIGA ligands (40 μM) for 2 h at 37 °C. The pregnenolone amounts were quantified by a competitive enzyme-linked immunosorbent assay. The data are expressed as percentage of pregnenolone production compared to the control, which was set to 100%, and represent the means ± SEM of three different determinations performed in duplicate. The statistical analysis was performed using one-way ANOVA and Bonferroni's post-test; $*$ $p < 0.05$, $**$ $p < 0.01$, $***$ $p < 0.001$ vs. the control.

Table 2. Pregnenolone production in C6 and U87MG cells. The data are expressed as percentage of pregnenolone production compared to the control, which was set to 100%, and represent the means ± SEM of three different determinations performed in duplicate. The statistical analysis was performed using one-way ANOVA and Bonferroni's post-test.

Compounds	Pregnenolone Production in C6 cells (% ± SEM)	p	Pregnenolone Production in U87MG cells (% ± SEM)	p
PIGA1128	148.4 ± 4.643	<0.01	190.0 ± 8.356	<0.001
PIGA1136	164.0 ± 2.096	<0.01	177.3 ± 7.963	<0.001
PIGA1130	155.0 ± 5.774	<0.01	145.8 ± 3.393	<0.01
PIGA1137	208.0 ± 4.933	<0.001	287.6 ± 11.98	<0.001
PIGA1138	245.0 ± 5.774	<0.001	298.7 ± 10.14	<0.001
PK 11195	138.5 ± 9.717	<0.05	143.8 ± 20.40	<0.01

2.3. The Best Performing PIGA Ligands in Terms of Steroidogenesis Stimulation Are Characterized by a Long Residence Time

Our very recent results have shown that the time over which a ligand interacts with TSPO directly affects its steroidogenic efficacy [46]. The time of the ligand-target interaction is a kinetic parameter known as Residence Time (RT), and it is calculated by the reciprocal of the dissociation rate constant (k_{off}). For unlabelled TSPO ligands, k_{off} is experimentally derived by the competition kinetic association assay [46]. The RT values of the most promising compounds were evaluated to relate the TSPO kinetic binding parameters and steroidogenic activity of the PIGA ligands. Consistent with our previous results, the 2-naphthyl derivative PIGA1138, which is characterized by a long RT (141 min) [46], was the best performing ligand in terms of its ability to stimulate steroidogenesis in both the C6 and U87MG cells. In contrast, the standard PK11195, which is characterized by a short RT (33 min) [46], showed a reduced ability to stimulate steroidogenesis compared to PIGA1138. The RT of PIGA1128 has been already determined (55 min) [46]; as expected, it showed an intermediate ability to stimulate steroidogenesis. The RTs of additional 2-naphthyl derivatives with promising steroidogenic ability in human U87MG cells (PIGA1136 and PIGA1137) were here examined. Kinetic experiments showed that the association rate constant (k_{on}), k_{off} and RT value for PIGA1136 were 2.52×10^7 $M^{-1} \cdot min^{-1}$, 0.0178 min^{-1} and 56 min, respectively. For PIGA1137, the k_{on}, k_{off} and RT values were 3.56×10^7 $M^{-1} \cdot min^{-1}$, 0.0185 min^{-1} and 54 min, respectively. For PIGA1136 and PIGA1137, the kinetically derived K_d values (k_{off}/k_{on}) were 0.71 and 0.52 nM, respectively. These values were in good agreement with the previously reported K_i values [52] obtained from competition binding experiments at equilibrium (the K_i values for PIGA1136 and PIGA1137 were 0.53 and 0.56 nM, respectively).

2.4. The Oxidative Metabolism Activity/Proliferation Index of PIGA1138 Is Related to Steroid Production

The OMAP index and the percentage of pregnenolone production could not be directly compared, as they were obtained in different experimental settings. However, the correlation analyses indicated a strong relationship between these two parameters. Indeed, Spearman's correlation analysis of the OMAP index and pregnenolone production in U87MG cells treated with micromolar concentrations of the compounds revealed a highly significant p value ($p = 0.0004$, Figure 3).

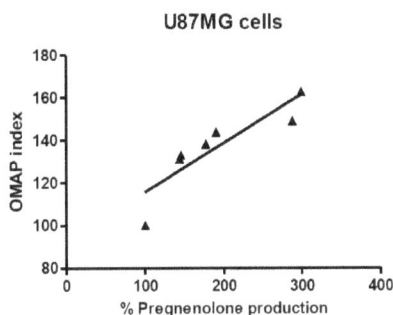

Figure 3. Spearman's correlation analyses of the Oxidative Metabolism Activity/Proliferation (OMAP) index and pregnenolone production in U87MG cells. The percentages of pregnenolone production obtained in the steroidogenesis experiments (U87MG cells were exposed to 40 μM PIGA ligand in saline medium for 2 h) were correlated to the OMAP indexes obtained in the metabolic experiments (U87MG cells were exposed to 1 μM PIGA ligand in serum-reduced medium for 48 h). The statistical analyses were performed using the Spearman r correlation, reporting a $p < 0.001$.

To deeply investigate the correlation between the ability of the most promising PIGA ligands to promote astrocyte survival and their neurosteroidogenic activity, the ligand-mediated metabolic activation of U87MG cells was evaluated in the presence and absence of DL-aminoglutethimide (AMG), an inhibitor of cytochrome P450 side chain cleavage (P450scc), the enzyme that catalyzes the first step of steroidogenesis. The assay was performed for PIGA1138, which was selected as the most representative compound based on its ability to effectively increase both the OMAP index and pregnenolone production. The cells were maintained in a serum-reduced growth condition; in this basal metabolic state, AMG alone did not affect the OMAP index (Figure 4).

The results showed that PIGA1138 increased the OMAP index of the U87MG cells in a concentration-dependent manner and the effects were completely counteracted by the co-treatment with AMG (Figure 4), clearly supporting the hypothesis that the PIGA ligand-induced increase in the OMAP index was mainly related to steroid production.

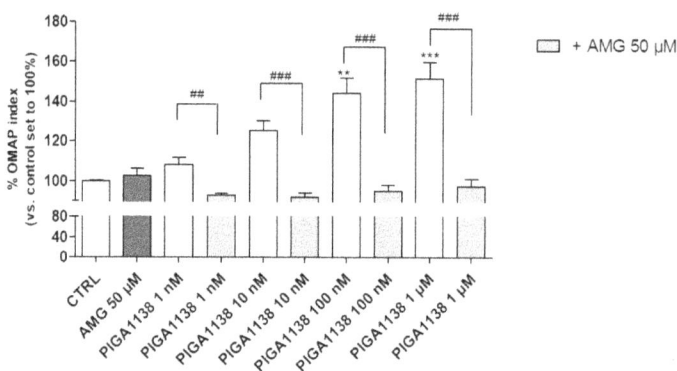

Figure 4. Influence of pregnenolone production on the U87MG OMAP index. U87MG cells were treated with different concentrations of PIGA ligands (1 nM–10 μM) in the absence or presence of AMG (50 μM) in serum-reduced media (1% FBS), and the viable cells were counted after 48 h of treatment using the MTS assay. The data are expressed as percentages of the OMAP index compared to the control, which was set to 100%, and represent the means ± SEM of three different experiments performed in duplicate. The statistical analysis was performed using one-way ANOVA and Bonferroni's post-test; ** $p < 0.01$, *** $p < 0.001$ vs. the control; ## $p < 0.01$, ### $p < 0.001$ vs. the respective treatment without DL-aminoglutethimide (AMG).

2.5. PIGA1138 Promoted the Activation of Oxidative Metabolism in Normal Human Astrocytes

Although the U87MG cell line is widely used as an astrocyte model in vitro, we verified the consistency of the obtained data by evaluating the effects of PIGA1138 and PK11195 on healthy normal human astrocytes. As first step, the TSPO expression levels were quantified using [^{3}H]PK11195 as a probe. In the whole membranes derived from normal human astrocytes, the [^{3}H]PK11195 binding reached saturation, showing a maximal binding capacity of 9.548 fmol/mg. In terms of the equilibrium dissociation constant, the experimentally derived [^{3}H]PK11195 binding affinity was 2.8 nM.

Then, the effects of PIGA1138 and PK11195 on healthy normal human astrocytes maintained under serum-reduced growth conditions were evaluated (Figure 5A). PK11195 did not increase the astrocytes' OMAP index. In contrast, PIGA1138 significantly promoted the astrocytes' well-being ($p < 0.001$) (Figure 5A), in accordance with the data obtained in the U87MG cells (Figure 1).

Finally, to verify if the serum-reduced conditions used in the experiments could affect the obtained results, parallel cell cultures maintained under normal growth conditions (20% serum medium) were treated with ligand and assessed. As shown in Figure 5B, PIGA1138 also increased the astrocytes' OMAP index under the normal growth culture conditions, whereas PK11195 did not affect the astrocytes' well-being. The results obtained for PIGA1138 are in accordance with those obtained in astrocytes grown under serum-reduced conditions, suggesting that the increase in steroid production exerted a positive effect when the cells were maintained in a housekeeping metabolic state and promoted the general well-being of the astrocytes under normal growth conditions.

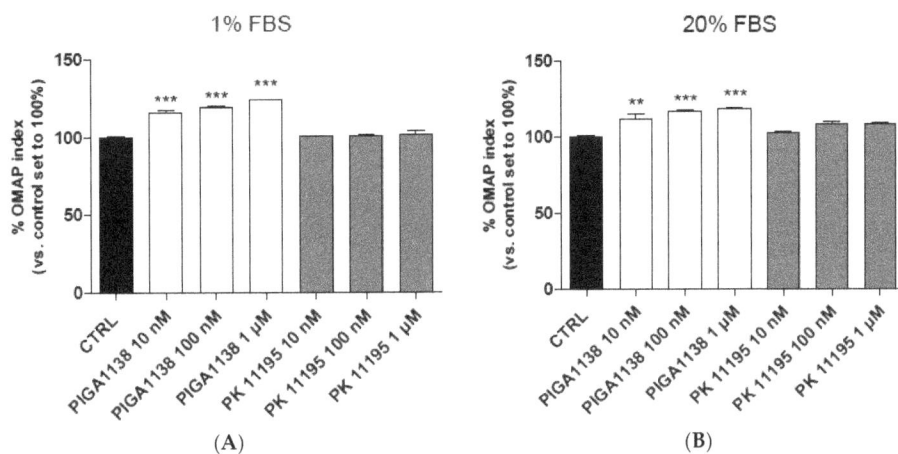

Figure 5. In vitro response of the human astrocytes to the PIGA1138 and PK11195 treatments. (**A**) Human astrocytes were treated with different concentrations of PIGA1138 and PK11195 (10 nM–1 μM) in serum-reduced media (1% FBS), and the MTS assay was performed after 48 h of treatment; (**B**) Human astrocytes were treated with different concentrations of PIGA1138 and PK11195 (10 nM–1· μM) in complete medium and the MTS assay was performed after 48 h of treatment. The data are expressed as percentages of metabolic activity compared to the control, which was set to 100%, and represent the means ± SEM of three different experiments performed in duplicate. ** $p < 0.01$, *** $p < 0.001$ vs. the control.

3. Discussion

In this study, TSPO PIGA ligand treatments showed a positive relation between steroidogenesis induction and the Oxidative Metabolism Activity/Proliferation Index in rat and human cell models. Notably, the data obtained in these astrocytic cell models were consistent with those acquired in healthy human astrocytes. Firstly, the ability of a number of PIGA ligands to stimulate the OMAP index was shown. Then, the ability of the ligands to induce steroidogenesis was evaluated and related with their TSPO residence time. It is likely that the metabolic stimulation was mediated by the modulation of steroid levels, as the positive effects of the compounds were completely counteracted by the treatment with the neurosteroid synthesis inhibitor AMG.

As previously reported in the nervous system, neuroactive steroids exert several direct regulatory activities on neurons and glial cells [59–66]; the role of TSPO in the release of neurosteroids led us to investigate the relation among the TSPO compound binding ability, OMAP index, and neurosteroid release in rat and human astrocytes. We selected a number of previously reported PIGA ligands with high affinities towards TSPO. The PIGA-induced increase in the OMAP index showed a good agreement with previous data reporting that three different TSPO ligands (triakontatetraneuropeptide, octadecaneuropeptide and Ro5-4864), at low concentrations, induce a dose-dependent increase in DNA synthesis in rat primary astrocytes by activating TSPO [67,68]. These results provide the evidence of a role for TSPO ligands in the control of glial cell proliferation. Indeed, although some forerunner studies have shown that low ligand concentrations inhibit astrocyte proliferation [69,70], most studies have shown a proliferative role for TSPO ligands. For instance, TSPO-related proliferation has been evaluated in C6 glioma cells in serum-free medium, as well as in a standard fibroblast cell line, showing that nanomolar concentrations of PK11195 and Ro5-4864 increased the growth rate and [³H]thymidine incorporation [71]. A significant increase in [³H]thymidine incorporation in human glioma cells following treatment with 10 nM PK11195 in serum-free media has been confirmed; the same study also showed an increase in mitochondrial mass and lipid fluidity [72]. It has been suggested that the changes in mitochondrial lipid metabolism might lead to mitochondrial biogenesis to support the increased metabolic requirements for cell division [72–74].

Concerning the well-known discrepancy between the functional efficacy and affinity of TSPO ligands [75], the PIGA compounds used here presented different functional effects on astrocytes in terms of the OMAP index, despite their similar TSPO affinities. Interestingly, the PIGA compounds that were able to interact with TSPO for a longer time (high RT) also presented higher pregnenolone synthesis induction and promoted better astrocyte well-being, in accordance with our previous data suggesting that the residence time is a predictive parameter for estimating steroidogenic activity [46]. Consistent with the present findings, it has been shown that primary astrocytes and C6 cells that were treated with nanomolar concentrations of PK11195 and Ro5-4864 exhibited an increase in the progesterone content in the medium that was 2–3-fold higher than the basal levels [76]. Similarly, the TSPO ligand AC-5216 increased the allopregnanolone level in meningioma cells [77], supporting the theory that TSPO ligands have a role in the regulation of steroid production. To our knowledge, this is the first study investigating steroidogenesis in a rat cellular model and in human glial cells in parallel, showing a good agreement between the data obtained in these two different cellular models.

Steroidogenesis stimulation is a widely proposed mechanism for the neuroprotective actions of TSPO ligands (for a review see [40]). In this line, we have recently demonstrated the involvement of steroidogenesis in the pro-survival properties of the PIGA ligands against cytotoxic insults, such as lipid peroxidation (induced by cellular glutathione depletion) and inflammatory responses (induced by LPS/IFN-γ cell exposure) [45]. Moreover, the PIGA-elicited modulation of the StAR protein levels has been recently demonstrated; in particular, we have shown that the stimulation of astroglial-derived cells with PIGAs leads to an increase of the 30 kDa intra-mitochondrial StAR, an indirect evidence of an increased cholesterol transfer into mitochondria [45].

It is well known that the effects of steroids are mediated by changes in cellular metabolism [78]. Here, we clearly showed that the positive effects of the PIGA ligands on the astrocytes were mediated by steroids, as they were completely prevented by the pre-treatment with the inhibitor of steroid synthesis, AMG. These data support the theory that the autocrine effects are due to neurosteroid release by the astrocytes themselves.

In conclusion, the importance of the induction of neurosteroidogenesis on astrocyte well-being was investigated in a human astrocyte model in vitro. Astrocytes play a pivotal role in the complex central nervous system network. The loss or the gain of astrocyte functions could be the basis of several pathological conditions [79–86]. In this respect, the positive effects of TSPO-stimulated neurosteroid release on astrocytes well-being were demonstrated. The development of molecules able to stimulate steroid release could represent a therapeutic strategy for central nervous system diseases characterized by astrocyte loss. Furthermore, these ligands may be exploited as pharmacological tools to deeply investigate the autocrine/paracrine roles of neurosteroids in the control of astrocyte metabolism.

4. Experimental Section

4.1. Materials

[^{3}H] PK11195 (Specific Activity, 85.7 µCi/nmol) was obtained from Perkin-Elmer Life Sciences (Perkin Elmer Italia, Monza, Italy). PK11195 and the protease inhibitors were purchased from Sigma-Aldrich (Sigma-Aldrich S.r.l., Milan, Italy). Dulbecco's Modified Eagle's Medium, fetal bovine serum, L-glutamine, penicillin, and streptomycin were purchased from Lonza (Milan, Italy). The enzyme-linked immunosorbent assay (ELISA) used to measure the pregnenolone levels was obtained from IBL (Hamburg, Germany). SU10603 and trilostane were gifts from Novartis Farma (Varese, Italy) and Susanne Zister (University of Dublin, Dublin, Ireland), respectively. All other chemical reagents were obtained from commercial sources.

Int. J. Mol. Sci. **2016**, *17*, 1028

4.2. Drugs

The compounds PIGA1228, PIGA1248, PIGA1175, PIGA1165, PIGA1174, PIGA1128, PIGA1136, PIGA1130, PIGA1137, PIGA1138, PIGA1226, PIGA1244, and PIGA1212 were synthesized according to the experimental procedure that we previously described [51,52]. Briefly, the appropriate 2-arylindoles, which were commercially available or easily obtained with a one-step Fischer indole synthesis, were reacted with oxalyl chloride at room temperature in anhydrous diethyl ether to produce the corresponding 2-arylindolylglyoxylyl chlorides. These compounds were then treated with the appropriate dialkylamine in dry toluene solution in the presence of triethylamine at room temperature to yield the target PIGA ligands [51,52].

4.3. Cell Culture

U87MG cells were purchased from the National Institute for Cancer Research of Genoa (Genoa Italy) and cultured in RPMI supplemented with 10% fetal bovine serum (FBS), 2 mM L-glutamine, 1% non-essential amino acids, penicillin (100 U/mL) and streptomycin (100 mg/mL) at 37 °C in 5% CO_2. C6 rat glioma cells were cultured in Dulbecco's Modified Eagle's Medium (DMEM) supplemented with 10% FBS, 2 mM L-glutamine, penicillin (100 U/mL) and streptomycin (100 mg/mL) at 37 °C in 5% CO_2. Human astrocytes were obtained from GIBCO (Life Technologies, Milan, Italy). Human astrocytes were cultured in DMEM supplemented with 10% FBS, 1% non-essential amino acids, 1% N-2 Supplement, penicillin (100 U/mL) and streptomycin (100 mg/mL) at 37 °C in 5% CO_2.

4.4. The Oxidative Metabolism Activity/Proliferation Index in the Astrocyte Cell Models

Rat C6 cells, human U87MG cells or human primary astrocytes were seeded in 96-well plates (10,000 cells/well) and maintained in their specific, complete culture media for 24 h. Then, the culture media were refreshed with serum-reduced media (1% FBS). After 16 h, the cells were treated with increasing concentrations of the PIGA ligands (ranging from 10 nM to 1 μM) for 48 h. In the experiments used to evaluate the specific contribution of PIGA ligand-induced steroid production, the U87MG cells were pretreated (1 h before the addition of the PIGA ligands) with AMG (50 μM), a potent inhibitor of steroidogenesis. After 48 h, the MTS reagent was added to the PIGA ligand-treated cells, and the colorimetric MTS conversion was quantified after 2 h by measuring the absorbance at 490 nm with a microplate reader (WallacVictor 2, 1420 Multilabel Counter, Perkin Elmer, MA, USA).

4.5. Pregnenolone Quantification

The amount of pregnenolone in the rat C6 and human U87MG astrocyte models were quantified as previously reported [46]. Briefly, C6 or U87MG cells were incubated with 40 μM TSPO PIGA ligands in saline medium (140 mM NaCl, 5 mM KCl, 1.8 mM $CaCl_2$, 1 mM $MgSO_4$, 10 mM glucose, 10 mM N-2-hydroxyethylpiperazine-N'-2-ethanesulfonic acid (HEPES)–NaOH, pH 7.4, and 0.1% bovine serum albumin) containing the inhibitors of pregnenolone metabolism trilostane (25 μM) and SU10603 (10 μM). After 2 h of incubation, the conditioned salt medium was collected and the amount of pregnenolone secreted into the medium was quantified by an ELISA.

4.6. RT Determination of the TSPO PIGA Ligands

The RT values of the TSPO PIGA ligands were calculated from the reciprocal of their dissociation rate constant (k_{off}). The k_{off} values were assessed by the competition kinetic association assay, as previously reported [46]. In particular, the TSPO-specific and TSPO-selective radioligand [^{3}H]PK11195 (approximately 20 nM, specific activity of 21.4 μCi/nmol) and PIGA ligand were simultaneously added to the final reaction volume (500 μL) containing a rat kidney membrane homogenate (30 μg of proteins) and assay buffer (50 mM Tris-HCl, pH 7.4). The incubation of the samples was terminated at various times by vacuum filtration. The PIGA ligands were solubilized with DMSO and tested at a concentration corresponding to a three-fold higher value than the respective value of the inhibition constant (K_i). The K_i values have been reported previously [52]. The nonspecific binding was determined in the presence of 1 μM PK11195. The experimentally derived data were analyzed using the "kinetics of competitive binding" using Prism 5.0 (GraphPad Software Inc., San Diego, CA, USA).

4.7. Radiolabel Binding Experiment in Human Astrocytes Using [^{3}H] PK11195

For crude membranes, confluent human astrocytes derived from a 175 cm^2 cell flask were harvested using phosphate-buffered saline (PBS), pH 7.4, supplemented with EDTA 0.04%. After the cells were collected by centrifugation (200× g for 5 min), the pellet was suspended in 10 mL of ice-cold buffer (5 mM Tris–HCl, pH 7.4 containing protease inhibitors (160 μg/mL benzamidine, 200 μg/mL bacitracin and 20 μg/mL trypsin inhibitor)) and homogenized with an Ultraturrax. Then, the homogenate was centrifuged at 48,000× g for 15 min at 4 °C and the supernatant was discarded. The obtained pellet was suspended in 10 mL of 50 mM Tris–HCl, pH 7.4 (assay buffer) containing the same amounts of protease inhibitors as described above, and the homogenate was pelleted by centrifugation (48,000× g for 15 min at 4 °C). The pellet was washed once with assay buffer and an additional centrifugation step was performed (48,000× g for 15 min at 4 °C). The resulting cell membrane pellet was suspended at a final concentration of 1 mg of protein/mL in assay buffer and used for the binding assays. The protein content of a 20 μL membrane suspension was measured by the Bradford method using the Bio-Rad Protein Assay reagent, according to the manufacturer's protocol, with bovine serum albumin (BSA) as the standard.

To determine the specific binding of [^{3}H]-PK11195 to the human astrocyte membrane suspensions, equilibrium radioligand binding assays were performed essentially as previously described [46,87]. Briefly, different aliquots of human astrocyte membranes (10–100 μg of proteins) were incubated with [^{3}H]-PK11195 (1.5 nM) in the presence (non-specific binding) or absence (total binding) of unlabelled PK11195 (1 μM) in a final volume of 500 mL of assay buffer for 90 min at 0 °C. For the saturation experiments, aliquots of human astrocyte membranes (20 μg of proteins) were incubated with eight different concentrations of [^{3}H]-PK11195 (0.5–30 nM) in duplicate using the conditions described above. In each assay, the final ethanol concentration in the incubation buffer was less than 1% and did not interfere with specific [^{3}H]-PK11195 binding.

4.8. Statistical Analysis

The data are reported as the means ± SEM of at least three independent experiments. All statistical analyses were performed using GraphPad 5.0 Prism Software (GraphPad Software, La Jolla, CA, USA). One-way ANOVA with Bonferroni's post-test and Spearman's correlation analyses were used to assess the statistical significance of the data. A p value ⩽ 0.05 was considered statistically significant.

Acknowledgments: Funding for this study was provided by the Italian Ministry of University and Scientific Research (PRIN-prot.2010W7YRLZ_005 and FIRB-prot.RBFR10ZJQT_002).

Author Contributions: Eleonora Da Pozzo designed and performed the experiments, analyzed the results, and wrote the manuscript; Chiara Giacomelli performed the experiments, analyzed the results and assisted with the writing of the manuscript; Barbara Costa analyzed the results and made a significant contribution to the writing of the manuscript; Chiara Cavallini performed residence time experiments; Sabrina Taliani, Elisabetta Barresi, Federico Da Settimo designed and furnished the PIGA ligands and assisted with the writing the final draft; Claudia Martini played a fundamental role as the supervisor of the project and provided important assistance in writing the article. All authors contributed to and have approved the final manuscript.

Conflicts of Interest: The authors declare no conflict of interest.

References

1. Rupprecht, R.; Holsboer, F. Neuroactive steroids: Mechanisms of action and neuropsychopharmacological perspectives. *Trends Neurosci.* **1999**, *9*, 410–416. [CrossRef]
2. Reddy, D.S. Neurosteroids: Endogenous role in the human brain and therapeutic potentials. *Prog. Brain Res.* **2010**, *186*, 113–137. [PubMed]

Int. J. Mol. Sci. **2016**, 17, 1028

3. Pfaff, D.W.; Gerlach, J.L.; McEwen, B.S.; Ferin, M.; Carmel, P.; Zimmerman, E.A. Autoradiographic localization of hormone-concentrating cells in the brain of the female rhesus monkey. *J. Comp. Neurol.* **1976**, *170*, 279–293. [CrossRef] [PubMed]
4. Buisson, B.; Bertrand, D. Steroid modulation of the nicotinic acetylcholine receptor. In *Neurosteroids: A New Regulatory Function in the Nervous System*, 1999th ed.; Baulieu, E.E., Robel, P., Schumacher, M., Eds.; Humana Press: Totowa, NJ, USA, 1999.
5. Gibbs, T.T.; Yaghoubi, N.; Weaver, C.E., Jr.; Park-Chung, M.; Russek, S.J.; Farb, D.H. Modulation of ionotropic glutamate receptors by neuroactive steroids. In *Neurosteroids: A New Regulatory Function in the Nervous System*, 1999th ed.; Baulieu, E.E., Robel, P., Schumacher, M., Eds.; Humana Press: Totowa, NJ, USA, 1999; pp. 167–190.
6. Majewska, M.D. Neurosteroid antagonists of the GABAA receptors. In *Neurosteroids: A New Regulatory Function in the Nervous System*, 1999th ed.; Baulieu, E.E., Robel, P., Schumacher, M., Eds.; Humana Press: Totowa, NJ, USA, 1999; pp. 155–166.
7. Bastianetto, S.; Ramassamy, C.; Poirier, J.; Quirion, R. Dehydroepiandrosterone (DHEA) protects hippocampal cells from oxidative stress-induced damage. *Mol. Brain Res.* **1999**, *6*, 35–41. [CrossRef]
8. Gee, K.W.; McCaule, L.D.; Lan, N.C. A putative receptor for neurosteroids on the GABAA receptor complex: The pharmacological properties and therapeutic potential of epalons. *Crit. Rev. Neurobiol.* **1995**, *9*, 207–227. [PubMed]
9. Ueda, H.; Yoshida, A.; Tokuyama, S.; Mizuno, K.; Maruo, J.; Matsuno, K.; Mita, S. Neurosteroids stimulate G protein-coupled sigma receptors in mouse brain synaptic membrane. *Neurosci. Res.* **2001**, *41*, 33–40. [CrossRef]
10. Schlichter, R.; Keller, A.F.; de Roo, M.; Breton, J.D.; Inquimbert, P.; Poisbeau, P. Fast nongenomic effects of steroids on synaptic transmission and role of endogenous neurosteroids in spinal pain pathways. *J. Mol. Neurosci.* **2006**, *28*, 33–51. [CrossRef]
11. Herd, M.B.; Belelli, D.; Lambert, J.J. Neurosteroid modulation of synaptic and extrasynaptic GABA(A) receptors. *Pharmacol. Ther.* **2007**, *116*, 20–34. [CrossRef] [PubMed]
12. Magnaghi, V. GABA and neuroactive steroid interactions in glia: New roles for old players? *Curr. Neuropharmacol.* **2007**, *5*, 47–64. [CrossRef] [PubMed]
13. Duenas, M.; Luquin, S.; Chowen, J.A.; Torres-Aleman, I.; Naftolin, F.; Garcia-Segura, L.M. Gonadal hormone regulation of insulin-like growth factor-I-like immunoreactivity in hypothalamic astroglia of developing and adult rats. *Neuroendocrinology* **1994**, *59*, 528–538. [CrossRef] [PubMed]
14. Kirschner, P.B.; Henshaw, R.; Weise, J.; Trubetskoy, V.; Finklestein, S.; Schulz, J.B.; Beal, M.F. Basic fibroblast growth factor protects against excitotoxicity and chemical hypoxia in both neonatal and adult rats. *J. Cereb. Blood Flow Metab.* **1995**, *15*, 619–623. [CrossRef] [PubMed]
15. Stone, D.J.; Rozovsky, I.; Morgan, T.E.; Anderson, C.P.; Hajian, H.; Finch, C.E. Astrocytes and microglia respond to estrogen with increased apoE mRNA in vivo and in vitro. *Exp. Neurol.* **1997**, *143*, 313–318. [CrossRef] [PubMed]
16. Flores, C.; Salmaso, N.; Cain, S.; Rodaros, D.; Stewart, J. Ovariectomy of adult rats leads to increased expression of astrocytic basic fibroblast growth factor in the ventral tegmental area and in dopaminergic projection regions of the entorhinal and prefrontal cortex. *J. Neurosci.* **1999**, *19*, 8665–8673. [PubMed]
17. Buchanan, C.D.; Mahesh, V.B.; Brann, D.W. Estrogen-astrocyte-luteinizing hormone-releasing hormone signaling: A role for transforming growth factor-β1. *Biol. Reprod.* **2000**, *62*, 1710–1721. [CrossRef] [PubMed]
18. Galbiati, M.; Martini, L.; Melcangi, R.C. Oestrogens, via transforming growth factor α, modulate basic fibroblast growth factor synthesis in hypothalamic astrocytes: In vitro observations. *J. Neuroendocrinol.* **2002**, *14*, 829–835. [CrossRef] [PubMed]
19. Kazanis, I.; Giannakopoulou, M.; Philippidis, H.; Stylianopoulou, F. Alterations in IGF-I, BDNF and NT-3 levels following experimental brain trauma and the effect of IGF-I administration. *Exp. Neurol.* **2004**, *186*, 221–234. [CrossRef] [PubMed]
20. Platania, P.; Seminara, G.; Aronica, E.; Troost, D.; Vincenza Catania, M.; Angela Sortino, M. 17β-estradiol rescues spinal motoneurons from AMPA-induced toxicity: A role for glial cells. *Neurobiol. Dis.* **2005**, *20*, 461–470. [CrossRef] [PubMed]
21. Mendez, P.; Cardona-Gomez, G.P.; Garcia-Segura, L.M. Interactions of insulin-like growth factor-I and estrogen in the brain. *Adv. Exp. Med. Biol.* **2005**, *567*, 285–303. [PubMed]

89

22. Dhandapani, K.M.; Brann, D.W. Role of astrocytes in estrogen-mediated neuroprotection. *Exp. Gerontol.* **2007**, *42*, 70–75. [CrossRef] [PubMed]

23. Cerciat, M.; Unkila, M.; Garcia-Segura, L.M.; Arevalo, M.A. Selective estrogen receptor modulators decrease the production of interleukin-6 and interferon-gamma-inducible protein-10 by astrocytes exposed to inflammatory challenge in vitro. *Glia* **2010**, *58*, 93–102. [CrossRef] [PubMed]

24. Xu, S.L.; Bi, C.W.; Choi, R.C.; Zhu, K.Y.; Miernisha, A.; Dong, T.T.; Tsim, K.W. Flavonoids induce the synthesis and secretion of neurotrophic factors in cultured rat astrocytes: A signaling response mediated by estrogen receptor. *Evid. Based Complement. Altern. Med.* **2013**, *2013*. [CrossRef] [PubMed]

25. Spence, R.D.; Wisdom, A.J.; Cao, Y.; Hill, H.M.; Mongerson, C.R.; Stapornkul, B.; Itoh, N.; Sofroniew, M.V.; Voskuhl, R.R. Estrogen mediates neuroprotection and anti-inflammatory effects during EAE through ERα signaling on astrocytes but not through ERβ signaling on astrocytes or neurons. *J. Neurosci.* **2013**, *33*, 10924–10933. [CrossRef] [PubMed]

26. Lee, E.; Sidoryk-Wêgrzynowicz, M.; Wang, N.; Webb, A.; Son, D.S.; Lee, K.; Aschner, M. GPR30 regulates glutamate transporter GLT-1 expression in rat primary astrocytes. *J. Biol. Chem.* **2012**, *287*, 26817–26828. [CrossRef] [PubMed]

27. Araújo, G.W.; Beyer, C.; Arnold, S. Oestrogen influences on mitochondrial gene expression and respiratory chain activity in cortical and mesencephalic astrocytes. *J. Neuroendocrinol.* **2008**, *20*, 930–941.

28. Irwin, R.W.; Yao, J.; Hamilton, R.T.; Cadenas, E.; Brinton, R.D.; Nilsen, J. Progesterone and estrogen regulate oxidative metabolism in brain mitochondria. *Endocrinology* **2008**, *149*, 3167–3175. [CrossRef] [PubMed]

29. Morohaku, K.; Pelton, S.H.; Daugherty, D.J.; Butler, W.R.; Deng, W.; Selvaraj, V. Translocator protein/peripheral benzodiazepine receptor is not required for steroid hormone biosynthesis. *Endocrinology* **2014**, *155*, 89–97. [CrossRef] [PubMed]

30. Tu, L.N.; Morohaku, K.; Manna, P.R.; Pelton, S.H.; Butler, W.R.; Stocco, D.M.; Selvaraj, V. Peripheral benzodiazepine receptor/translocator protein global knock-out mice are viable with no effects on steroid hormone biosynthesis. *J. Biol. Chem.* **2014**, *289*, 27444–27454. [CrossRef] [PubMed]

31. Banati, R.B.; Middleton, R.J.; Chan, R.; Hatty, C.R.; Kam, W.W.; Quin, C.; Graeber, M.B.; Parmar, A.; Zahra, D.; Callaghan, P.; et al. Positron emission tomography and functional characterization of a complete PBR/TSPO knockout. *Nat. Commun.* **2014**, *5*. [CrossRef] [PubMed]

32. Papadopoulos, V.; Lecanu, L. Translocator protein (18 kDa) TSPO: An emerging therapeutic target in neurotrauma. *Exp. Neurol.* **2009**, *219*, 53–57. [CrossRef] [PubMed]

33. Frye, C.A. Neurosteroids' effects and mechanisms for social, cognitive, emotional, and physical functions. *Psychoneuroendocrinology* **2009**, *34*, S143–S161. [CrossRef] [PubMed]

34. Fan, J.; Campioli, E.; Midzak, A.; Culty, M.; Papadopoulos, V. Conditional steroidogenic cell-targeted deletion of TSPO Unveils a Crucial role in Viability and Hormone-Dependent Steroid Formation. *Proc. Natl. Acad. Sci. USA* **2015**, *112*, 7261–7266. [CrossRef] [PubMed]

35. Mellon, S.H.; Deschepper, C.F. Neurosteroid biosynthesis: Genes for adrenal steroidogenic enzymes are expressed in the brain. *Brain Res.* **1993**, *629*, 283–292. [CrossRef]

36. King, S.R.; Ginsberg, S.D.; Ishii, T.; Smith, R.G.; Parker, K.L.; Lamb, D.J. The steroidogenic acute regulatory protein is expressed in steroidogenic cells of the day-old brain. *Endocrinology* **2004**, *145*, 4775–4780. [CrossRef] [PubMed]

37. Papadopoulos, V.; Baraldi, M.; Guilarte, T.R.; Knudsen, T.B.; Lacapère, J.J.; Lindemann, P.; Norenberg, M.D.; Nutt, D.; Weizman, A.; Zhang, M.R.; et al. Translocator protein (18 kDa): New nomenclature for the peripheral-type benzodiazepine receptor based on its structure and molecular function. *Trends Pharmacol. Sci.* **2006**, *27*, 402–409. [CrossRef] [PubMed]

38. Lacapère, J.J.; Papadopoulos, V. Peripheral-type benzodiazepine receptor: Structure and function of a cholesterol-binding protein in steroid and bile acid biosynthesis. *Steroids* **2003**, *68*, 569–585. [CrossRef]

39. Da Pozzo, E.; Costa, B.; Martini, C. Translocator protein (TSPO) and neurosteroids: Implications in psychiatric disorders. *Curr. Mol. Med.* **2012**, *12*, 426–442. [CrossRef] [PubMed]

40. Da Pozzo, E.; Giacomelli, C.; Barresi, E.; Costa, B.; Taliani, S.; da Settimo, F.; Martini, C. Targeting the 18 kDa translocator protein: Recent perspectives for neuroprotection. *Biochem. Soc. Trans.* **2015**, *43*, 559–565. [CrossRef] [PubMed]

41. Veiga, S.; Azcoitia, I.; Garcia-Segura, L.M. Ro5-4864, a peripheral benzodiazepine receptor ligand, reduces reactive gliosis and protects hippocampal hilar neurons from kainic acid excitotoxicity. *J. Neurosci. Res.* **2005**, *80*, 129–137. [CrossRef] [PubMed]

42. Veiga, S.; Carrero, P.; Pernia, O.; Azcoitia, I.; Garcia-Segura, L.M. Translocator protein 18 kDa is involved in the regulation of reactive gliosis. *Glia* **2007**, *55*, 1426–1436. [CrossRef] [PubMed]

43. Girard, C.; Liu, S.; Cadepond, F.; Adams, D.; Lacroix, C.; Verleye, M.; Gillardin, J.M.; Baulieu, E.E.; Schumacher, M.; Schweizer-Groyer, G. Etifoxine improves peripheral nerve regeneration and functional recovery. *Proc. Natl. Acad. Sci. USA* **2008**, *105*, 20505–20510. [CrossRef] [PubMed]

44. Zhao, Y.Y.; Yu, J.Z.; Li, Q.Y.; Ma, C.G.; Lu, C.Z.; Xiao, B.G. TSPO-specific ligand vinpocetine exerts a neuroprotective effect by suppressing microglial inflammation. *Neuron Glia Biol.* **2011**, *7*, 187–197. [CrossRef] [PubMed]

45. Santoro, A.; Mattace Raso, G.; Taliani, S.; da Pozzo, E.; Simorini, F.; Costa, B.; Martini, C.; Laneri, S.; Sacchi, A.; Cosimelli, B.; et al. TSPO-ligands prevent oxidative damage and inflammatory response in C6 glioma cells by neurosteroid synthesis. *Eur. J. Pharm. Sci.* **2016**, *88*, 124–131. [CrossRef] [PubMed]

46. Costa, B.; da Pozzo, E.; Giacomelli, C.; Barresi, E.; Taliani, S.; da Settimo, F.; Martini, C. TSPO ligand residence time: A new parameter to predict compound neurosteroidogenic efficacy. *Sci. Rep.* **2016**, *6*. [CrossRef] [PubMed]

47. Chen, J.H.; Tsou, T.C.; Chiu, I.M.; Chou, C.C. Proliferation inhibition, DNA damage, and cell-cycle arrest of human astrocytoma cells after acrylamide exposure. *Chem. Res. Toxicol.* **2010**, *23*, 1449–1458. [CrossRef] [PubMed]

48. Li, Y.; Cheng, D.; Cheng, R.; Zhu, X.; Wan, T.; Liu, J.; Zhang, R. Mechanisms of U87 astrocytoma cell uptake and trafficking of monomeric versus protofibril Alzheimer's disease amyloid-β proteins. *PLoS ONE* **2014**, *9*, e99939. [CrossRef] [PubMed]

49. Maresca, B.; Spagnuolo, M.S.; Cigliano, L. Haptoglobin modulates β-amyloid uptake by U-87 MG astrocyte cell line. *J. Mol. Neurosci.* **2015**, *56*, 35–47. [CrossRef] [PubMed]

50. Munoz, L.; Kavanagh, M.E.; Phoa, A.F.; Heng, B.; Dzamko, N.; Chen, E.J.; Doddareddy, M.R.; Guillemin, G.J.; Kassiou, M. Optimisation of LRRK2 inhibitors and assessment of functional efficacy in cell-based models of neuroinflammation. *Eur. J. Med. Chem.* **2015**, *95*, 29–34. [CrossRef] [PubMed]

51. Da Settimo, F.; Simorini, F.; Taliani, S.; La Motta, C.; Marini, A.M.; Salerno, S.; Bellandi, M.; Novellino, E.; Greco, G.; Cosimelli, B.; et al. Anxiolytic-like effects of N,N-dialkyl-2-phenylindol-3-ylglyoxylamides by modulation of translocator protein promoting neurosteroid biosynthesis. *J. Med. Chem.* **2008**, *51*, 5798–5806. [CrossRef] [PubMed]

52. Barresi, E.; Bruno, A.; Taliani, S.; Cosconati, S.; da Pozzo, E.; Salerno, S.; Simorini, F.; Daniele, S.; Giacomelli, C.; Marini, A.M.; et al. Deepening the topology of the Translocator Protein binding site by novel N,N-dialkyl-2-arylindol-3-ylglyoxylamides. *J. Med. Chem.* **2015**, *58*, 6081–6092. [CrossRef] [PubMed]

53. Kugler, W.; Veenman, L.; Shandalov, Y.; Leschiner, S.; Spanier, I.; Lakomek, M.; Gavish, M. Ligands of the mitochondrial 18 kDa translocator protein attenuate apoptosis of human glioblastoma cells exposed to erucylphosphohomocholine. *Cell Oncol.* **2008**, *30*, 435–450. [PubMed]

54. Banfalvi, G. *Cell Cycle Synchronization. Methods and Protocols*; Humana Press: Totowa, NJ, USA, 2011.

55. Chen, M.; Huang, J.; Yang, X.; Liu, B.; Zhang, W.; Huang, L.; Deng, F.; Ma, J.; Bai, Y.; Lu, R.; Huang, B.; Gao, Q.; Zhuo, Y.; Ge, J. Serum starvation induced cell cycle synchronization facilitates human somatic cells reprogramming. *PLoS ONE* **2012**, *7*, e28203. [CrossRef] [PubMed]

56. Takahashi, S.; Abe, T.; Gotoh, J.; Fukuuchi, Y. Substrate-dependence of reduction of MTT: A tetrazolium dye differs in cultured astroglia and neurons. *Neurochem. Int.* **2002**, *40*, 441–448. [CrossRef]

57. Dunigan, D.D.; Waters, S.B.; Owen, T.C. Aqueous soluble tetrazolium/formazan MTS as an indicator of NADH- and NADPH-dependent dehydrogenase activity. *Biotechniques* **1995**, *19*, 640–649. [PubMed]

58. Mosmann, T. Rapid colorimetric assay for cellular growth and survival: Application to proliferation and cytotoxicity assays. *J. Immunol. Methods* **1983**, *65*, 55–63. [CrossRef]

59. Papadopoulos, V.; Guarneri, P.; Kreuger, K.E.; Guidotti, A.; Costa, E. Pregnenolone biosynthesis in C6-2B glioma cell mitochondria: Regulation by a mitochondrial diazepam binding inhibitor receptor. *Proc. Natl. Acad. Sci. USA* **1992**, *89*, 5113–5117. [CrossRef] [PubMed]

60. Azcoitia, I.; Sierra, A.; Veiga, S.; Garcia-Segura, L.M. Aromatase expression by reactive astroglia is neuroprotective. *Ann. N. Y. Acad. Sci.* **2003**, *1007*, 298–305. [CrossRef] [PubMed]

61. Groeneveld, G.J.; van Muiswinkel, F.L.; Sturkenboom, J.M.; Wokke, J.H.; Bär, P.R.; van den Berg, L.H. Ovariectomy and 17β-estradiol modulate disease progression of a mouse model of ALS. *Brain Res.* **2004**, *1021*, 128–131. [CrossRef] [PubMed]

62. Garcia-Ovejero, D.; Azcoitia, I.; Doncarlos, L.L.; Melcangi, R.C.; Garcia-Segura, L.M. Glia-neuron crosstalk in the neuroprotective mechanisms of sex steroid hormones. *Brain Res. Rev.* **2005**, *48*, 273–286. [CrossRef] [PubMed]

63. Conejo, N.M.; González-Pardo, H.; Cimadevilla, J.M.; Argüelles, J.A.; Díaz, F.; Vallejo-Seco, G.; Arias, J.L. Influence of gonadal steroids on the glial fibrillary acidic protein-immunoreactive astrocyte population in young rat hippocampus. *J. Neurosci. Res.* **2005**, *79*, 488–494. [CrossRef] [PubMed]

64. Barreto, G.; Veiga, S.; Azcoitia, I.; Garcia-Segura, L.M.; Garcia-Ovejero, D. Testosterone decreases reactive astroglia and reactive microglia after brain injury in male rats: Role of its metabolites, oestradiol and dihydrotestosterone. *Eur. J. Neurosci.* **2007**, *25*, 3039–3046. [CrossRef] [PubMed]

65. Choi, C.I.; Lee, Y.D.; Gwag, B.J.; Cho, S.I.; Kim, S.S.; Suh-Kim, H. Effects of estrogen on lifespan and motor functions in female hSOD1 G93A transgenic mice. *J. Neurol. Sci.* **2008**, *268*, 40–47. [CrossRef] [PubMed]

66. Garcia-Segura, L.M. *Hormones and Brain Plasticity*; Oxford University Press: New York, NY, USA, 2009.

67. Arevalo, M.A.; Santos-Galindo, M.; Bellini, M.J.; Azcoitia, I.; Garcia-Segura, L.M. Actions of estrogens on glial cells: Implications for neuroprotection. *Biochim. Biophys. Acta* **2010**, *1800*, 1106–1112. [CrossRef] [PubMed]

68. Gandolfo, P.; Patte, C.; Thoumas, J.L.; Leprince, J.; Vaudry, H.; Tonon, M.C. The endozepine ODN stimulates [3H]thymidine incorporation in cultured rat astrocytes. *Neuropharmacology* **1999**, *38*, 725–732. [CrossRef]

69. Gandolfo, P.; Patte, C.; Leprince, J.; Régo, J.L.; Mensah-Nyagan, A.G.; Vaudry, H.; Tonon, M.C. The triakontatetraneuropeptide (TTN) stimulates thymidine incorporation in rat astrocytes through peripheral-type benzodiazepine receptors. *J. Neurochem.* **2000**, *75*, 701–707. [CrossRef] [PubMed]

70. Neary, J.T.; Jorgensen, S.L.; Oracion, A.M.; Bruce, J.H.; Norenberg, M.D. Inhibition of growth factor-induced DNA synthesis in astrocytes by ligands of peripheral-type benzodiazepine receptors. *Brain Res.* **1995**, *675*, 27–30. [CrossRef]

71. Bruce, J.H.; Ramirez, A.M.; Lin, L.; Oracion, A.; Agarwal, R.P.; Norenberg, M.D. Peripheral-type benzodiazepines inhibit proliferation of astrocytes in culture. *Brain Res.* **1991**, *564*, 167–170. [CrossRef]

72. Ikezaki, K.; Black, K.L. Stimulation of cell growth and DNA synthesis by peripheral benzodiazepine. *Cancer Lett.* **1990**, *49*, 115–120. [CrossRef]

73. Miccoli, L.; Oudard, S.; Beurdeley-Thomas, A.; Dutrillaux, B.; Poupon, M.F. Effect of 1-(2-chlorophenyl)-N-methyl-N-(1-methylpropyl)-3-isoquinoline carboxamide (PK11195), a specific ligand of the peripheral benzodiazepine receptor, on the lipid fluidity of mitochondria in human glioma cells. *Biochem. Pharmacol.* **1999**, *58*, 715–721. [CrossRef]

74. Shiraishi, T.; Black, K.L.; Ikezaki, K.; Becker, D.P. Peripheral benzodiazepine induces morphological changes and proliferation of mitochondria in glioma cells. *J. Neurosci. Res.* **1991**, *30*, 463–474. [CrossRef] [PubMed]

75. Black, K.L.; Shiraishi, T.; Ikezak, K.; Tabuchi, K.; Becker, D.P. Peripheral benzodiazepine stimulates secretion of growth hormone and mitochondrial proliferation in pituitary tumour GH3 cells. *Neurol. Res.* **1994**, *16*, 74–80. [PubMed]

76. Scarf, A.M.; Auman, K.M.; Kassiou, M. Is there any correlation between binding and functional effects at the translocator protein (TSPO) (18 kDa)? *Curr. Mol. Med.* **2012**, *12*, 387–397. [CrossRef] [PubMed]

77. Alho, H.; Varga, V.; Krueger, K.E. Expression of mitochondrial benzodiazepine receptor and its putative endogenous ligand diazepam binding inhibitor in cultured primary astrocytes and C-6 cells: Relation to cell growth. *Cell Growth Differ.* **1994**, *5*, 1005–1014. [PubMed]

78. Gao, Z.W.; Huang, J.B.; Lin, Q.; Qin, Q.; Liang, Y.J.; Zhou, L.; Luo, M. The effects of PK11195 on meningioma was associated with allopregnanolone biosynthesis, which was mediated by translocator protein 18 kDa. *Cancer Biomark.* **2016**, *16*, 65–69. [CrossRef] [PubMed]

79. Hechter, O.; Halkerston, I.D. Effects of steroid hormones on gene regulation and cell metabolism. *Annu. Rev. Physiol.* **1965**, *27*, 133–162. [CrossRef] [PubMed]

80. Takuma, K.; Baba, A.; Matsuda, T. Astrocyte apoptosis: Implications for neuroprotection. *Prog. Neurobiol.* **2004**, *72*, 111–127. [CrossRef] [PubMed]

81. Wagner, B.; Natarajan, A.; Grünaug, S.; Kroismayr, R.; Wagner, E.F.; Sibilia, M. Neuronal survival depends on EGFR signaling in cortical but not midbrain astrocytes. *EMBO J.* **2006**, *25*, 752–762. [CrossRef] [PubMed]

82. Bauer, J.; Elger, C.E.; Hans, V.H.; Schramm, J.; Urbach, H.; Lassmann, H.; Bien, C.G. Astrocytes are a specific immunological target in Rasmussen's encephalitis. *Ann. Neurol.* **2007**, *62*, 67–80. [CrossRef] [PubMed]

83. Ricci, G.; Volpi, L.; Pasquali, L.; Petrozzi, L.; Siciliano, G. Astrocyte-neuron interactions in neurological disorders. *J. Biol. Phys.* **2009**, *35*, 317–336. [CrossRef] [PubMed]

84. Wingerchuk, D.M. Neuromyelitis optica spectrum disorders. *Continuum (Minneap. Minn.)* **2010**, *16*, 105–121. [CrossRef] [PubMed]

85. Escartin, C.; Rouach, N. Astroglial networking contributes to neurometabolic coupling. *Front. Neuroenerg.* **2013**, *5*. [CrossRef] [PubMed]

86. Najjar, S.; Pearlman, D.M.; Alper, K.; Najjar, A.; Devinsky, O. Neuroinflammation and psychiatric illness. *J. Neuroinflamm.* **2013**, *10*. [CrossRef] [PubMed]

87. Phatnani, H.; Maniatis, T. Astrocytes in neurodegenerative disease. *Cold Spring Harb. Perspect. Biol.* **2015**, *7*. [CrossRef] [PubMed]

International Journal of
Molecular Sciences

MDPI

Article

Synthesis and Evaluation of Tricarbonyl ^{99m}Tc-Labeled 2-(4-Chloro)phenyl-imidazo[1,2-*a*] pyridine Analogs as Novel SPECT Imaging Radiotracer for TSPO-Rich Cancer

Ji Young Choi [1,2], Rosa Maria Iacobazzi [3,4], Mara Perrone [3], Nicola Margiotta [5], Annalisa Cutrignelli [3], Jae Ho Jung [1], Do Dam Park [1], Byung Seok Moon [1], Nunzio Denora [3,*], Sang Eun Kim [1,2,6] and Byung Chul Lee [1,6,*]

[1] Department of Nuclear Medicine, Seoul National University College of Medicine, Seoul National University Bundang Hospital, Seongnam 13620, Korea; cjy0929@snu.ac.kr (J.Y.C.); jaehoboa@paran.com (J.H.J.); ddp2194@naver.com (D.D.P.); bsmoon@snu.ac.kr (B.S.M.); kse@snu.ac.kr (S.E.K.)

[2] Department of Transdisciplinary Studies, Graduate School of Convergence Science and Technology, Seoul National University, Seoul 16229, Korea

[3] Department of Pharmacy–Drug Sciences, University of Bari "Aldo Moro", Bari 70125, Italy; rosamaria.iacobazzi@uniba.it (R.M.I.); mara.perrone@uniba.it (M.P.); annalisa.cutrignelli@uniba.it (A.C.)

[4] Istituto Tumori IRCCS "Giovanni Paolo II", Flacco, St. 65, Bari 70124, Italy

[5] Department of Chemistry, University of Bari "Aldo Moro", Bari 70125, Italy; nicola.margiotta@uniba.it

[6] Center for Nanomolecular Imaging and Innovative Drug Development, Advanced Institutes of Convergence Technology, Suwon 16229, Korea

* Correspondence: nunzio.denora@uniba.it (N.D.); leebc@snu.ac.kr (B.C.L.); Tel.: +39-080-5442-767 (N.D.); +82-31-7872-956 (B.C.L.)

Academic Editor: Mateus Webba da Silva
Received: 4 June 2016; Accepted: 1 July 2016; Published: 7 July 2016

Abstract: The 18-kDa translocator protein (TSPO) levels are associated with brain, breast, and prostate cancer progression and have emerged as viable targets for cancer therapy and imaging. In order to develop highly selective and active ligands with a high affinity for TSPO, imidazopyridine-based TSPO ligand (CB256, **3**) was prepared as the precursor. ^{99m}Tc- and Re-CB256 (**1** and **2**, respectively) were synthesized in high radiochemical yield (74.5% $\pm$ 6.4%, decay-corrected, $n = 5$) and chemical yield (65.6%) by the incorporation of the $[^{99m}Tc(CO)_3(H_2O)_3]^+$ and $(NEt_4)_2[Re(CO)_3Br_3]$ followed by HPLC separation. Radio-ligand **1** was shown to be stable (>99%) when incubated in human serum for 4 h at 37 °C with a relatively low lipophilicity (log$D = 2.15 \pm 0.02$). The rhenium-185 and -187 complex **2** exhibited a moderate affinity ($K_i = 159.3 \pm 8.7$ nM) for TSPO, whereas its cytotoxicity evaluated on TSPO-rich tumor cell lines was lower than that observed for the precursor. In vitro uptake studies of **1** in C6 and U87-MG cells for 60 min was found to be 9.84% $\pm$ 0.17% and 7.87% $\pm$ 0.23% ID, respectively. Our results indicated that ^{99m}Tc-CB256 can be considered as a potential new TSPO-rich cancer SPECT imaging agent and provides the foundation for further in vivo evaluation.

Keywords: translocator protein; TSPO; Tricarbonyltechnetium-99m; ^{99m}Tc(CO)$_3$; TSPO-rich tumors; SPECT

1. Introduction

The 18-kDa translocator protein (TSPO) is a mitochondrial five transmembrane protein, which is principally located in the outer mitochondrial membrane and associated with a wide number of biological processes including cell proliferation, apoptosis, steroidogenesis, and immunomodulation [1–4]. Moreover, elevated TSPO levels are well documented in oncology and have been correlated with tumor proliferation, invasion, and metastasis including brain, colorectal, liver, breast, oral cavity, and prostate carcinomas [5–7]. Thus, TSPO is a suitable imaging target for both inflammatory neurodegenerative diseases and cancer because it is highly expressed in activated microglial cells and surrounding TSPO-rich tumors, but absent in normal healthy tissues, except for kidney, heart, and gonads. Over the years, TSPO-specific ligands have been widely investigated and shown to be valuable tools for targeting the progression of pathologies associated with overexpression of TSPO. Various TSPO ligands are known from many different structural classes such as isoquinoline carboxamides (e.g., PK 11195), benzodiazepines (e.g., Ro-54864), phenoxyarylacetamides (e.g., DAA1106), aryloxyanilides (e.g., PBR28) and 2-phenyl-imidazo[1,2-*a*]pyridine acetamides (e.g., alpidem) [8–12]. Although a wide number of TSPO ligands have been developed for positron emission tomography (PET) and single photon emission computed tomography (SPECT) imaging, few ^{99m}Tc-labeled ligands have been reported so far [13–16]. Among the radionuclides, ^{99m}Tc is desirable due to its ideal physical properties (γ-emission of 141 keV, $t_{1/2}$ = 6 h) for imaging purposes [17,18]. Moreover, ^{99m}Tc is readily obtained by daily elution from [99]Mo/^{99m}Tc-generator and thus, it is very convenient and suitable for routine clinical use. In addition, the tricarbonyl technetium-99m ([99m]Tc-(CO)$_3$) unit has been shown to be useful for introducing ^{99m}Tc into biomolecules because of its high chemical stability and small size. In order to take advantage of molecular imaging techniques with [99m]Tc-(CO)$_3$, many low-molecular weight [99m]Tc(I)-complex with tridentate ligands have been developed and used for the preparation of [99m]Tc-(CO)$_3$-labeled radiotracers [19–24].

In our previous studies, potent and selective imidazopyridine-based TSPO ligands—which could carry both a cytostatic platinum species and a rhenium complex as the precursor for introducing the [99m]Tc-(CO)$_3$ unit—were reported [25–30]. Among the imidazopyridine-based TSPO ligands investigated so far, 2-(8-(2-(bis(pyridin-2-yl)methyl)amino)acetamido)-2-(4-chlorophenyl)*H* imidazo [1,2-*a*] pyridin-3-yl)-*N*,*N*-dipropylacetamide (CB256, **3**) contains the TSPO-targeting moiety and a metal ion anchor, namely di-(2-picolyl)amine, for a bifunctional chelate approach [29].

Our goal was to exploit the imidazopyridine-based ligand CB256 for incorporating the tricarbonyl ^{99m}Tc radioisotope to obtain a new TSPO-selective imaging agent. In addition, Re-CB256 was prepared as a model of [186/188]Re-CB256, a potential TSPO-targeted internal radiation therapy agent.

2. Results and Discussion

2.1. Synthesis of Re-CB256 and ^{99m}Tc-CB256

The precursor **3** (CB256) was prepared from the TSPO ligand CB86 and bromoacetyl bromide in the presence of triethylamine, followed by *N*-alkylation with di-(2-picolyl)amide, according to a previously described method [29]. The coordination potential of CB256 towards Pt(II) and Re(I) metal ions was already exploited by some of us and, in particular, two dinuclear Pt/Re and Re/Re complexes were prepared, indicating that the introduction of the di-(2-picolyl)amine moiety allows the coordination of a metal ion such as Pt^{2+} or Re$^+$ [30]. Unlike our previous results of the dinuclear Re complex, only the homologous ^{99m}Tc complex was generated in the radiolabeling reaction due to the low concentration of technetium. The cold Re-CB256 (**2**) was prepared to identify the chemical characteristics of ^{99m}Tc-CB256 based on the similar chemical properties between Tc and Re complex. The coordination of Re to the imidazopyridine residue reduced the affinity of CB256 towards TSPO, hence, in this investigation, we have sought to coordinate a single metal ion to the di-(2-picolyl)amine chelate residue to obtain a diagnostic drug (^{99m}Tc in compound **1**) or a model of a therapeutic drug (using cold Re in compound **2**).

As shown in Scheme 1, the "cold" rhenium complex **2** was prepared by treating the TSPO ligand **3** in methanol at 65 °C with (NEt$_4$)$_2$[ReBr$_3$(CO)$_3$]. The reaction was monitored by HPLC until the precursor peak (t_R = 14.5 min) disappeared. The desired Re(CO)$_3$ core, coordinated to CB256 (**2**), was obtained in good yield (60%–71%) and revealed a HPLC retention time of 22 min. Compound **2** was characterized by ^{1}H-NMR (Figure 1) and ^{13}C-NMR spectroscopy and by HRMS (ESI).

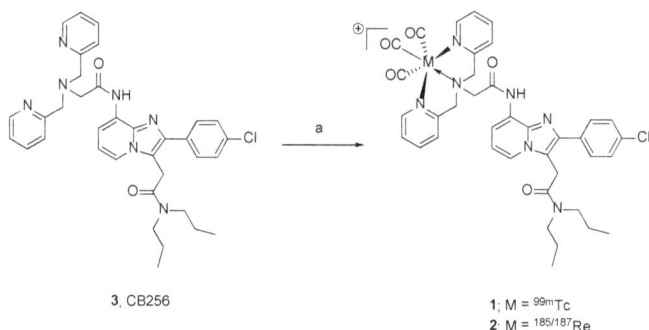

Scheme 1. Synthesis of ^{99m}Tc- and Re-CB256. Reagents and conditions: (a) $[^{99m}Tc(CO)_3(H_2O)_3]^+$, MeOH–H$_2$O, 65 °C, 0.5 h or (NEt$_4$)$_2$[Re(CO)$_3Br_3$], MeOH, 65 °C, 1 h.

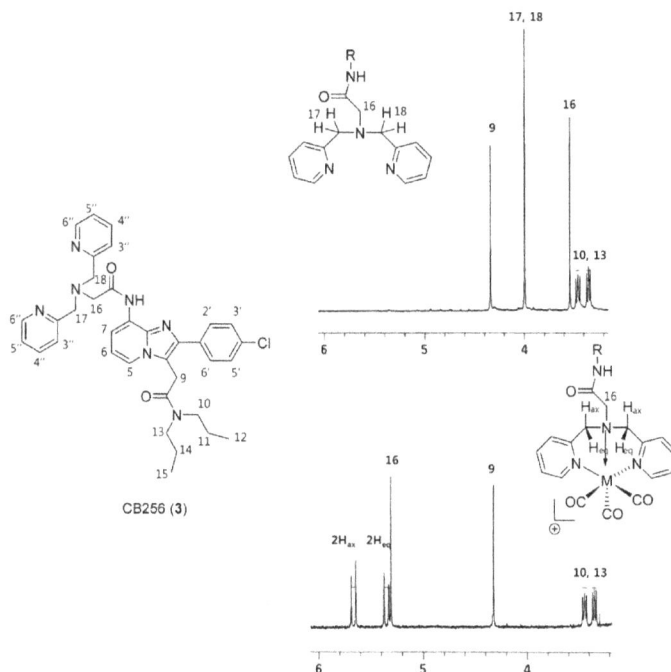

Figure 1. Section of the ^{1}H-NMR spectra between 6.0 and 3.2 ppm of the precursor **3** (**upper**) and of the "cold" rhenium complex **2** (**bottom**).

The coordination of the ^{99m}Tc radioisotope to the di-(2-picolyl)amine moiety of the TSPO-ligand **3** was obtained in aqueous media by using the *fac*-$[^{99m}Tc(H_2O)_3(CO)_3]^+$ synthon, which can be readily generated from ^{99m}TcO$_4{}^-$ and CO gas in the presence of NaBH$_4$ [20,31].

The radiochemical yield and purity of **1** were 74.5% $\pm$ 6.4% (decay-corrected) and >95%, respectively. Compound **1** was characterized by HPLC (Figure 2) by comparison with the chromatogram of **2**. The retention times for **1** and **2** were found to be 22.5 and 22 min, respectively.

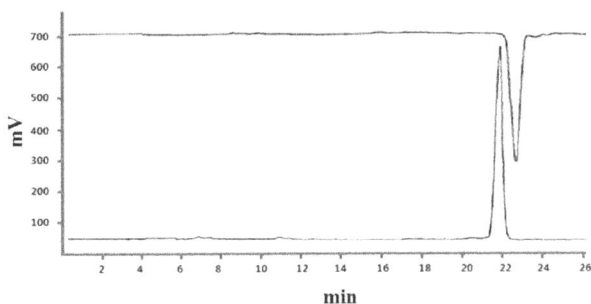

Figure 2. HPLC chromatograms of [99mTc]-CB256 (**1**, red) and Re-CB256 (**2**, black). Xterra RP-18; 20%–90% acetonitrile-water; flow rate 3 mL/min.

2.2. In Vitro Stability Studies and Partition Coefficient of ^{99m}Tc-CB256 (**1**)

The percentage of **1** remaining in solution after 4 h of incubation in human serum at 37 °C was 99% as calculated by radio-TLC scanner, indicating a high in vitro stability of the radiotracer. The partition coefficient of **1** ($LogD = 2.15 \pm 0.02$ vs. 1.08 ± 0.02 for **3**) indicated a relatively low lipophilicity compared to that of fluorine-substituted imidazo [1,2-*a*]pyridine acetamide analogs (i.e., 3.00 ± 0.03 for [18F]CB251) prepared in our previous investigations [32]. However, the lower lipophilicity did not prevent the uptake of **1** in tumor cells (see Section 2.3).

2.3. In Vitro Cell Uptake Assay of ^{99m}Tc-CB256 (**1**)

The in vitro uptake of **1** by tumor cells was measured in two different cancer cell lines overexpressing the TSPO receptor, namely C6 rat glioma and U87-MG human glioblastioma cell lines. The results are shown in Figure 3 and indicate that the uptake of **1** was time-dependent and reached almost the highest level after 60 min of incubation ($9.84\% \pm 0.17\%$ and $7.87\% \pm 0.23\%$ ID in C6 and U87-MG cells, respectively). This result is a direct consequence of the lipophilicity of compound **1**. In blocking experiments conducted on U87-MG cells in the presence of the TSPO ligand PK 11195, the cell uptake of **1** was markedly decreased throughout the experimental period and the observed relative uptake reduction was 63.5%. These displacement studies indicate that the uptake of **1** in the tumor cells was selectively and specifically mediated by TSPO, and support the potential use of compound **1** as TSPO marker for SPECT diagnosis.

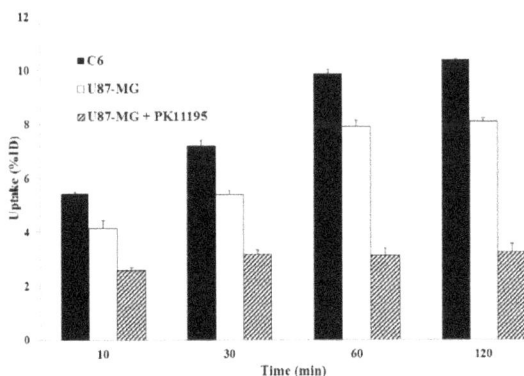

Figure 3. Uptake kinetic of ^{99m}Tc-CB256 (**1**) into C6 and U87-MG cells in the presence or absence of PK 11195. Data are expressed as percentage injected dose (% ID, mean with S.D. *n* = 4). Closed bar: uptake of **1** in C6 cells; open bar: uptake of **1** in U87-MG cells; right-handed striped bar: uptake of **1** in U87-MG cells in the presence of 300 µM PK 11195.

2.4. In Vitro Cell Binding Affinity of CB256 (3) and Re-CB256 (2)

The affinity for TSPO of compounds **3** and **2** was evaluated by measuring their ability to displace the reference compound [^{3}H]-PK 11195 from the membrane extracts of C6 glioma cells. The results show that the free TSPO ligand **3** has an appreciable affinity for TSPO (148 nM), which, however, is lower than that of the reference compound PK 11195 (9 nM). This result is in agreement with that previously reported [30], and the reduced affinity of compound **3** could be explained by the steric bulk generated by the dipicolylaminic moiety at position 8 of the imidazopyridine nucleus, which is crucial for the interaction of the ligand with mitochondrial TSPO [29,30]. As expected, compound **2** showed an affinity (159 nM) comparable to that of the free ligand (Table 1). In fact, as already reported, the coordination of a metal ion (Pt(II)) to the tridentate bis-(2-picolyl)amine residue did not significantly alter the TSPO affinity of CB256, while metalation at the imidazopyridine moiety greatly reduced the affinity for TSPO [30]. Even though the affinity of compounds **2** and **3** are lower to that of the reference compound PK 11195, these binding values can be still good for biological applications.

Table 1. Affinities (K_i/nM) of Re-CB256 (**2**) for TSPO from rat C6 glioma cells membranes. Corresponding values for PK 11195 and CB256 (**3**) are also reported for comparison.

Compound	K_i (nM) for TSPO
PK 11195	9.10 ± 1.2
CB256 (**3**)	148.2 ± 11.3
Re-CB256 (**2**)	159.3 ± 8.70

2.5. In Vitro Cytotoxicity Assays of CB256 (3) and Re-CB256 (2)

Table 2 summarizes the cytotoxicity of **2** and **3** against HepG2, MCF7, and U87 cancer cells exposed for a period of 72 h. Our previous investigation has shown that compound **3** is extremely effective toward C6 glioma cells [29]. The high cytotoxicity of **3** was correlated with its ability to produce double-strand lesions on DNA after coordination of a biometal, such as Cu(I) [29]. In the present study, compound **3** was found to be cytotoxic against HepG2, MCF7, and U87 cancer cells, confirming the above-mentioned evidence. On the contrary, as expected, compound **2** has much lower cytotoxicity. The lower cytotoxicity of Re-CB256 with respect to uncoordinated CB256 can be explained by its inability to coordinate a biometal (CuI) and therefore to act as a double-strand breaker of DNA.

Table 2. Cytotoxicity of CB256 (**3**) and Re-CB256 (**2**) toward HepG2, MCF7 and U87 cancer cell lines.

Compound	IC$_{50}$ (μM) [a]		
	HepG2	MCF7	U87
CB256 (**3**)	30 ± 5	38 ± 3	35 ± 2
Re-CB256 (**2**)	>50 (67%) [b]	>50 (56%) [b]	>50 (59%) [b]

[a] Cells were seeded at a density of ~5000 cells per well into 96-well plates. Following overnight incubation, cells were treated with a range of drug concentrations (from 0.01 to 50 μM) and incubated at 37 °C under a humidified atmosphere with 5% CO$_2$ for a period of 72 h. Data are the mean values $\pm$ SD of three independent experiments performed in triplicate; [b] In parenthesis the percentage of cell viability at highest tested concentration (50 μM).

3. Experimental Section

3.1. Materials and Methods

All commercial reagents and solvents were used without further purification unless otherwise specified. Reagents and solvents were purchased from Sigma-Aldrich and TCI. ^{1}H- and ^{13}C-NMR spectra were recorded on a Varian at 400-MR (400 MHz) spectrometer (Agilent Technologies, Santa Clara, CA, USA) at ambient temperature. Chemical shifts were reported in parts per million (ppm, δ units). Electrospray mass spectrometry (ESI-MS) was performed on a LC/MS spectrometer

(Agilent 6130 Series, Agilent Technologies). HPLC was carried out on a Thermo Separation Products System (Fremont, CA, USA) equipped with a semi-preparative column (Waters, Xterra RP-C18, 10 μm, 10 × 250 mm) and equipped with a UV detector (wavelength set at 254 nm) and a γ-ray detector (Bioscan, Poway, CA, USA). HPLC-grade solvents (J. T. Baker, Phillipsburg, NJ, USA) were used for HPLC purification after membrane filtering (Whatman, Maidstone, UK, 0.22 μm). The column was eluted with a solvent mixture of acetonitrile-water (0.1% trifluoroacetic acid) using a gradient condition. The HPLC eluent started with 20% acetonitrile-water (0.1% trifluoroacetic acid) and the ratio was increased with a solvent mixture of 90% acetonitrile-water (0.1% trifluoroacetic acid) over 30 min at a flow rate of 3 mL/min. TLC was performed on Merck F254 silica plates and radio-TLC was analyzed on a Bioscan radio-TLC scanner (Washington, DC, USA). All radioactivities were measured using a VDC-505 activity calibrator from Veenstra Instruments (Joure, The Netherlands). In vitro incubation was carried out at 37 °C using a block heater (Digi-Block Laboratory Device Inc., Holliston, MA, USA). $Na^{99m}TcO_4$ was eluted on a daily basis from $^{99}Mo/^{99m}Tc$ generators (Samyoung Unitech, Seoul, Korea). The organometallic precursor $(NEt_4)_2[ReBr_3(CO)_3]$ and the radioactive precursor $[^{99m}Tc(CO)_3(H_2O)_3]^+$ were prepared as previously reported [29,31,33].

3.2. Synthesis of Re-CB256 (2)

A solution of **3** (2 mg, 3 μmol) in methanol (1 mL) was treated with $(NEt_4)_2[ReBr_3(CO)_3]$ (2.3 mg, 3 μmol). The reaction mixture was stirred at 65 °C for 1 h. The solvent was removed under reduced pressure and then the product was separated by a semi-preparative HPLC system. The fraction of **2** was collected at 22 min as yellow solid: m.p. 190.3–210.4 °C; ^1H-NMR (400 MHz, acetone-D_6) δ 9.02 (d, J = 5.2 Hz, 2H, $H_{6''}$), 8.27 (d, J = 6.8 Hz, 1H, H_7), 8.10 (t, J = 7.6 Hz, 2H, $H_{3''}$), 8.07 (d, J = 6.0 Hz, 1H, H_5), 7.81 (d, J = 8.0 Hz, 2H, $H_{2'}/H_{6'}$), 7.72–7.69 (m, 2H, $H_{4''}$), 7.55 (d, J = 6.8 Hz, 2H, $H_{3'}/H_{5'}$), 7.53 (m, 2H, $H_{5''}$), 6.99 (t, J = 7.2 Hz, 1H, H_6), 5.69 (d, J = 17.2 Hz, 2H, $H_{17(ax)}/H_{18(ax)}$), 5.37 (d, J = 17.2 Hz, 2H, $H_{17(eq)}/H_{18(eq)}$), 5.31 (s, 2H, H_{16}), 4.32 (s, 2H, H_9), 3.44 (t, J = 7.6 Hz, 2H, H_{10}), 3.34 (t, J = 7.6 Hz, 2H, H_{13}), 1.73–1.67 (m, 4H, H_{11}/H_{14}), 0.92–0.84 (m, 6H, H_{12}/H_{15}); ^{13}C-NMR (100 MHz, acetone-D_6) δ 197.1 (overlap with CO and CONH), 196.7, 169.4, 168.7, 163.0, 153.6, 142.2, 135.1, 131.5, 130.3, 127.5, 125.4, 122.2, 120.0, 114.2, 106.6, 71.7, 70.0, 51.0, 49.0, 23.6, 22.3, 12.3, 12.1; MS (ESI) m/z 894.2 (M$^+$, 100%), 892.2 (53%), 895.2 (40%). HRMS (ESI) m/z $C_{38}H_{38}O_5N_7ClRe$ calcd: 894.2175; found: 894.2155.

3.3. Synthesis of ^{99m}Tc-CB256 (1)

A solution of $[^{99m}Tc(H_2O)_3(CO)_3]^+$ in saline (250 μL, approximately 44 MBq) was added to a solution of **3** (1 mg, 1.5 μmol) dissolved in methanol (250 μL). The reaction mixture was stirred at 65 °C for 20 min. After the reaction time, the mixture was cooled in an ice-bath and diluted with 10 mL of water. This solution was loaded into a C18 Sep-Pak cartridge, washed with 5 mL of water, and eluted with 1.5 mL of acetonitrile. The combined solvent fractions were removed by a stream of nitrogen gas. The product was purified by a semi-preparative HPLC system. The radiochemically pure **1** eluted off with a retention time of 22.5 min, and the radiochemical yield, calculated from a homemade $[^{99m}Tc(H_2O)_3(CO)_3]^+$ solution in saline, was 74.5% ± 6.4% (decay-corrected). The obtained **1** was diluted with excess water, passed through a C18 Sep-Pak cartridge and washed with water (5 mL). The desired product was eluted by ethanol (1.5 mL) and exchanged to 10% ethanol-saline for in vitro experiments. The identity was confirmed by coinjection with authentic compound **2** as shown in Figure 2.

3.4. In Vitro Stability Study

The stability of **1** was assayed by monitoring the Radio-TLC profile and determining its radiochemical purity. Human serum was prepared from human whole blood by centrifuging at 3500 rpm for 5 min. An aliquot (3.7 MBq) of **1** in 10% ethanol-saline (0.1 mL) was added to human serum (0.5 mL) and incubated at 37 °C for 4 h. At the indicated time points (10, 30, 60, 120, and 240 min), the sample was taken and then added to acetonitrile (0.1 mL). After vortexing (20 s), the mixture was centrifuged at 3500 rpm for 5 min. The obtained supernatant was analyzed by radio-TLC using methanol-dichloromethane (1:9, R_f = 0.6 for **1**) as the developing solvents.

3.5. LogD Determination

The Log$D$ value was measured by mixing a solution of **1** in 5% ethanol-saline (10 μL, approximately 0.74 MBq) with sodium phosphate buffer (0.15 M, pH 7.4, 5 mL) and *n*-octanol (5 mL) in a test tube. After vortexing for 1 min, each tube was then stored for 3 min at room temperature and the phases were separated. Samples of each phase (100 μL) were counted for radioactivity. Log$D$ is expressed as the logarithm of the ratio of the counts from *n*-octanol versus that of the sodium phosphate buffer.

3.6. In Vitro Cell Uptake Assay of ^{99m}Tc-CB256 (1)

Rat C6-glioma cells (C6) and human glioblastoma U87-MG cells were purchased from the American Type Culture Collection (ATCC). C6 cells were cultured in Dulbecco's Modified Eagle's Medium (DMEM) containing high glucose (WelGENE), supplemented with 10% heat-inactivated fetal bovine serum (FBS; GIBCO) and antibiotics (100 units/mL penicillin G and 10 μg/mL reptomycin; GIBCO) at 37 °C in a humidified 5% CO_2 atmosphere. U87-MG cells were cultured in DMEM supplemented with 4 mM L-glutamine, 4500 mg/L glucose, 1 mM sodium pyruvate, 1.5 g/mL sodium bicarbonate, 10% FBS, and antibiotics-antimycotic at 37 °C in a humidified 5% CO_2 atmosphere. Both cells (1 × 10^6 cells in 0.1 mL medium per test tubes) were incubated with ^{99m}Tc-CB256 (**1**) (0.74 MBq in 0.1 mL 10% ethanol-saline) at 10, 30, 60, and 120 min. In an inhibition study for specificity, ^{99m}Tc-CB256 involved with 300 μM PK 11195 was performed in U87-MG cells. After incubation, the cells were quickly washed twice (<15 s) with 1 mL of ice-cold phosphate buffer saline and centrifuged at 3500 rpm for 5 min. Each supernatant was removed for counting, and the remaining sample containing cells was measured by a γ counter (1480 WIZARD, Perkin-Elmer). ^{99m}Tc uptake was expressed as the percentage injected dose (% ID).

3.7. In Vitro Cell-Binding Assays

Binding affinity and selectivity to the 18-kDa translocator protein TSPO and to CBR were assessed using in vitro receptor-binding assays. These experiments were carried out as previously described [28].

3.8. Cytotoxicity Assays

Cytotoxicity assays were carried out against HepG2, MCF7, and U87 cancer cells seeded at a density of 5000 cells/well. All tested compounds were dissolved in DMSO prior to their dilution with cell culture medium to the predetermined experimental concentrations (eight concentrations ranging from 0.01 to 50 μM), with the final DMSO concentration never exceeding 1%. Cytotoxicity (IC$_{50}$) values for the tested compounds were determined using the 3-(4,5-dimethylthiazol-2-yl)-2,5-diphenyltetrazolium bromide (MTT) assay. Briefly, the cells were seeded in a 96-well plate and incubated at 37 °C for 72 h with the tested compounds. Then, 10 μL of 5 mg/mL MTT were added to each well and the plates were incubated for an additional 4 h at 37 °C. Subsequently, cells were lysed by addition of 150 μL of 50% (*v*/*v*) DMSO and 50% (*v*/*v*) ethanol solution, and the absorbance of each individual well was measured using a microplate reader at 570 nm (Wallac Victor3, 1420 Multilabel Counter, Perkin-Elmer (manufactured for WALLAC Oy, Turku, Finland)). The reported values are the average of triplicate measurements performed in at least three separate experiments.

4. Conclusions

A ^{99m}Tc-labeled imidazopyridine-based bifunctional chelate ligand (**1**) was prepared in one step by coordination of the tricarbonyl ^{99m}Tc core to the di-(2-picolyl)amine residue, with good radiochemical yield. The resulting complex (**1**) showed high stability in vitro. The affinity toward TSPO of **2** proved that the tricarbonyl rhenium moiety did not alter the TSPO affinity of CB256 (**3**). The low cytotoxicity of **1** further demonstrates that if the dipicolylamine moiety is coordinated to a metal ion—in the present case the tricarbonyl Re-core—it is not able to bind endogenous biometals to exert its DNA cleavage activity and cause double-strand DNA lesions. In vitro studies on TSPO-rich tumor cells suggest that radiolabeled **1** may have potential to act as a useful SPECT radiotracer for the evaluation of TSPO-overexpressing tissues, and provides the foundation for further in vivo biological evaluation.

Acknowledgments: This study was supported by grants from the Korea Health Technology R&D Project through the Korea Health Industry Development Institute (KHIDI), funded by the Ministry of Health & Welfare, Republic of Korea (HI14C1072, HI16C0947) and the National Research Foundation of Korea (NRF) grant funded by Ministry of Science, ICT and Future Planning, Republic of Korea (NRF-2014R1A2A2A01007980). This study was also supported by grant from the Seoul National University Bundang Hospital Research Fund (14-2014-006). The University of Bari (Bari, Italy) and the Inter-University Consortium for Research on the Chemistry of Metal Ions in Biological Systems (C.I.R.C.M.S.B.) are gratefully acknowledged.

Author Contributions: Byung Chul Lee and Nunzio Denora conceived and designed the experiment; Ji Young Choi, Jae Ho Jung, Do Dam Park, Byung Seok Moon, and Rosa Maria Iacobazzi, Mara Perrone, Nicola Margiotta, Annalisa Cutrignelli performed the synthesis, radio-labeling and cell experiments; Byung Chul Lee, Sang Eun Kim, and Nunzio Denora analyzed the data; Ji Young Choi, Byung Chul Lee, and Nicola Margiotta, Nunzio Denora wrote the paper.

Conflicts of Interest: The authors declare no conflict of interest.

Abbreviations

TSPO	18-kDa mitochondrial translocator protein
PET	positron emission computed tomography
SPECT	single photon emission computed tomography
HPLC	high performance liquid chromatography
ESI-MS	electrospray mass spectrometry
HRMS	high resolution mass spectrometry
CBR	central benzodiazephine receptor
U87-MG	U87-malignant glioma
% ID	% injected dose
NMR	nuclear magnetic resonance
TLC	thin layer chromatography
S.D.	standard deviation

References

1. Papadopoulos, V.; Baraldi, M.; Guilarte, T.R.; Knudsen, T.B.; Lacapère, J.J.; Lindemann, P.; Norenberg, M.D.; Nutt, D.; Weizman, A.; Zhang, M.R.; et al. Translocator protein (18 kDa): New nomenclature for the peripheral-type benzodiazepine receptor based on its structure and molecular function. *Trends Pharmacol. Sci.* **2006**, *27*, 402–409. [CrossRef] [PubMed]

2. Rupprecht, R.; Papadopoulos, V.; Rammes, G.; Baghai, T.C.; Fan, J.; Akula, N.; Groyer, G.; Adams, D.; Schumacher, M. Translocator protein (18 kDa) (TSPO) as a therapeutic target for neurological and psychiatric disorders. *Nat. Rev. Drug Discov.* **2010**, *9*, 971–988. [CrossRef] [PubMed]

3. Maaser, K.; Grabowski, P.; Sutter, A.P.; Höpfner, M.; Foss, H.D.; Stein, H.; Berger, G.; Gavish, M.; Zeitz, M.; Scherübl, H. Overexpression of the peripheral benzodiazepine receptor is a relevant prognostic factor in stage III colorectal cancer. *Clin. Cancer Res.* **2002**, *8*, 3205–3209. [PubMed]

4. Corsi, L.; Geminiani, E.; Baraldi, M. Peripheral Benzodiazepine Receptor (PBR) new insight in cell proliferation and cell differentiation review. *Curr. Clin. Pharmacol.* **2008**, *3*, 38–45. [CrossRef] [PubMed]

5. Austin, C.J.D.; Kahlert, J.; Kassiou, M.; Rendina, L.M. The translocator protein (TSPO): A novel target for cancer chemotherapy. *Int. J. Biochem. Cell Biol.* **2013**, *45*, 1212–1216. [CrossRef] [PubMed]

6. Zheng, J.; Boisgard, R.; Siquier-Pernet, K.; Decaudin, D.; Dollé, F.; Tavitian, B. Differential expression of the 18 kDa translocator protein (TSPO) by neoplastic and inflammatory cells in mouse tumors of breast cancer. *Mol. Pharm.* **2011**, *8*, 823–832. [CrossRef] [PubMed]

7. Han, Z.; Slack, R.S.; Li, W.; Papadopoulos, V. Expression of peripheral benzodiazepine receptor (PBR) in human tumors: Relationship to breast, colorectal, and prostate tumor progression. *J. Recept. Signal Transduct.* **2003**, *2*, 225–238. [CrossRef] [PubMed]

8. Le Fur, G.; Perrier, M.L.; Vaucher, N.; Imbault, F.; Flamier, A.; Benavides, J.; Uzan, A.; Renault, C.; Dubroeucq, M.C.; Guérémy, C. Peripheral benzodiazepine binding sites: Effect of PK11195, 1-(2-chlorophenyl)-*N*-(1-methylpropyl)-3-isoquinolinecarboxamide I. In vitro studies. *Life Sci.* **1983**, *32*, 1839–1847. [CrossRef]

9. Marangos, P.L.; Pate, J.; Boulenger, J.P.; Clark-Rosenberg, R. Characterization of peripheral-type benzodiazepine binding sites in brain using [³H]Ro 5-4864. *Mol. Pharmacol.* **1982**, *22*, 26–32. [PubMed]

10. Romeo, E.; Auta, J.; Kozikowski, A.P.; Ma, D.; Papadopoulos, V.; Puia, G.; Costa, E.; Guidotti, A. 2-Aryl-3-indoleacetamides (FGIN-1): A new class of potent and specific ligands for the mitochondrial DBI receptor (MDR). *J. Pharmacol. Exp. Ther.* **1992**, *262*, 971–978. [PubMed]

11. Imaizumi, M.; Briard, E.; Zoghbi, S.S.; Gourley, J.P.; Hong, J.; Fujimura, Y.; Pike, V.W.; Innis, R.B.; Fujita, M. Brain and whole-body imaging in nonhuman primates of [¹¹C]PBR28, a promising PET radioligand for peripheral benzodiazepine receptors. *NeuroImage* **2008**, *39*, 1289–1298. [CrossRef] [PubMed]

12. Denora, N.; Laquintana, V.; Pisu, M.G.; Dore, R.; Murru, L.; Latrofa, A.; Trapani, G.; Sanna, E. 2-Phenyl-imidazo[1,2-*a*]pyridine compounds containing hydrophilic groups as potent and selective ligands for peripheral benzodiazepine receptors: Synthesis, binding affinity and electrophysiological studies. *J. Med. Chem.* **2008**, *51*, 6876–6888. [CrossRef] [PubMed]

13. Cappelli, A.; Mancini, A.; Sudati, F.; Valenti, S.; Anzini, M.; Belloli, S.; Moresco, R.M.; Matarrese, M.; Vaghi, M.; Fabro, A.; et al. Synthesis and biological characterization of novel 2-quinolinecarboxamide ligands of the peripheral benzodiazepine receptors bearing technetium-99m or rhenium. *Bioconjug. Chem.* **2008**, *19*, 1143–1153. [CrossRef] [PubMed]

14. Ching, A.S.C.; Kuhnast, B.; Damont, A.; Roeda, D.; Tavitian, B.; Dollé, F. Current paradigm of the 18-kDa translocator protein (TSPO) as a molecular target for PET imaging in neuroinflammation and neurodegenerative diseases. *Insights Imaging* **2012**, *3*, 111–119. [CrossRef] [PubMed]

15. Turkheimer, F.E.; Rizzo, G.; Bloomfield, P.S.; Howes, O.; Zanotti-Fregonara, P.; Bertoldo, A.; Veronese, M. The methodology of TSPO imaging with positron emission tomography. *Biochem. Soc. Trans.* **2015**, *43*, 586–592. [CrossRef] [PubMed]

16. Vivash, L.; O'Brien, T.J. Imaging microglial activation with TSPO PET: Lighting up neurologic diseases? *J. Nucl. Med.* **2016**, *57*, 165–168. [CrossRef] [PubMed]

17. Schwochau, K. Technetium radiopharmaceuticals. Fundamentals, synthesis, structure, and development. *Angew. Chem. Int. Ed.* **1994**, *33*, 2258–2267. [CrossRef]

18. Jurisson, S.; Berning, D.; Jia, W.; Ma, D.S. Coordination compounds in nuclear medicine. *Chem. Rev.* **1993**, *93*, 1137–1156. [CrossRef]

19. Schibli, R.; Netter, M.; Scapozza, L.; Birringer, M.; Schelling, P.; Dumas, C.; Schoch, J.; Schubiger, P.A. First organometallic inhibitors for human thymidine kinase: Synthesis and in vitro evaluation of rhenium(I)- and technetium(I)-tricarbonyl complexes of thymidine. *J. Oranomet. Chem.* **2003**, *668*, 67–74. [CrossRef]

20. Lee, B.C.; Kim, D.H.; Lee, J.H.; Sung, H.J.; Choe, Y.S.; Chi, D.Y.; Lee, K.H.; Choi, Y.; Kim, B.T. ⁹⁹ᵐTc(CO)₃-15-[*N*-(Acetyloxy)-2-picolylamino]pentadecanoic acid: A potential radiotracer for evaluation of fatty acid metabolism. *Bioconjug. Chem.* **2007**, *18*, 1332–1337. [CrossRef] [PubMed]

21. Lee, B.C.; Moon, B.S.; Kim, J.S.; Jung, J.H.; Park, H.S.; Katzenellenbogen, J.A.; Kim, S.E. Synthesis and biological evaluation of RGD peptides with the ⁹⁹ᵐTc/¹⁸⁸Re chelated iminodiacetate core: Highly enhanced uptake and excretion kinetics of theranostics against tumor angiogenesis. *RSC Adv.* **2013**, *3*, 782–792. [CrossRef]

22. Alves, S.; Correia, J.D.; Gano, L.; Rold, T.L.; Prasanphanich, A.; Haubner, R.; Rupprich, M.; Alberto, R.; Decristoforo, C.; Santos, I.; et al. In vitro and in vivo evaluation of a novel ⁹⁹ᵐTc(CO)₃-pyrazolyl conjugate of cyclo-(Arg-Gly-Asp-d-Tyr-Lys). *Bioconjug. Chem.* **2007**, *18*, 530–537. [CrossRef] [PubMed]

23. North, A.J.; Hayne, D.J.; Schieber, C.; Price, K.; White, A.R.; Crouch, P.J.; Rigopoulos, A.; O'Keefe, G.J.; Tochon-Danguy, H.; Scott, A.M.; et al. Toward hypoxia-selective rhenium and technetium tricarbonyl complexes. *Inorg. Chem.* **2015**, *54*, 9594–9610. [CrossRef] [PubMed]
24. Nayak, D.K.; Halder, K.K.; Baishya, R.; Sen, T.; Mitra, P.; Debnath, M.C. Tricarbonyltechnetium(I) and tricarbonylrhenium(I) complexes of amino acids: Crystal and molecular structure of a novel cyclic dimeric Re(CO)$_3$-amino acid complex comprised of the OON donor atom set of the tridentate ligand. *Dalton Trans.* **2013**, *42*, 13565–13575. [CrossRef] [PubMed]
25. Margiotta, N.; Ostuni, R.; Ranaldo, R.; Denora, N.; Laquintana, V.; Trapani, G.; Liso, G.; Natile, G. Synthesis and characterization of a platinum(II) complex tethered to a ligand of the peripheral benzodiazepine receptor. *J. Med. Chem.* **2007**, *50*, 1019–1027. [CrossRef] [PubMed]
26. Margiotta, N.; Denora, N.; Ostuni, R.; Laquintana, V.; Anderson, A.; Johnson, S.W.; Trapani, G.; Natile, G. Platinum(II) complexes with bioactive carrier ligands having high affinity for the translocator protein. *J. Med. Chem.* **2010**, *53*, 5144–5154. [CrossRef] [PubMed]
27. Piccinonna, S.; Margiotta, N.; Denora, N.; Iacobazzi, R.M.; Pacifico, C.; Trapani, G.; Natile, G. A model radiopharmaceutical agent targeted to translocator protein 18 kDa (TSPO). *Dalton Trans.* **2013**, *42*, 10112–10115. [CrossRef] [PubMed]
28. Piccinonna, S.; Denora, N.; Margiotta, N.; Laquintana, V.; Trapani, G.; Natile, G. Synthesis, characterization, and binding to the translocator protein (18 kDa, TSPO) of a new rhenium complex as a model of radiopharmaceutical agents. *Z. Anorg. Allg. Chem.* **2013**, *639*, 1606–1612. [CrossRef]
29. Denora, N.; Margiotta, N.; Laquintana, V.; Lopedota, A.; Cutrignelli, A.; Losacco, M.; Franco, M.; Natile, G. Synthesis, characterization, and in vitro evaluation of a new TSPO selective bifunctional chelate ligand. *ACS Med. Chem. Lett.* **2014**, *5*, 685–689. [CrossRef] [PubMed]
30. Margiotta, N.; Denora, N.; Piccinonna, S.; Laquintana, V.; Lasorsa, F.M.; Franco, M.; Natile, G. Synthesis, characterization, and in vitro evaluation of new coordination complexes of platinum(II) and rhenium(I) with a ligand targeting the translocator protein (TSPO). *Dalton Trans.* **2014**, *43*, 16252–16264. [CrossRef] [PubMed]
31. Alberto, R.; Schibli, R.; Egli, A.; Schubiger, A.P. A Novel Organometallic Aqua Complex of Technetium for the Labeling of Biomolecules: Synthesis of [^{99m}Tc(OH$_2$)$_3$(CO)$_3$]$^+$ from [^{99m}TcO$_4$]$^-$ in Aqueous Solution and Its Reaction with a Bifunctional Ligand. *J. Am. Chem. Soc.* **1998**, *120*, 7987–7988. [CrossRef]
32. Perrone, M.; Moon, B.S.; Park, H.S.; Laquintana, V.; Jung, J.H.; Cutrignelli, A.; Lopedota, A.; Franco, M.; Kim, S.E.; Lee, B.C.; et al. A novel pet imaging probe for the detection and monitoring of translocator protein 18 kDa expression in pathological disorders. *Sci. Rep.* **2016**, *6*. [CrossRef] [PubMed]
33. Schibli, R.; Katti, K.V.; Volkert, W.A.; Barnes, C.L. Novel coordination behavior of *fac*-[ReBr$_3$(CO)$_3$]$_2$- with 1,3,5-triaza-7-phosphaadamantane (PTA). Systematic investigation on stepwise replacement of the halides by PTA ligand. Phase transfer studies and X-ray crystal structure of [NEt$_4$][ReBr$_2$((PTA)(CO)$_3$], [ReBr(PTA)$_2$(CO)$_3$], and [Re(PTA)$_3$(CO)$_3$]PF$_6$. *Inorg. Chem.* **1998**, *37*, 5306–5312.

International Journal of
Molecular Sciences

MDPI

Article

Regulation of Translocator Protein 18 kDa (TSPO) Expression in Rat and Human Male Germ Cells

Gurpreet Manku [1,2] and Martine Culty [1,2,3,]*

[1]   The Research Institute of the McGill University Health Centre, McGill University, Montreal, QC H4A 3J1,
     Canada; gurpreet.manku@mail.mcgill.ca
[2]   Departments of Medicine, McGill University, Montreal, QC H4A 3J1, Canada
[3]   Pharmacology & Therapeutics, McGill University, Montreal, QC H3G 1Y6, Canada
*   Correspondence: martine.culty@mcgill.ca; Tel.: +1-514-934-1934 (ext. 43752); Fax: +1-514-933-8784

Academic Editors: Giovanni Natile and Nunzio Denora
Received: 10 August 2016; Accepted: 30 August 2016; Published: 6 September 2016

Abstract: Translocator protein 18 kDa (TSPO) is a high affinity cholesterol- and drug-binding protein highly expressed in steroidogenic cells, such as Leydig cells, where it plays a role in cholesterol mitochondrial transport. We have previously shown that TSPO is expressed in postnatal day 3 rat gonocytes, precursors of spermatogonial stem cells. Gonocytes undergo regulated phases of proliferation and migration, followed by retinoic acid (RA)-induced differentiation. Understanding these processes is important since their disruption may lead to the formation of carcinoma in situ, a precursor of testicular germ cell tumors (TGCTs). Previously, we showed that TSPO ligands do not regulate gonocyte proliferation. In the present study, we found that TSPO expression is downregulated in differentiating gonocytes. Similarly, in F9 embryonal carcinoma cells, a mouse TGCT cell line with embryonic stem cell properties, there is a significant decrease in TSPO expression during RA-induced differentiation. Silencing TSPO expression in gonocytes increased the stimulatory effect of RA on the expression of the differentiation marker *Stra8*, suggesting that TSPO exerts a repressive role on differentiation. Furthermore, in normal human testes, TSPO was located not only in Leydig cells, but also in discrete spermatogenic phases such as the forming acrosome of round spermatids. By contrast, seminomas, the most common type of TGCT, presented high levels of TSPO mRNA. TSPO protein was expressed in the cytoplasmic compartment of seminoma cells, identified by their nuclear expression of the transcription factors OCT4 and AP2G. Thus, TSPO appears to be tightly regulated during germ cell differentiation, and to be deregulated in seminomas, suggesting a role in germ cell development and pathology.

Keywords: gonocytes; spermatogonia; translocator protein 18 kDa (TSPO); differentiation; human testis; germ cells; seminoma

1. Introduction

Translocator protein 18 kDa (TSPO; previously known as peripheral benzodiazepine receptor PBR) is a high affinity drug- and cholesterol-binding protein strongly expressed in steroidogenic cells, including testicular Leydig cells, where it is located at the outer mitochondrial membrane (OMM) [1–3]. TSPO plays an essential role in cholesterol transfer from the outer to the inner mitochondrial membrane, where CYP11A1 catalyzes pregnenolone formation from cholesterol, the first step of the steroidogenic cascade [1]. Although present mainly as monomers, TSPO can form polymers upon hormonal stimulation and oxidative stress, and is also found as polymers in tumorigenic cells [2,4]. In addition to steroidogenic tissues, TSPO is expressed in multiple tissues and cells, where it is involved in various processes, including cellular proliferation, apoptosis, transport, and differentiation [4–7]. While the role of TSPO in steroidogenic cells has been well documented, its role in non-steroidogenic cells is not well understood.

We previously made the serendipitous observation that TSPO was present in neonatal rat gonocytes (also known as pre- and pro-spermatogonia), the precursor cells of type A spermatogonia, including cells of the first spermatogenic wave and spermatogonial stem cells (SSCs) [8]. In gonocytes, TSPO expression was nuclear, and its drug ligand did not regulate gonocyte proliferation [8]. TSPO was also found in pachytene spermatocytes and dividing spermatogonia in adult rat testis [8]. In a recent study, we found that TSPO transcripts are also abundant in adult mouse spermatogonia, pachytene spermatocytes and round spermatids, while the protein is observed in mouse sperm [9].

Adequate gonocyte development is critical for the establishment of a SSC reservoir that will support the production of sperm throughout adulthood [10]. These transitional cells undergo proliferation and migration to the basement membrane of the seminiferous cords, prior to their differentiation [10–12]. We have previously shown that gonocyte proliferation is stimulated by the coordinated action of platelet-derived growth factor (PDGF)-BB and 17β-estradiol, and requires ERK1/2 pathway activation [13,14]. We also found that gonocyte differentiation is stimulated by all-trans retinoic acid (RA) and involves the activation of PDGFR, SRC and JAK2/STAT5 pathways, highlighting the complexity of this process [15,16]. Another important regulatory process is the apoptosis of gonocytes that failed to migrate and/or differentiate, taking place at the end of the first postnatal week [17]. Studies have suggested that improper gonocyte development may lead to infertility and testicular germ cell tumors (TGCTs), a reproductive pathology on the rise in the past decades [18,19]. Thus, better understanding of gonocyte development should elucidate the mechanisms leading to germline stem cell formation, as well as the origins of TGCTs.

In the present study, we examined whether TSPO is altered during differentiation in gonocytes and in the F9 embryonal carcinoma cell line, a mouse TGCT that has retained embryonic stem cell properties, as well as somatic and germ cell markers [16]. TSPO's potential role in RA-induced gonocyte differentiation was examined by silencing its expression. TSPO expression profiles were also determined in human adult normal testis and TGCTs. Together, these studies suggest that TSPO is regulated during germ cell differentiation and in germ cell tumors.

2. Results

2.1. Changes in TSPO Expression during Gonocyte Differentiation and in Spermatogonia

We have previously shown that TSPO is expressed in the nucleus of PND3 gonocytes but that binding of its drug ligand does not affect gonocyte proliferation. Here, the comparison of *Tspo* mRNA levels between purified PND3 gonocytes and the more differentiated PND8 spermatogonia showed that gonocytes expressed significantly higher amounts of *Tspo* compared to spermatogonia (Figure 1A). TSPO protein levels were also higher in gonocytes, where it appeared nuclear, than in spermatogonia, where it was found in the cytoplasm (Figure 1B). At both ages, the low purity cell preparations contained somatic cells that were either TSPO-positive (likely contaminant Leydig cells) or TSPO-negative (likely myoid and Sertoli cells) (Figure 1B).

We then treated isolated gonocytes with RA, previously shown to induces gonocyte differentiation [15,16], and found that RA treatment induced a significant downregulation (30% decrease) of *Tspo* mRNA expression, concomitant with an eight-fold increase in the differentiation marker *Stra8* (Figure 1C,D). TSPO protein levels were also decreased in RA-treated gonocytes (Figure 1E). These data suggested that TSPO is actively down-regulated during gonocyte differentiation.

2.2. TSPO Expression in Differentiating F9 Mouse Embryonal Carcinoma Cells

We have previously used F9 embryonal carcinoma cells as a model to study signalling pathways involved in RA-induced differentiation [15,16]. Here, we found that TSPO mRNA expression was significantly down-regulated in F9 cells treated with RA, similarly to what was observed in gonocytes (Figure 2A). TSPO protein, which was mainly found as 36 kDa dimers in F9 cells, was also greatly depleted in RA-treated F9 cells, suggesting an active down-regulation of TSPO during F9 cell differentiation (Figure 2B).

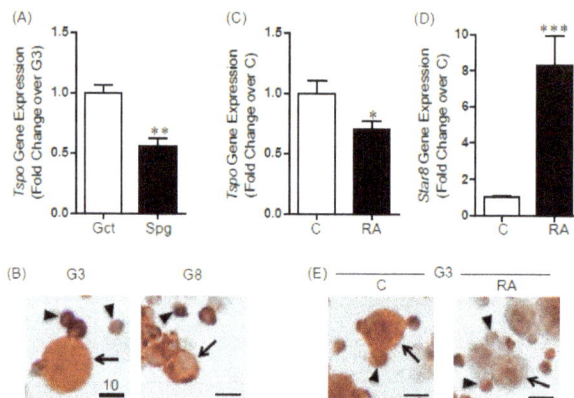

Figure 1. Translocator protein 18 kDa (TSPO) expression in differentiating rat gonocytes and in spermatogonia. (**A**) *Tspo* mRNA expression was determined by quantitative-PCR (qPCR) analysis in high purity PND3 gonocytes (Gct) and PND8 spermatogonia (Spg). Results shown are from 3–4 independent germ cell preparations and are plotted as mean ± SEM. ** *p*-value < 0.01; (**B**) immunocytochemical (ICC) analysis of TSPO protein expression in low purity isolated PND3 gonocytes (G3) and PND8 spermatogonia (G8). Arrows indicate germ cells; arrowheads indicate somatic cells. Scale bar = 10 μm; (**C,D**) mRNA expression of *Tspo* (**C**) and *Stra8* (**D**) in gonocytes treated with or without retinoic acid (RA, 10^{-6} M) for 24 h in medium supplemented with 2.5% FBS. Results shown are from 5 independent germ cell preparations with each treatment done in duplicate wells, and are plotted as mean ± SEM. * *p*-value < 0.05; *** *p*-value < 0.001; (**E**) ICC analysis of TSPO protein expression in low purity isolated gonocytes treated with or without RA (10^{-6} M) for 72 h. Arrows, arrowheads and scales are as in (**B**). (**B,E**) Representative cells are shown.

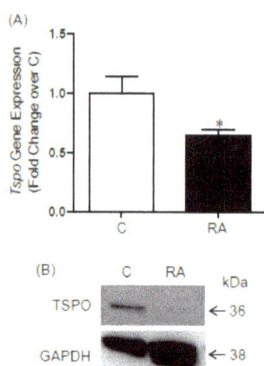

Figure 2. Changes in TSPO expression in differentiating F9 embryonal carcinoma cells. (**A**) *Tspo* mRNA expression in F9 cells treated with or without RA (10^{-7} M) for 24 h in medium supplemented with 10% FBS. Results shown are from 5 independent experiments, with each treatment done in duplicate wells, and are plotted as mean ± SEM. * *p*-value < 0.05; (**B**) immunoblot analysis of TSPO protein expression in F9 cells treated with or without RA (10^{-7} M) for 72 h. TSPO 36 kDa dimer band is observed. GAPDH: Loading control. A representative immunoblot is shown.

2.3. Effects of TSPO Knockdown on Gonocyte Differentiation and on the Expression of Germ Cell Specific Genes

To examine the potential role of TSPO in gonocyte differentiation, we knocked down its expression using siRNA transfection with Lipofectamine. Treatment with siRNA significantly reduced *Tspo* mRNA expression by 90% in isolated gonocytes after 24-h treatment, confirming knockdown (Figure 3A). As shown by immunoblot analysis, a successful TSPO knockdown was also seen at the protein level, affecting 18 kDa monomers, as well as 36 kDa dimers and 54 kDa trimers in cells treated with siRNA, compared to those treated with DsiRNA (Figure 3B), similar to mock samples (not shown).

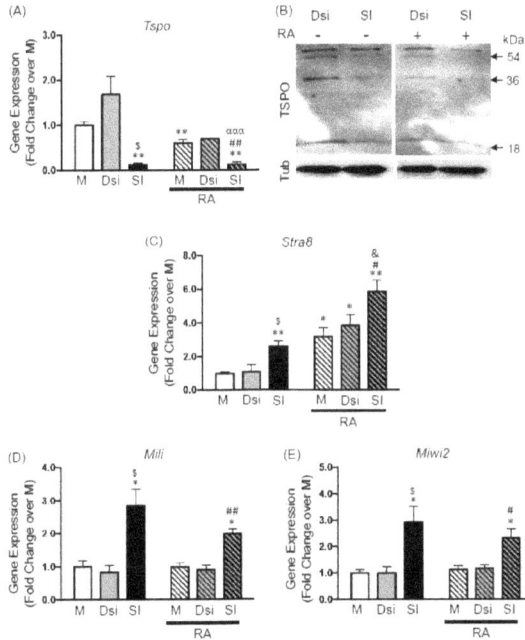

Figure 3. Effect of TSPO knockdown on gene expression in gonocytes. Silencing experiments were performed as described in the method section, by treating isolated gonocytes with either medium (mock; M), non-targeting DsiRNA (Dsi), or TSPO siRNA (SI) for 24 h (RNA analysis) or 48 h (protein analysis), followed by an additional 24 h incubation with or without RA (10^{-6} M). (**A**) *Tspo* mRNA expression. Results shown are from 3 independent germ cell preparations (each condition done in duplicate) and are plotted as mean ± SEM. * p-value < 0.05; and ** p-value < 0.01 compared to M; $ p-value < 0.05 compared to Dsi; ## p-value < 0.01 compared to M + RA; $\alpha\alpha\alpha$ p < 0.001 compared to Dsi + RA; (**B**) immunoblot analysis of TSPO protein expression. Representative gels are shown. TSPO monomers (18 kDa), dimers (36 kDa) and trimers (54 kDa) are indicated by arrows. Tubulin (Tub) = loading control; (**C–E**) mRNA expression of *Stra8* (**C**), *Mili* (**D**), and *Miwi2* (**E**). Results shown are from 3 independent germ cell preparations (each done in duplicate) expressed as means ± SEM. * p-value < 0.05; and ** p-value < 0.01 compared to M; $ p-value < 0.05 compared to Dsi; # p-value < 0.05 and ## p-value < 0.01 compared to M + RA; & p-value < 0.05 compared to SI.

RA-induced differentiation in mock and DsiRNA samples was confirmed by significant three- to four-fold increases in the mRNA expression of the differentiation marker *Stra8* [15,16,20] (Figure 3C). RA treatment significantly decreased *Tspo* mRNA expression by 40%, compared to levels found in control mock cells (Figure 3A); similarly to the results obtained with non-transfected gonocytes (Figure 1C). RA treatment also induced a decreasing trend in TSPO mRNA levels in cells treated with DsiRNA. Looking at protein levels, RA treatment decreased TSPO bands in DsiRNA—treated cells

(Figure 3B). Furthermore, in siRNA knockdown cells, RA treatment for 24 h induced more pronounced decreases in TSPO protein bands than with siRNA alone (Figure 3B). Taken together, these results suggested that one of the effects of RA in differentiating gonocytes is to repress TSPO expression.

Next, we examined the effects of silencing TSPO expression on *Stra8* expression in gonocytes, in the absence and presence of RA. Interestingly, TSPO silencing induced a significant three-fold increase in *Stra8* mRNA levels in the absence of added RA (beyond the small RA amounts provided by 2.5% FBS), suggesting that removing TSPO promoted the expression of this differentiation gene in gonocytes (Figure 3C). The addition of 10^{-6} M RA in TSPO knockdown cells further doubled *Stra8* expression in comparison to its levels in TSPO knockdown cells and in control mock cells (Figure 3C). These data suggest that TSPO may play a repressive role in gonocyte differentiation, and that Stra8 is involved in removing TSPO to facilitate cell differentiation.

We then analyzed the mRNA expression profiles of *Mili* and *Miwi2*, two RNA binding proteins of the PIWI family critical for maintaining the genomic integrity of early germ cells, including gonocytes, by preventing transposon expression, and for the progression of spermatogenesis [21,22]. Interestingly, silencing TSPO significantly increased by three-fold the expression of both *Mili* and *Miwi2* transcripts (Figure 3E,F). RA treatment in mock, DsiRNA or TSPO knockdown cells had no effect on these genes, suggesting that these genes are not regulated by RA in differentiating gonocytes, while TSPO appears to play a repressive role in the expression of these genes.

2.4. TSPO Expression in Normal Human Adult Testicular Tissues

Considering TSPO expression in subsets of adult rat male germ cells, we next examined whether TSPO is also expressed in adult human spermatogenic cells. TSPO mRNA was examined by qPCR analysis in frozen biopsies from three patients with normal spermatogenesis, and TSPO protein expression were determined in testes paraffin sections from three other patients (Table 1). TSPO was highly expressed in the interstitial area, corresponding to its expected localization in Leydig cells (Figure 4A). Interestingly, TSPO was also expressed in round spermatids, more specifically, in what appeared to be the forming acrosomes (Figure 4A). To confirm that TSPO co-localized with nascent acrosomes, we performed immunofluorescence analysis on normal human adult testicular tissue biopsies using both an anti-TSPO antibody, and PNA, a peanut agglutinin lectin commonly used as acrosomal marker [23]. As shown in the magnified panels of Figure 4B, there was clear co-localization of TSPO and acrosomal immunofluorescent signals in distinct sections of tubules, indicating a stage-specific pattern of TSPO protein expression in human germ cells.

Table 1. Human normal and tumoral testicular tissue information.

Patient	Pathology	Age	Tissue Type
		Samples for qPCR analysis	
1	Normal		Normal spermatogenesis, Vasectomy patient
2	Normal		Normal spermatogenesis, Vasectomy patient
3	Normal		Normal spermatogenesis, Vasectomy patient
4	Seminoma		Left testis, Seminoma, Age 36, Stage I
5	Seminoma		Right testis, Seminoma, Age 26, Stage I
6	Seminoma		Right testis, Seminoma, Age 50, Stage I
		Samples for IF analysis	
1	Normal	73	Normal testicular parenchyma
2	Normal	57	Normal testicular parenchyma
3	Normal	73	Normal testicular parenchyma
4	Seminoma	NA	100% Seminoma
5	Seminoma	NA	100% Seminoma
6	Seminoma	38	100% Seminoma
7	Seminoma	NA	100% Seminoma

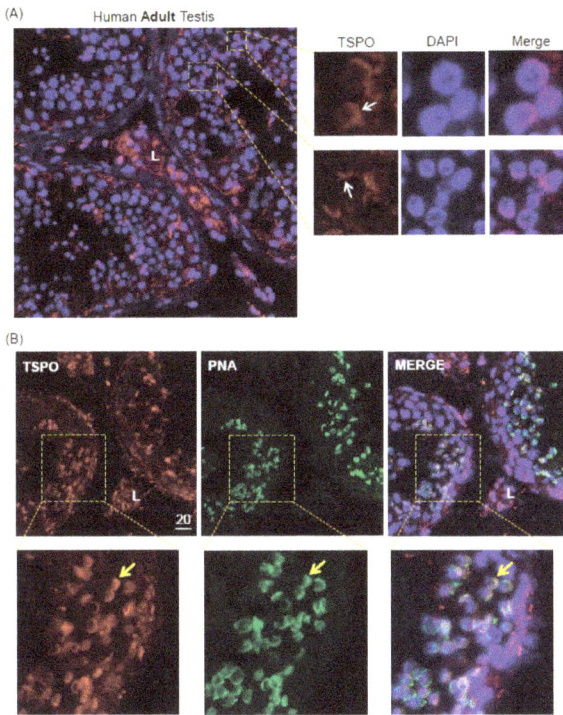

Figure 4. TSPO protein expression in normal human adult testis. Immunofluorescence (IF) analysis was performed on paraffin sections. (**A**) Patient 1. **Red**: TSPO protein expression; **Blue**: DAPI nuclear staining. Leydig cells are indicated by "L". Areas showing expression in round spermatids (acrosomal formation) are shown in enlarged panels. Representative images are shown; (**B**) Patient 2. **Red**: TSPO protein expression; **Green**: PNA labeling; **Blue**: DAPI nuclear staining. Examples of co-localization between TSPO and PNA are indicated by yellow arrow in enlarged panels. Representative images are shown. Scale bar: 20 μm.

2.5. TSPO Expression in Seminomas

Because of TSPO expression in gonocytes, and the suspected link between impaired gonocytes development and TGCT formation, we next examined whether TSPO might also be expressed in seminoma, the most common type of TGCTs. A comparison of *TSPO* transcript levels between normal testes and seminoma biopsies showed that seminomas significantly expressed higher levels of *TSPO* mRNA than normal testicular tissues (Figure 5A). TSPO protein expression was assessed in paraffin sections from several seminoma cases, in parallel with pluripotency genes OCT3/4 (OCT4; POU5F1) and AP2γ (TFAP2C), considered as seminoma markers [24,25]. Interestingly, TSPO was expressed in OCT4- and AP2G-positive cells, where it was localized in the cytoplasmic compartment, forming a ring around tumor cell nuclei, while the two transcription factors presented nuclear stainings (Figure 5B,C). The observation of specimen from different patients also highlighted zones of heterogeneity in some seminomas, with the presence of large cells with irregularly shaped nuclei, resembling embryonal carcinoma cells rather than seminoma cells (Figure 5C). In this case, TSPO appeared as a diffuse and weak signal spread around the nuclei, rather than as the well-defined cytoplasmic ring found in more rounded seminoma cells. In some specimens, a very strong TSPO signal was observed in OCT4-negative and AP2G-negative cells, much smaller in size than the tumor cells, likely corresponding to Leydig cells (Figure 5B,C). Taken together, these data showed that TSPO is expressed in seminoma cells, although at lower level than in the presumptive Leydig cells found within some tumors.

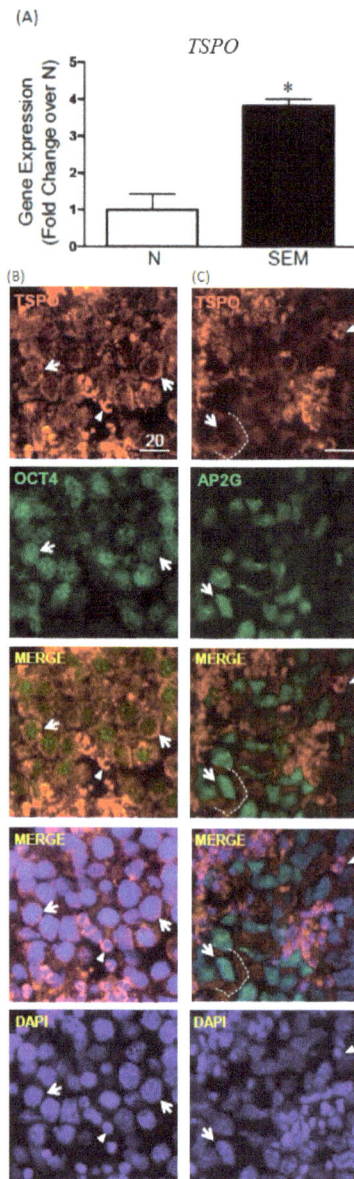

Figure 5. TSPO mRNA and protein expression in human seminomas. (**A**) *TSPO* mRNA expression in normal human adult testicular tissue and seminoma biopsies. n = 3–4, * p-value < 0.05; (**B**) TSPO (**red**) and OCT4 (**green**) protein expression in seminomas, visualized by IF in sections from Patient 4. Blue: DAPI nuclear staining. Representative images are shown; (**C**) TSPO (**red**) and AP2γ (**green**) protein expression in seminomas examined by IF in Patient 6. Representative images shown. (**B,C**) Arrows indicate representative seminoma cells; arrowheads indicate TSPO-positive small somatic cells (likely Leydig cells). Scale bars in Figure 4B,C are 20 μm.

3. Discussion

TSPO is a ubiquitous and multifunctional protein [4–7], shown by various pharmacological and molecular means to play a critical role in steroidogenesis [26,27]. While TSPO main location in testis is in the mitochondria of steroidogenic Leydig cells, we have found in previous studies that TSPO is also expressed in rat and mouse male germ cells in an age and spermatogenic cycle dependent manner [8,9]. The goal of the present study was to examine if TSPO plays a role in gonocyte differentiation, an essential process at the basis of spermatogenesis, and to examine its patterns of expression in healthy human testicular germ cells and in seminomas, which are thought to arise from disrupted gonocyte differentiation [18].

Our initial finding that gonocytes express higher levels of TSPO than PND8 spermatogonia led us to speculate that TSPO expression is down-regulated during gonocyte differentiation. Moreover, the change in TSPO subcellular localization from nuclear in gonocytes to cytoplasmic in spermatogonia is reminiscent of other factors translocating in spermatogenic cells from one compartment to another in function of their activation status. For example, FOXO1 has been shown to translocate from cytoplasm to nucleus between PND1 and 3, a period corresponding to the transition from quiescence to mitosis, and to remain nuclear in SSC [28]. TSPO nuclear localization has been reported before, particularly in breast cancer cells, where it is involved in cell proliferation [4]. However, this does not seem to be the case in gonocytes, since adding a TSPO binding drug to gonocytes in presence or absence of proliferation factors did not have an effect [8].

To examine whether TSPO is down-regulated during gonocyte differentiation, isolated PND3 gonocytes were treated with the differentiation factor RA. Our data confirmed that RA exerts a repressive effect on the expression of TSPO mRNA and protein during gonocyte differentiation, a process characterized by an increase in Stra8 expression. Interestingly, TSPO promoter has been shown to contain putative binding sites for RARβ and RARγ [7]. Together with our previous data showing that PND3 gonocytes express high levels of RARγ [16], this suggests that RARγ activation may be involved in TSPO down-regulation.

Similar effects were observed in F9 mouse embryonal carcinoma cells, a cell line issued from a mouse testicular teratoma, which shares a number of pluripotency markers with embryonic stem cells as well as gonocytes [15,16,29,30]. In F9 cells, which express genes from both the somatic and germline lineages, RA treatment can induce the expression of markers of somatic (e.g., collagen IV) and germ cell (e.g., Stra8) fates, depending of the culture conditions and downstream pathway activated [16,30]. All TGCTs, including embryonal carcinoma, are believed to derive from a common precursor, the carcinoma in situ, itself resulting from the failed differentiation of gonocytes [18]. Therefore, one cannot exclude at present that RA-induced TSPO repression in F9 cells may correspond to a function retained from gonocytes, which was not altered during the process of carcinogenesis.

To better understand TSPO role in gonocyte differentiation, we knocked down *Tspo* in gonocytes treated or not with RA. Surprisingly, silencing TSPO expression with siRNA was sufficient to significantly increase *Stra8* expression in gonocytes, in basal conditions with low levels of RA provided by FBS. Moreover, TSPO knock down potentiated RA-mediated increases in *Stra8* transcripts, implying that TSPO represses Stra8 expression. Among its multiple roles, TSPO has been reported to be positively regulated during adipocyte differentiation and homeostasis, with its knockdown impairing the release of adipokines, glucose uptake, and adipogenesis [31,32]. We have previously shown that gonocyte differentiation involves the activation of platelet-derived growth factor receptors and downstream SRC, JAK2, and STAT5 pathways [16] and an active ubiquitin proteasome system [20]. In contrast to these positive regulators of gonocyte differentiation, the present results suggest that TSPO acts as a repressor on gonocyte differentiation, and that one of RA effects is to eliminate this block. This is reminiscent of the downregulation by RA of Nanos2 in fetal and neonatal gonocytes, where this RNA binding protein was shown to repress Stra8 expression and prevent premature entry of germ cells into meiosis [33]. It would be interesting to examine in future studies whether TSPO ligands with agonist properties might exert a repressing effect on gonocyte differentiation, as suggested by our silencing results.

We then analyzed the mRNA expression of two germ cell markers, the PIWI-interacting RNA binding proteins Mili, and Miwi2, both required for spermatogenesis and meiotic progression. MILI is expressed from gonocytes to early spermatocytes, and has been shown to be essential for spermatogonial stem cell self-renewal and differentiation [21]. MIWI2 is strongly expressed in gonocytes, where it is guiding the re-methylation of transposons and preventing their activation, following the global erasure of DNA methylation marks in primordial germ cells [22]. Besides their role in transposon methylation, these genes have also been implicated in histone modifications and post-transcriptional regulation [34]. While the expression patterns of the two genes did not appear to be regulated by RA, TSPO knockdown led to increases in both *Mili* and *Miwi2* transcripts, suggesting TSPO involvement in the down-regulation of these genes, independent of differentiation signals. While preserving the genomic integrity of germ cells is critical for species survival, allowing for some degree of transposon mobility has played a role in evolution and adaptability of species to environmental changes. Considering the multiple roles of TSPO, the fact that it is highly conserved throughout evolution from bacteria to human [35], and its nuclear localization in gonocytes, one could envision a role for TSPO in maintaining the balance between transposon mobility and preserving genomic integrity in gonocytes. Further experiments will be required to test this hypothesis.

Next, we examined whether TSPO is also expressed in human normal and pathological germ cells, which had not been reported in the literature. Our study showed that normal testicular tissues express TSPO in discreet phases of spermatogenic cells, similarly to our previous findings in rat and mouse testis [8,9]. In human testis, besides its expected high expression in Leydig cells, TSPO was clearly expressed in the forming acrosomes of round spermatids, as confirmed by co-localization with PNA, a peanut agglutinin lectin, commonly used as an acrosomal marker [23]. This is a novel finding, suggesting that TSPO is involved in the formation of specialized organelles in germ cells. Interestingly, TSPO has been previously reported to be expressed in the Golgi apparatus of liver cells, suggesting that it might play a similar role in somatic cells [36].

Finally, in our search of TSPO role in germ cells, we examined its expression patterns in seminoma, the most common testicular germ cell tumor type in human, proposed to originate from abnormal gonocytes [18]. The comparative study of several normal and seminoma samples revealed a significant upregulation in *TSPO* mRNA levels in seminomas. Immunofluorescent analysis of TSPO and two pluripotency transcription factors OCT3/4 and AP2γ used as seminoma markers [24,25] confirmed TSPO protein localization in seminoma cells. Several studies have reported changes in TSPO expression and potential role in various cancers. For example, TSPO expression is upregulated in glioma tumor cells, compared to normal brain [37]. TSPO has also been implicated in lung cancer development, as cigarette smoke exposure was found to alter TSPO protein, leading the way for the initiation and progression of lung cancer [38]. Studies have shown that TSPO is highly expressed in estrogen-receptor (ER) negative breast tumors, representing a potential target for the development of new therapies for this subset of breast cancer [39]. Moreover, studies have shown that that TSPO expression correlates positively with the invasiveness and/or malignancy of breast, colorectal and prostate cancers [4,40–42]. Thus, our study revealed one more type of cancer presenting TSPO dysregulation. Determining whether TSPO alterations reflect its involvement in testicular tumorigenesis process, or if they indicate the retention of an early germ cell marker in deficient germ cells remains to be determined.

Taken together, our study shows that TSPO is down regulated during gonocyte differentiation, in which it might play a repressive role. Moreover, our expression studies in human normal testes confirmed that TSPO is expressed in subsets of adult germ cells, suggesting a function in acrosome formation, while the analysis of tumor samples revealed its mRNA up-regulation and protein localization in seminoma cells. Further studies are needed to determine the exact role of TSPO in normal spermatogenic cell development, from gonocyte to more mature germ cells, and in testicular cancer.

4. Materials and Methods

4.1. Animals

Newborn male Sprague Dawley rats were obtained from Charles Rivers Laboratories (Saint-Constant, QC, Canada). PND3 and PND8 pups were euthanized and handled according to protocols approved by the McGill University Health Centre Animal Care Committee and the Canadian Council on Animal Care (McGill Animal Care protocol #2015-7582; approved July 2015).

4.2. Germ Cell Isolation

Gonocytes were isolated from PND3 rat testes as previously described [13,43]. Briefly, testes from 40 rats were isolated and decapsulated. Gonocytes were isolated by sequential enzymatic digestion and differential plating overnight in RPMI 1640 medium (Invitrogen, Burlington, ON, Canada) with 5% fetal bovine serum (FBS) (Invitrogen), 2% penicillin/streptomycin (CellGro, Manassas, VA, USA), and 1% amphotericin B (CellGro). The next morning, non-adherent germ cells were further separated using a 2%–4% bovine serum albumin (BSA) (Roche Diagnostics, Indianapolis, IN, USA) gradient. Gonocytes were identified by morphology and by their larger size compared to Sertoli/myoid cells. Fractions containing the most gonocytes were pooled, centrifuged, and collected with a purity of at least 85% for RNA analysis. Samples with lower purity were used for immunocytochemical analysis. Spermatogonia were isolated from the testes of 10 PND8 pups per preparation, using the same method used for gonocyte isolation described above [20,43]. All experiments were performed using a minimum of three independent gonocyte or spermatogonia preparations.

4.3. Gonocyte Treatment

Isolated gonocytes were cultured in RPMI media (Invitrogen) containing 2.5% fetal bovine serum (FBS), 2% penicillin/streptomycin (CellGro) and 1% amphotericin B (CellGro) at 37 °C and 3.5% CO_2. Gonocytes were plated at a density of 8000–10,000 cells/well in 24-well plates. The cells were treated with or without all-trans retinoic acid (RA) (10^{-6} M) (Sigma, Oakville, ON, Canada) for 24 h (RNA analysis) or 72 h (protein analysis). Within each experiment, duplicate wells were used for each treatment condition.

4.4. TSPO Knockdown Using Silencing RNA

For silencing experiments, once gonocytes were collected from the BSA gradient, they were transfected with 100 nM of a mixture of three pre-designed siRNA sequences (Table 2) (IDT, San Jose, CA, USA) using Lipofectamine RNAiMAX (Invitrogen) and Opti-MEM transfection medium (Invitrogen). RPMI culture media without antibiotics, FBS or amphotericin B was used to plate the cells. A red fluorescent dye and a non-targeting universal negative control DsiRNA were transfected at 10 nM each, and served as positive and negative controls, respectively (IDT). Cells were transfected for 24 h (mRNA silencing analysis) and 48 h (protein silencing analysis), and then treated with or without RA (10^{-6} M) (Sigma) for an additional 24 h, in medium supplemented with 2.5% FBS.

Table 2. Translocator protein 18 kDa (TSPO) siRNA sequences. Three siRNA pairs (#1–3) were used.

#	Sense Strand	Antisense Strand
1	5′-CCUACAUAAUCUGGAAAGAGCUGGG-3′	5′-CCCAGCUCUUUCCAGAUUAUGUAGGAG-3′
2	5′-GCAGUUGCAAUCACUAUGUCUCAAT-3′	5′-AUUGAGACAUAGUGAUUGCAACUGCUG-3′
3	5′-CACCUAGCCAUCAGGAAUGCAGCCC-3′	5′-AUUGAGACAUAGUGAUUGCAACUGCUG-3′

4.5. F9 Mouse Embryonal Carcinoma Cell Culture

F9 cells were maintained in DMEM medium (Invitrogen) containing 10% FBS (Invitrogen) at 37 °C and 5.0% CO_2. Cells were plated on gelatine-coated culture dishes on day 1 and treated on day 2 with or without RA (10^{-7} M) (Sigma) for 24 h and 72 h for mRNA and protein analysis, respectively.

4.6. Human Specimens

Frozen and paraffin-embedded tissues from human adult normal testes and seminoma specimens, together with their pathology reports, were obtained from the Departments of Pathology and Surgery at the McGill University Health Centre, Montreal, and at La Paz University Hospital (Universidad Autonoma de Madrid, Madrid, Spain), with informed consent forms, following institutional ethics review board requirements, as described before [44]. Six non-cancer samples presenting normal seminiferous tubules and spermatogenesis, as well as seven seminoma specimens were examined (see Table 1 for details).

4.7. RNA Extraction and cDNA Synthesis

Total RNA was extracted from cell pellets or frozen biopsies using the PicoPure RNA isolation kit (Arcturus, Mountain View, CA, USA) and digested with DNase I (Qiagen, Valencia, CA, USA). For quantitative PCR (qPCR) analysis, cDNA was synthesized from the extracted RNA by using the single strand cDNA transcriptor synthesis kit (Roche Diagnostics) following the manufacturer's instructions.

4.8. Quantitative Real Time PCR (qPCR)

QPCR was performed using a LightCycler 480 with a SYBR Green PCR Master Mix kit (Roche Diagnostics). The primer sets used were designed using the Roche primer design software (Roche Diagnostics) and are listed in Table 3. QPCR cycling conditions were as follows: Initial step at 95 °C followed by 45 cycles at 95 °C for 10 s, 61 °C for 10 s, and 72 °C for 10 s. The direct detection of PCR products was monitored by measuring the increase in fluorescence caused by the binding of SYBR Green dye to double-stranded DNA. The comparative threshold cycle (C_t) method was used to analyze the data and 18S rRNA was used for data normalization, as it was previously shown to be an adequate housekeeping gene for both rat germ cells and human seminomas [20,44]. Assays were performed in triplicate. All experiments were performed using a minimum of three independent sample preparations, with each condition performed in duplicate, and the mean ± SEM are shown.

Table 3. Quantitative real time PCR primers. Lower case character letters in sequences were added by the primer designing program and are not part of the gene sequences.

Species	Gene	Forward Sequence	Reverse Sequence
Rat	*Tspo*	cgcaATGGGAGCCTACTTTGTGcG	GCCAGGAGGGTTTCTGCAAG
Rat	*Stra8*	TGCTTTTGATGTGGCGAGCT	GCGCTGATGTTAGACAGACGCT
Rat	*Mili*	cggaaGACATCCAGTACAGAGTCATTCcG	GGGTCTCTTCTTGCTCGCTGA
Rat	*Miwi2*	AGCTGCACAGGTATTCCGCTAGAA	ATGGCAGACAGGACTTCGTCGATT
Rat	*18s*	cgggTGCTCTTAGCTGAGTGTCCcG	CTCGGGCCTGCTTTGAACAC
Mouse	*Tspo*	CCCGCTTGCTGTACCCTTACC	CACCGCATACATAGTAGTTGAGCAC
Mouse	*18s*	CGGAATCTTAATCATGGCCTCAGTTC	ACCGCAGCTAGGAATAATGGAAT
Human	*TSPO*	GGTGGATCTCCTGCTGGTCA	CGACCAATACGCAGTAGTTGAGTGTG
Human	*18S*	cggacTGTGATGCCCTTAGATGTCcG	GTAGGGTAGGCACACGCTGAG

4.9. Immunoblot Analysis

F9 cells or gonocytes were solubilized in Laemmli buffer and protein concentrations were determined using the BCA protein assay kit (Thermo Scientific, Burlington, ON, Canada). As previously described [20], proteins (30 µg per lane) were separated on pre-casted 4%–20% tris-glycine gels (Invitrogen). Gels were transferred to a polyvinylidene fluoride (PVDF) membrane (Bio-Rad, Mississauga, ON, Canada). After blocking with 5% milk, membranes were probed with an anti TSPO antibody (lab-made; [8]) to determine protein expression. GAPDH and tubulin (Trevigen, Gaithersburg, MD, USA) were used as loading controls. After an overnight incubation at 4 °C, bound antibodies were detected using anti-biotin and horseradish peroxidase (HRP) coupled goat anti-rabbit secondary antibodies (Cell Signaling, Danvers, MA, USA) and ECL-enhanced chemiluminescence (GE Healthcare, Mississauga, ON, Canada). Images were taken using the LAS-4000 gel documentation system (Fujifilm, Mississauga, ON, Canada).

4.10. Immunocytochemistry (ICC)

Microscopic slides were prepared using isolated germ cells and treated gonocytes collected by cytospin centrifugation, and processed for ICC as previously described [20,43]. In brief, slides were treated with Dako Target Retrieval solution (DAKO, Burlington, ON, Canada), a 30% hydrogen peroxide methanol solution, and then non-specific protein interactions were blocked with PBS (Invitrogen) containing 10% goat serum (Vector Laboratories, Burlington, ON, Canada), 1% BSA (Roche Diagnostics) and 0.02% Triton X100 (Promega; Madison, WI, USA) for one hour; and then incubated with anti TSPO primary antibody (lab-made) diluted in PBS containing 1% BSA (Roche Diagnostics) and 0.02% Triton X100 (Promega) overnight at 4 °C. The slides were then incubated with a biotin-conjugated goat anti-rabbit secondary antibody (BD Pharmingen, San Diego, CA, USA) diluted in PBS-1% BSA for one hour at room temperature. Immunoreactivity was detected using a combination of streptavidin-peroxidase (Invitrogen) and AEC single use solution (Invitrogen). Slides were counter-stained with hematoxylin (Sigma Aldrich), coated with Crystal Mount (Electron Microscopy Sciences, Hatfield, PA, USA) and dried, and then cover-slipped using Permaflour (Thermo Scientific) and glass coverslips (Fisher Scientific, Toronto, ON, Canada). The slides were then viewed using a BX40 Olympus microscope (Olympus, Center Valley, PA, USA) coupled to a DP70 Olympus digital camera (Olympus).

4.11. Immunofluorescence (IF)

Paraffin-embedded human adult normal testicular tissues and seminoma specimens were processed for IF as previously described [44], following a protocol similar to that in the ICC section above, except that slides were dewaxed and rehydrated using Citrosolv (Fisher Scientific) and Trilogy solution (Cell Marque IVD, Rocklin, CA, USA), prior treatment with Dako Target Retrieval solution. Sections were incubated overnight with a combination of either TSPO antibody (Epitomics, Burlingame, CA, USA) and OCT 3/4 antibody (Santa Cruz, Dallas, TX, USA) diluted in PBS containing 1% BSA and 0.02% Triton X100 overnight at 4 °C, or TSPO antibody and AP2γ antibody (Santa Cruz), or TSPO antibody alone. Next, slides were incubated with a fluorescent goat anti-rabbit Alexa Fluor 546 secondary antibody and a fluorescent goat anti-mouse Alexa Fluor 488 secondary antibody (Invitrogen) diluted in PBS (Invitrogen) containing 1% BSA (Roche Diagnostics) for one hour at room temperature. Some slides were then stained for 30 min with a lectin from *Arachis hypogaea* (peanut) PNA (Sigma) to stain the acrosomes. Nuclear staining for all slides was done using nuclear DAPI (Invitrogen) staining for 5 min. Slides were mounted with PermaFluor Mountant (Thermo Scientific) and cover-slipped. Slides were then examined using a BX40 Olympus microscope (Olympus) coupled to a DP70 Olympus digital camera (Olympus).

4.12. Statistical Analysis

Statistical analysis was performed using an unpaired two-tail Student's *t*-test using statistical analysis functions in the GraphPad Prism 5.0 program (GraphPad Inc., San Diego, CA, USA). All experiments were performed with a minimum of three independent experiments, each including duplicate wells per condition. A *p*-value less than 0.05 was considered statistically significant.

Acknowledgments: We would like to thank Joanna Bouihol, Dahee Kim, and Annie Boisvert for their outstanding technical assistance.

Author Contributions: Gurpreet Manku participated to the design of the project, data analysis, writing of the manuscript, and performed all experiments; Martine Culty designed the project, participated to data analysis and manuscript writing.

Conflicts of Interest: The authors declare no conflict of interest.

References

1. Papadopoulos, V.; Baraldi, M.; Guilarte, T.R.; Knudsen, T.B.; Lacapère, J.J.; Lindemann, P.; Norenberg, M.D.; Nutt, D.; Weizman, A.; Zhang, M.R.; et al. Translocator protein (18 kDa): New nomenclature for the peripheral-type benzodiazepine receptor based on its structure and molecular function. *Trends Pharmacol. Sci.* **2006**, *27*, 402–409. [CrossRef] [PubMed]
2. Delavoie, F.; Li, H.; Hardwick, M.; Robert, J.C.; Giatzakis, C.; Peranzi, G.; Yao, Z.X.; Maccario, J.; Lacapere, J.J.; Papadopoulos, V. In vivo and in vitro peripheral-type benzodiazepine receptor polymerization: Functional significance in drug ligand and cholesterol binding. *Biochemistry* **2003**, *42*, 4506–4519. [CrossRef] [PubMed]
3. Midzak, A.; Rone, M.; Aghazadeh, Y.; Culty, M.; Papadopoulos, V. Mitochondrial protein import and the genesis of steroidogenic mitochondria. *Mol. Cell. Endocrinol.* **2011**, *336*, 70–79. [CrossRef] [PubMed]
4. Hardwick, M.; Fertikh, D.; Culty, M.; Li, H.; Vidic, B.; Papadopoulos, V. Peripheral-type benzodiazepine receptor (TSPO) in human breast cancer: Correlation of breast cancer cell aggressive phenotype with TSPO expression, nuclear localization, and TSPO-mediated cell proliferation and nuclear transport of cholesterol. *Cancer Res.* **1999**, *59*, 831–842. [PubMed]
5. Corsi, L.; Geminiani, E.; Avallone, R.; Baraldi, M. Nuclear location-dependent role of peripheral benzodiazepine receptor (PBR) in hepatic tumoral cell lines proliferation. *Life Sci.* **2005**, *76*, 2523–2533. [CrossRef] [PubMed]
6. Veenman, L.; Papadopoulos, V.; Gavish, M. Channel-like functions of the 18 kDa Translocator protein (TSPO): Regulation of apoptosis and steroidogenesis as part of the host-defense response. *Curr. Pharm. Des.* **2007**, *13*, 2385–2405. [CrossRef] [PubMed]
7. Batarseh, A.; Papadopoulos, V. Regulation of translocator protein 18 kDa (TSPO) expression in health and disease states. *Mol. Cell. Endocrinol.* **2010**, *327*, 1–12. [CrossRef] [PubMed]
8. Manku, G.; Wang, Y.; Thuillier, R.; Rhodes, C.; Culty, M. Developmental expression of the Translocator protein 18 kDa (TSPO) in testicular germ cells. *Curr. Mol. Med.* **2012**, *12*, 467–475. [CrossRef] [PubMed]
9. Culty, M.; Liu, Y.; Manku, G.; Chan, W.Y.; Papadopoulos, V. Expression of steroidogenesis-related genes in murine male germ cells. *Steroids* **2012**, *103*, 105–114. [CrossRef] [PubMed]
10. Culty, M. Gonocytes, the forgotten cells of the germ cell lineage. *Birth Defects Res. C Embryo Today* **2009**, *876*, 1–26. [CrossRef] [PubMed]
11. Culty, M. Gonocytes, from the fifties to the present: Is there a reason to change the name? *Biol. Reprod.* **2013**, *89*, 46. [CrossRef] [PubMed]
12. Manku, G.; Culty, M. Mammalian gonocyte and spermatogonia differentiation: Recent advances and remaining challenges. *Reproduction* **2015**, *149*, R139–R157. [CrossRef] [PubMed]
13. Li, H.; Papadopoulos, V.; Vidic, B.; Dym, M.; Culty, M. Regulation of rat testis gonocyte proliferation by platelet-derived growth factor and estradiol: Identification of signalling mechanisms involved. *Endocrinology* **1997**, *138*, 1289–1298. [CrossRef] [PubMed]
14. Thuillier, R.; Mazer, M.; Manku, G.; Boisvert, A.; Wang, Y.; Culty, M. Interdependence of PDGF and estrogen signaling pathways in inducing neonatal rat testicular gonocytes proliferation. *Biol. Reprod.* **2010**, *82*, 825–836. [CrossRef] [PubMed]
15. Wang, Y.; Culty, M. Identification and distribution of a novel platelet-derived growth factor receptor beta variant: Effect of retinoic acid and involvement in cell differentiation. *Endocrinology* **2007**, *148*, 2233–2250. [CrossRef] [PubMed]
16. Manku, G.; Wang, Y.; Merkbaoui, V.; Boisvert, A.; Ye, X.; Blonder, J.; Culty, M. Role of Retinoic Acid and Platelet-Derived Growth Factor Receptor crosstalk in the regulation of neonatal gonocyte and embryonal carcinoma cell differentiation. *Endocrinology* **2015**, *1*, 346–359. [CrossRef] [PubMed]
17. Manku, G.; Culty, M. Dynamic changes in the expression of apoptosis-related genes in differentiating gonocytes and in seminomas. *Asian J. Androl.* **2015**, *7*, 403–414. [CrossRef] [PubMed]
18. Skakkebaek, N.E.; Berthelsen, J.G.; Giwercman, A.; Muller, J. Carcinoma-in-situ of the testis: Possible origin from gonocytes and precursor of all types of germ cell tumours except spermatocytoma. *Int. J. Androl.* **1987**, *10*, 19–28. [CrossRef] [PubMed]
19. Huyghe, E.; Matsuda, T.; Thonneau, P. Increasing incidence of testicular cancer worldwide: A review. *J. Urol.* **2003**, *170*, 5–11. [CrossRef] [PubMed]

20. Manku, G.; Wing, S.S.; Culty, M. Expression of the Ubiquitin Proteasome System in Neonatal Rat Gonocytes and Spermatogonia: Role in Gonocyte Differentiation. *Biol. Reprod.* **2012**, *87*, 1–18. [CrossRef] [PubMed]

21. Unhavaithaya, Y.; Hao, Y.; Beyret, E.; Yin, H.; Kuramochi-Miyagawa, S.; Nakano, T.; Lin, H. MILI, a PIWI-interacting RNA-binding protein, is required for germ line stem cell self-renewal and appears to positively regulate translation. *J. Biol. Chem.* **2009**, *284*, 6507–6519. [CrossRef] [PubMed]

22. Carmell, M.A.; Girard, A.; van de Kant, H.J.; Bourc'his, D.; Bestor, T.H.; de Rooij, D.G.; Hannon, G.J. MIWI2 is essential for spermatogenesis and repression of transposons in the mouse male germline. *Dev. Cell* **2007**, *12*, 503–514. [CrossRef] [PubMed]

23. Kallajoki, M.; Virtanen, I.; Suominen, J. The fate of acrosomal staining during the acrosome reaction of human spermatozoa as revealed by a monoclonal antibody and PNA-lectin. *Int. J. Androl.* **1986**, *9*, 181–194. [CrossRef] [PubMed]

24. Looijenga, L.H.; Stoop, H.; Biermann, K. Testicular cancer: Biology and biomarkers. *Virchows Arch.* **2014**, *464*, 301–313. [CrossRef] [PubMed]

25. Pauls, K.; Jager, R.; Weber, S.; Wardelmann, E.; Koch, A.; Buttner, R.; Schorle, H. Transcription factor AP-2gamma, a novel marker of gonocytes and seminomatous germ cell tumors. *Int. J. Cancer* **2005**, *115*, 470–477. [CrossRef] [PubMed]

26. Papadopoulos, V. On the role of the translocator protein (18-kDa) TSPO in steroid hormone biosynthesis. *Endocrinology* **2014**, *155*, 15–20. [CrossRef] [PubMed]

27. Fan, J.; Campioli, E.; Midzak, A.; Culty, M.; Papadopoulos, V. Conditional steroidogenic cell-targeted deletion of TSPO unveils a crucial role in viability and hormone-dependent steroid formation. *Proc. Natl. Acad. Sci. USA* **2015**, *112*, 7261–7266. [CrossRef] [PubMed]

28. Goertz, M.J.; Wu, Z.; Gallardo, T.D.; Hamra, F.K.; Castrillon, D.H. Foxo1 is required in mouse spermatogonial stem cells for their maintenance and the initiation of spermatogenesis. *J. Clin. Investig.* **2011**, *121*, 3456–3466. [CrossRef] [PubMed]

29. Rochette-Egly, C.; Chambon, P. F9 embryo carcinoma cells: A cell autonomous model to study the functional selectivity of RARs and RXRs in retinoid signaling. *Histol. Histopathol.* **2001**, *16*, 909–922. [PubMed]

30. Kwon, J.T.; Jin, S.; Choi, H.; Kim, J.; Jeong, J.; Kim, J.; Kim, Y.; Cho, B.N.; Cho, C. Identification and characterization of germ cell genes expressed in the F9 testicular teratoma stem cell line. *PLoS ONE* **2014**, *9*, e103837. [CrossRef] [PubMed]

31. Campioli, E.; Batarseh, A.; Li, J.; Papadopoulos, V. The endocrine disruptor mono-(2-ethylhexyl) phthalate affects the differentiation of human liposarcoma cells (SW 872). *PLoS ONE* **2011**, *6*, e28750. [CrossRef] [PubMed]

32. Li, J.; Papadopoulos, V. Translocator protein (18 kDa) as a pharmacological target in adipocytes to regulate glucose homeostasis. *Biochem. Pharmacol.* **2015**, *97*, 99–110. [CrossRef] [PubMed]

33. Barrios, F.; Filipponi, D.; Pellegrini, M.; Paronetto, M.P.; Di Siena, S.; Geremia, R.; Rossi, P.; de Felici, M.; Jannini, E.A.; Dolci, S. Opposing effects of retinoic acid and FGF9 on Nanos2 expression and meiotic entry of mouse germ cells. *J. Cell Sci.* **2010**, *123*, 871–880. [CrossRef] [PubMed]

34. Peng, J.C.; Lin, H. Beyond transposons: The epigenetic and somatic functions of the Piwi-piRNA mechanism. *Curr. Opin. Cell Biol.* **2013**, *25*, 190–194. [CrossRef] [PubMed]

35. Fan, J.; Lindemann, P.; Feuilloley, M.G.; Papadopoulos, V. Structural and functional evolution of the translocator protein (18 kDa). *Curr. Mol. Med.* **2012**, *12*, 369–386. [CrossRef] [PubMed]

36. O'Beirne, G.B.; Woods, M.J.; Williams, D.C. Two subcellular locations for peripheral-type benzodiazepine acceptors in rat liver. *Eur. J. Biochem.* **1990**, *188*, 131–138. [CrossRef] [PubMed]

37. Janczar, K.; Su, Z.; Raccagni, I.; Anfosso, A.; Kelly, C.; Durrenberger, P.F.; Gerhard, A.; Roncaroli, F. The 18 kDa mitochondrial translocator protein in gliomas: From the bench to bedside. *Biochem. Soc. Trans.* **2015**, *43*, 579–585. [CrossRef] [PubMed]

38. Gavish, M.; Cohen, S.; Nagler, R. Cigarette smoke effects on TSPO and VDAC expression in a cellular lung cancer model. *Eur. J. Cancer Prev.* **2016**, *25*, 361–367. [CrossRef] [PubMed]

39. Wu, X.; Gallo, K.A. The 18 kDa translocator protein (TSPO) disrupts mammary epithelial morphogenesis and promotes breast cancer cell migration. *PLoS ONE* **2013**, *8*, e71258. [CrossRef] [PubMed]

40. Galiegue, S.; Casellas, P.; Kramar, A.; Tinel, N.; Simony-Lafontaine, J. Immunohistochemical assessment of the peripheral benzodiazepine receptor in breast cancer and its relationship with survival. *Clin. Cancer Res.* **2004**, *10*, 2058–2064. [CrossRef] [PubMed]

41. Han, Z.; Slack, R.S.; Li, W.; Papadopoulos, V. Expression of peripheral benzodiazepine receptor (PBR) in human tumors: Relationship to breast, colorectal, and prostate tumor progression. *J. Recept. Signal Transduct. Res.* **2003**, *23*, 225–238. [CrossRef] [PubMed]

42. Batarseh, A.; Barlow, K.D.; Martinez-Arguelles, D.B.; Papadopoulos, V. Functional characterization of the human translocator protein (18 kDa) gene promoter in human breast cancer cell lines. *Biochim. Biophys. Acta* **2012**, *1819*, 38–56. [CrossRef] [PubMed]

43. Manku, G.; Mazer, M.; Culty, M. Neonatal testicular gonocytes isolation and processing for immunocytochemical analysis. *Methods Mol. Biol.* **2012**, *825*, 17–29. [PubMed]

44. Manku, G.; Hueso, A.; Brimo, F.; Chan, P.; Gonzalez-Peramato, P.; Jabado, N.; Riazalhosseini, Y.; Gayden, T.; Bourgey, M.; Culty, M. Changes in claudin expression profiles during gonocyte differentiation and high expression in seminomas. *Andrology* **2016**, *4*, 95–110. [CrossRef] [PubMed]

International Journal of
Molecular Sciences

MDPI

Article

Effect of the CRAC Peptide, VLNYYVW, on mPTP Opening in Rat Brain and Liver Mitochondria

Tamara Azarashvili [1,*], Olga Krestinina [1], Yulia Baburina [1], Irina Odinokova [1], Vladimir Akatov [1], Igor Beletsky [1], John Lemasters [2] and Vassilios Papadopoulos [3]

[1] Institute of Theoretical and Experimental Biophysics, Russian Academy of Sciences, Institutskaya Str., Pushchino, Moscow Region 142290, Russia; krestinina@rambler.ru (O.K.); byul@rambler.ru (Y.B.); odinokova@rambler.ru (I.O.); akatov.vladimir@gmail.com (V.A.); ipbeletsky@gmail.com (I.B.)

[2] Departments of Drug Discovery & Biomedical Sciences and Biochemistry & Molecular Biology, Medical University of South Carolina, DD504 Drug Discovery Bldg., 70 President St., MSC 140, Charleston, SC 29425, USA; JJLemasters@musc.edu

[3] The Research Institute of the McGill University Health Center, and Departments of Medicine, Biochemistry, Pharmacology and Therapeutics, McGill University, 2155 Guy Street, Montreal, QC H3H 2R9, Canada; vassilios.papadopoulos@mcgill.ca

* Correspondence: tamara.azarashvili@gmail.com; Tel.: +7-496-773-9182; Fax: +7-496-733-0553

Academic Editors: Giovanni Natile and Nunzio Denora
Received: 25 September 2016; Accepted: 7 December 2016; Published: 13 December 2016

Abstract: The translocator protein (TSPO; 18 kDa) is a high-affinity cholesterol-binding protein located in the outer membrane of mitochondria. A domain in the C-terminus of TSPO was characterized as the cholesterol recognition/interaction amino acid consensus (CRAC). The ability of the CRAC domain to bind to cholesterol led us to hypothesize that this peptide may participate in the regulation of mitochondrial membrane permeability. Herein, we report the effect of the synthetic CRAC peptide, VLNYYVW, on mitochondrial permeability transition pore (mPTP) opening. It was found that the CRAC peptide alone prevents the mPTP from opening, as well as the release of apoptotic factors (cytochrome c, AIF, and EndoG) in rat brain mitochondria (RBM). Co-incubation of CRAC, together with the TSPO drug ligand, PK 11195, resulted in the acceleration of mPTP opening and in the increase of apoptotic factor release. VLNYYVW did not induce swelling in rat liver mitochondria (RLM). 3,17,19-androsten-5-triol (19-Atriol; an inhibitor of the cholesterol-binding activity of the CRAC peptide) alone and in combination with the peptide was able to stimulate RLM swelling, which was Ca^{2+}- and CsA-sensitive. Additionally, a combination of 19-Atriol with 100 nM PK 11195 or with 100 μM PK 11195 displayed the opposite effect: namely, the addition of 19-Atriol with 100 μM PK 11195 in a suspension of RLM suppressed the Ca^{2+}-induced swelling of RLM by 40%, while the presence of 100 nM PK 11195 with 19-Atriol enhanced the swelling of RLM by 60%. Taken together, these data suggest the participation of the TSPO's CRAC domain in the regulation of permeability transition.

Keywords: mitochondria; translocator protein (TSPO); cholesterol recognition/interaction amino acid consensus (CRAC); permeability transition

1. Introduction

Cholesterol, a constituent of biological membranes, regulates the physical states of membrane phospholipid bilayers and membrane fluidity, and it determines membrane permeability. It is not uniformly distributed in cell membranes, as the highest concentration is demonstrated in the plasma membrane, whereas mitochondria have the lowest concentrations [1]. Proteins play an important role in cholesterol distribution. There are segments of integral membrane proteins that are located at the membrane interface and facilitate interactions with cholesterol-binding proteins or that have partitioned into cholesterol-rich domains, characterized by the presence of a "cholesterol recognition amino acid consensus" sequence, otherwise known as the CRAC motif (CRAC domain). A CRAC motif is defined as a sequence pattern, -L/V-(X)(1–5)-Y-(X)$_{(1–5)}$-R/K-, in which (-X-)$_{(1–5)}$ represents between one and five residues of any amino acid [2–4]. Nuclear magnetic resonance (NMR) spectroscopy of the CRAC motif has demonstrated that the side chains of the motif generate a groove that is capable of accommodating a cholesterol molecule, with the central tyrosine playing a critical role in cholesterol binding [3].

The first protein studied with a CRAC motif was the peripheral-type benzodiazepine receptor, now known as the translocator protein (TSPO) [4]. TSPO is an 18 kDa hydrophobic protein and a tryptophan-rich sensory protein oxygen sensor; it is defined as a multi-spanning membrane protein consisting of five transmembrane alpha-helices [5], and it is an extra-mitochondrial C-terminal containing a cholesterol-binding domain [3], an intra-mitochondrial N-terminal, two extra-mitochondrial loops, and two intra-mitochondrial loops. The CRAC domain of the TSPO is located at the C-terminus of the protein. TSPO transfers cholesterol (its endogenous ligand) across the membrane [6,7]. In this process, the CRAC domain is critical for cholesterol binding. Additionally, TSPO binds with high affinity to a variety of distinct chemical drugs (synthetic exogenous ligands), including the isoquinoline carboxamide, PK 11195, and others [7]. TSPO is found in all examined tissues and it is involved in various cell functions, such as steroidogenesis, cell proliferation, mitochondrial respiration, and apoptosis [8,9]. In mammalian cells, TSPO is primarily located in the outer mitochondrial membrane and it is concentrated at the outer–inner membrane contact sites. Deletion of the C-terminus of recombinant mammalian TSPO, severely reduces cholesterol uptake, although PK 11195 binding is retained [10], suggesting the existence of distinct drug ligand- and cholesterol-binding sites in TSPO [11–13].

Mitochondrial membrane cholesterol is known to affect the permeability of the outer mitochondrial membrane. In this relation, it is worth noting that the major outer membrane protein, VDAC (which determines outer membrane permeability [14,15]) also binds cholesterol and is probably involved in its distribution between the inner and outer membranes of the mitochondria [16]. TSPO co-localizes with VDAC [17]. This tight physical interaction between both proteins indicates that TSPO may possibly regulate VDAC function, particularly, modulating cholesterol binding to VDAC. Besides, adenine nucleotide carrier (ANT) was found to participate in the distribution of cholesterol. It should be reminded that both VDAC and ANT are considered as the main regulators of the function of the permeability transition pore.

The CRAC-like motif was determined in the BAX sequence. BAX is a pro-apoptotic member of the Bcl-2 protein family that resides in an inactive state in the cytoplasm of normal cells. Oligomer BAX forms pores in the outer mitochondrial membrane [18]. Cholesterol modulates oligomerization and the insertion of BAX into the membranes. The existence of the CRAC domain in BAX may promote the protein incorporation into the cholesterol-enriched membrane [19]. The CRAC motif is also found in transmembrane domains of connexin43 (LLIQWYIY), which is present in mitochondria, where it might interact with TSPO [20]. Interestingly, proteins shown to bind cholesterol, e.g., VDAC and TSPO, or having the CRAC motif, e.g., TSPO and Bax, participate in the regulation of permeability of the outer mitochondrial membrane.

CRAC peptides have been synthesized and examined in artificial bilayer lipid membranes (BLM). Epand et al showed that the cholesterol-binding peptide, LWYIK, was able to stimulate the formation of cholesterol-rich domains in BLM, which is composed of phosphatidylcholine and cholesterol [21]. The CRAC motif is a primary structure pattern used to identify regions that may be responsible for preferential cholesterol binding in many proteins [22]. However, at the moment nothing is known about the effect of the CRAC peptide on mitochondrial function and in particular on the initiation of the mitochondrial permeability transition pore (mPTP) opening.

To understand the mechanism of action of the CRAC peptide in the cells, a novel ligand, 3,17,19-androsten-5-triol (19-Atriol), was recently identified, which has the ability to inhibit cholesterol binding at the CRAC motif [23]. 19-Atriol binds to a synthetic CRAC peptide and inhibits steroidogenesis in MA-10 mouse Leydig tumor cells, as well as in R2C rat Leydig tumor cells, at low micromolar concentrations. In addition, 19-Atriol suppresses PK 11195-stimulated steroidogenesis, with activity in the high nanomolar range. The binding of 19-Atriol to the CRAC domain did not perturb PK 11195 binding to TSPO. TSPO has been implicated in mitochondrial permeability transition and in cell death [23,24], so 19-Atriol was supposed to facilitate cell death through CRAC domain binding [25]. Collectively, the data described above suggest that the CRAC domain is critical for the TSPO function in cholesterol transport, and 19-Atriol might operate at the level of mitochondrial cholesterol transfer. At the moment, the effect of 19-Atriol on mitochondrial function has not been examined. Therefore, taking into consideration that the initiation of mPTP opening is considered as an initial stage of programmed cell death, the aim of the present work was to examine the effect of a synthetic CRAC peptide (VLNYYVW), 19-Atriol, as well as their combined effect on mPTP opening in purified rat brain non-synaptic mitochondria (RBM) and in rat liver mitochondria (RLM). Both RBM and RLM preparations were used to compare and contrast to previous work we performed on mPTP induction by TSPO drug ligands [20,23].

2. Results

2.1. Effect of the CRAC Peptide (VLNYYVW) and the Combined Effect of the CRAC Peptide with PK 11195 on Calcium Capacity in RBM

The effect of the synthetic peptide-a fragment of the TSPO cholesterol-binding consensus, VLNYYVW, on the mPTP function of rat non-synaptic brain mitochondria was examined. The functional state of mitochondria was determined by simultaneous measurement of the membrane potential ($\Delta\psi$) with a TPP$^+$-selective electrode, and Ca^{2+} release was measured with a Ca^{2+} electrode. Calcium pulses (50 µM each) were added to the mitochondria to reach a threshold calcium concentration for mPTP opening. The mitochondria maintained a high $\Delta\Psi_m$ level in the presence of succinate and rotenone.

As shown in Figure 1A, the first three pulses of Ca^{2+} (added to RBM) induced a decrease in $\Delta\Psi_m$ following its restoration. At the same time, Ca^{2+} rapidly accumulated in the mitochondrial matrix until it reached 150 µM; however, the fourth addition of calcium (reaching a total of 200 µM Ca2$^+$) caused Ca^{2+} release within approximately four minutes (250 s) after the last calcium addition, indicating that pore opening had initiated. Next, we checked whether a peptide with a sequence analogous to the cholesterol-binding consensus of TSPO (CRAC peptide, VLNYYVW) had the ability to change the threshold Ca^{2+} load needed for mPTP opening. The addition of 100 µg of VLNYYVW alone does not induce the opening of mPTP. Furthermore, pore opening was not observed after the fourth calcium pulses to the mitochondrial suspension (Figure 1B).

In this case, the mitochondria were more resistant, since calcium release and membrane depolarization were not observed. Then, we tested whether PK 11195, which is known to be among the most specific drugs binding to TSPO, can alter the effect of added VLNYYVW. Earlier, we reported that 100 nM PK 11195 was able to suppress mPTP opening in calcium-overloaded mitochondria, while 100 µM PK 11195 stimulated it [23]. Therefore, we used 100 µM PK 11195 to test the combined effect of PK11195 with VLNYYVW. Figure 1C shows that, in the presence of 100 µM PK 11195 alone in RBM suspension, the third addition of calcium immediately initiated pore opening. Thus, 150 µM Ca^{2+} was enough to initiate mPTP opening in the presence of 100 µM PK 11195. Next, we tested whether the VLNYYVW peptide was able to decrease the stimulating effects of PK 11195. In Figure 1D, the combined effect of 100 µg VLNYYVW and 100 µM PK 11195 on initiating of mPTP opening in RBM is shown. It was found that, taken together, these compounds significantly stimulate calcium release after two calcium additions (100 µM Ca^{2+}), demonstrating that 100 µg VLNYYVW in combination with 100 µM PK 11195 is able to decrease the threshold calcium concentration by two times and accelerate pore opening. It was observed that CsA (mPTP blocker) was able to prevent acceleration of mPTP opening caused by VLNYYVW and 100 µM PK 11195 (data not shown). Figure 1E presents the comparative summary data regarding the effect of VLNYYVW and PK 11195 on calcium capacity and membrane potential under mPTP opening. In the presence of 100 μM PK 11195, the calcium capacity in calcium-overloaded RBM decreased by two times. However, PK 11195, when added to the mitochondrial suspension in combination with VLNYYVW, results in a reduction in RBM's calcium capacity by almost three times. Thus, the VLNYYVW peptide (100 μg) alone was able to delay mPTP opening, while the peptide combined with PK 11195 strengthened the effect of the drug. These data allowed us to postulate that cooperation of the CRAC peptide with PK 11195 might decrease calcium retention and accelerate mPTP opening.

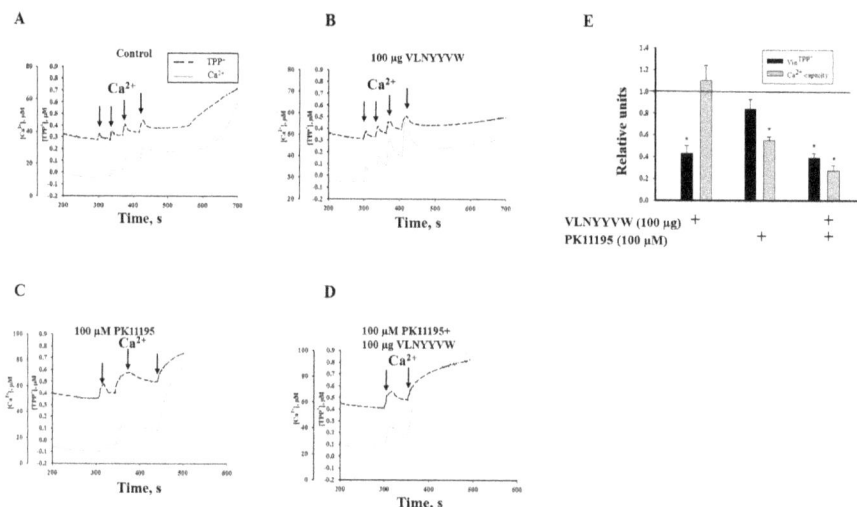

Figure 1. Effect of the VLNYYVW peptide and the translocator protein (TSPO) ligand, PK 11195, on Ca^{2+}-induced mPTP opening in RBM (rat brain non-synaptic mitochondria). The arrows indicate where $CaCl_2$ (50 μM) was added to the mitochondrial suspension. RBM were incubated in standard medium, as described in Materials and Methods. A value of $p < 0.05$ was considered to be significant (asterisk indicates $p < 0.05$). (**A**) Control RBM; (**B**) VLNYYVW (100 μg)-treated RBM; (**C**) PK 11195 (100 μM)-treated RBM; (**D**) The combined effect of VLNYYVW and PK 11195; (**E**) The quantitative characteristics of mitochondrial parameters. Ca^{2+} capacity and the rate of TPP+ influx, calculated as described in [23].

2.2. Effect of 19-Atriol (an Inhibitor of Cholesterol-Binding TSPO) on the Swelling of RLM

A novel ligand, 19-Atriol, was recently identified; it inhibits cholesterol binding at the TSPO CRAC motif, which is responsible for binding cholesterol and facilitating its translocation from the outer to inner mitochondrial membrane [25]. We used 19-Atriol to determine the possible participation of 19-Atriol in mPTP and its relationship with the CRAC peptide, VLNYYVW. Since Ca^{2+}-induced swelling is a parameter of mPTP function, we examined the effect of 19-Atriol on mitochondrial swelling by measuring this swelling as a decrease of absorbance at 540 nm. For that, we used RLM, which swell better, and isolation of mitochondria from the liver allowed us to obtain a sufficient amount of mitochondria for testing RLM swelling under different conditions and for the detection of the membrane potential and Ca^{2+}-induced Ca^{2+} release from RLM. First, we tested the effect of different concentrations of 19-Atriol in the range of 5–100 μM (5, 10, 50, and 100 μM) on various mPTP parameters, such as the Ca^{2+} release rate, membrane depolarization, and the time of calcium retention before pore opening (the lag-phase period). These parameters were measured in a chamber with installed selective electrodes, as in Figure 1. It was found that only 50 and 100 μM 19-Atriol had an effect on pore opening; they increased calcium release and depolarization, and shortened the lag-time period prior to calcium release, thus inducing pore opening (Figure 2A).

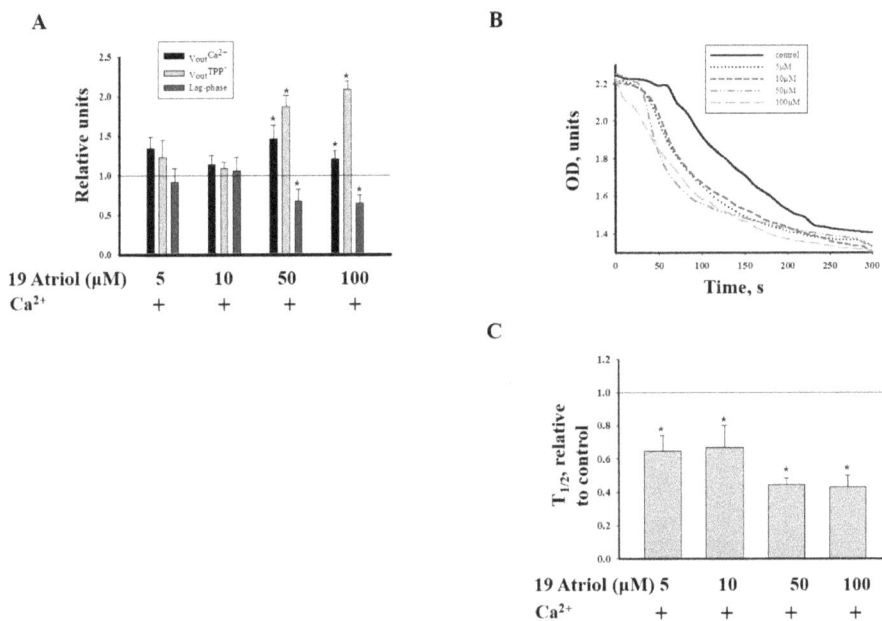

Figure 2. Concentration dependence of 19-Atriol on the parameters of RLM (rat liver mitochondria) functions. (**A**) Quantitative characteristics of parameters of RLM functions; (**B**) The high-amplitude swelling of RLM in the presence of different concentrations of 19-Atriol; (**C**) The average results of the half-time ($T_{1/2}$) to reach the maximal swelling of RLM. Lines in (**A**) and (**C**) show swelling of RLM under control conditions. This parameter was calculated as described in [26]. A value of $p < 0.05$ was considered to be significant (asterisk indicates $p < 0.05$).

To further assess the effects of 19-Atriol on mPTP, we examined the effect of different concentrations of 19-Atriol on Ca^{2+}-induced swelling of RLM (Figure 2B). Swelling was initiated by the addition of 200 µM Ca^{2+} to RLM (at a protein concentration of 0.5 mg/mL) incubated in standard medium (see Materials and Methods). Ca^{2+}-induced swelling of RLM was initiated in the presence of 5, 10, 50, and 100 µM 19-Atriol. Ca^{2+}-induced swelling of RLM (shown in Figure 2B) was accelerated by 5 and 10 µM 19-Atriol, and the effect was more pronounced in the presence of 50 and 100 µM of 19-Atriol. The diagram in Figure 2C shows the average results for the half-time ($T_{1/2}$) required to reach 50% of the maximal swelling of RLM. The parameter, $T_{1/2}$, strongly decreased in the presence of 50 µM and 100 µM of 19-Atriol, resulting in a two-fold reduction of the $T_{1/2}$ parameter when compared to the control (Ca^{2+}-induced swelling). Next, we examined the effect of 19-Atriol in cooperation with CRAC and with PK 11195 on Ca^{2+} accumulation and Ca^{2+}-induced mPTP opening in RLM.

Figure 3 demonstrates the effect of VLNYYVW and PK 11195 on the calcium capacity and calcium retention in 19-Atriol-treated RLM.

Figure 3 shows that in control RLM, pore opening was induced after the addition of the third calcium pulse (150 M Ca^{2+} in sum; curve 1) and following a lag-phase of 200 seconds. 19-Atriol was able to initiate mPTP opening at a lower calcium concentration (following the addition of 100 µM calcium), demonstrating a decrease in the threshold calcium concentration by 30% in the presence of 50 µM 19-Atriol, as compared with the control. The lag-phase was found to be 130 s (curve 2).

Curves 3–6 in Figure 3 show that 100 μM Ca^{2+} was enough to induce mPTP opening in RLM although its ability to retain calcium was rather different. Curve 3 in Figure 3 demonstrates that Ca^{2+} release was stimulated in the presence of the VLNYYVW peptide only and the lag-phase period was decreased by 30% (100 s). The effect of the CRAC peptide on mPTP opening in RLM was found to be the opposite of that seen with RBM. The presence of 100 μM PK 11195 alone caused acceleration of mPTP opening, shortening the lag-phase by 50%, compared to the 19-Atriol effect only (curve 4). Induction of mPTP opening in RLM in the presence of 19-Atriol with PK 11195, as well as in the presence of the combination of 19-Atriol and the VLNYYVW peptide is shown in curves 5 and 6. It was observed, that calcium retention (the lag-time period before pore opening) was decreased in the presence of 19-Atriol with VLNYYVW in comparison with the CRAC peptide alone, while in the presence of 19-Atriol with PK 11195 the lag-phase was prolonged by 30% in comparison with the presence of 19-Atriol alone. Since shortening of the lag-phase means accelerating mPTP opening, the results indicate that the VLNYYVW peptide itself, and combination of the peptide with 19-Atriol, significantly stimulates mPTP opening in RLM.

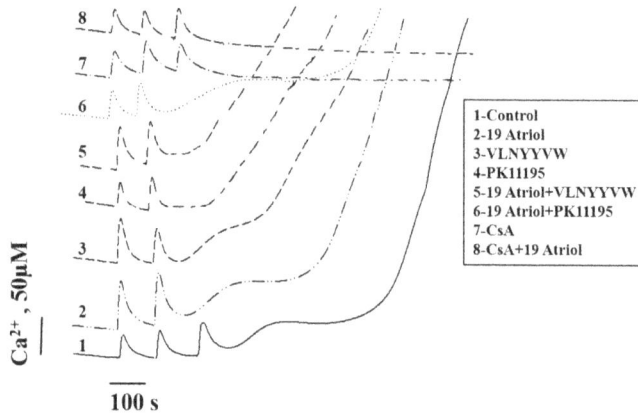

Figure 3. Effect of 19-Atriol, VLNYYVW, and PK 11195, as well as 19-Atriol in combination with VLNYYVW and PK 11195, on Ca^{2+} accumulation and Ca^{2+} release in RLM. Traces 7 and 8 indicate Ca^{2+} accumulation and release in the presence of CsA (an mPTP blocker).

19-Atriol, in combination with 100 μM PK 11195, slows down calcium release and prevents pore opening, demonstrating the opposite effect, particularly since 19-Atriol, or 100 μM PK 11195 alone, facilitates mPTP opening. Because the definition of the classical mPTP is that of a Ca^{2+}-induced and CsA–sensitive pore, we used a specific mPTP opening blocker to confirm the relationship of the effects reported above to mPTP. Curves 7 and 8 demonstrate the blocking of mPTP opening by CsA in control calcium-overloaded RLM, as well as in 19-Atriol-treated RLM. The same results have been obtained in experiments in the presence of CsA, performed under all conditions shown in curves 3–6. Calcium release was not found under all these conditions tested (not shown). The results obtained demonstrate that mPTP opening, induced in RLM by 19-Atriol and by 19-Atriol in combination with the CRAC peptide, were Ca^{2+}- and CsA-sensitive, suggesting the potential participation of the CRAC domain in mPTP function.

2.3. Combined Effect of 19-Atriol and the CRAC Peptide on Ca^{2+}-Induced RLM Swelling

We examined the combined effect of 19-Atriol with the CRAC peptide on Ca^{2+}-induced RLM swelling (as an additional parameter of mPTP function) to obtain additional evidence on the relationships between the effects of the drugs on mPTP. Figure 4A shows that Ca^{2+}-induced RLM swelling is significantly accelerated in the presence of 19-Atriol (trace 2), and it was completely prevented in the presence of CsA (trace 8). A decrease in RLM swelling was found in the presence of the CRAC peptide, VLNYYVW (trace 3), however, RLM swelling in the presence of both 19-Atriol and VLNYYVW was stimulated.

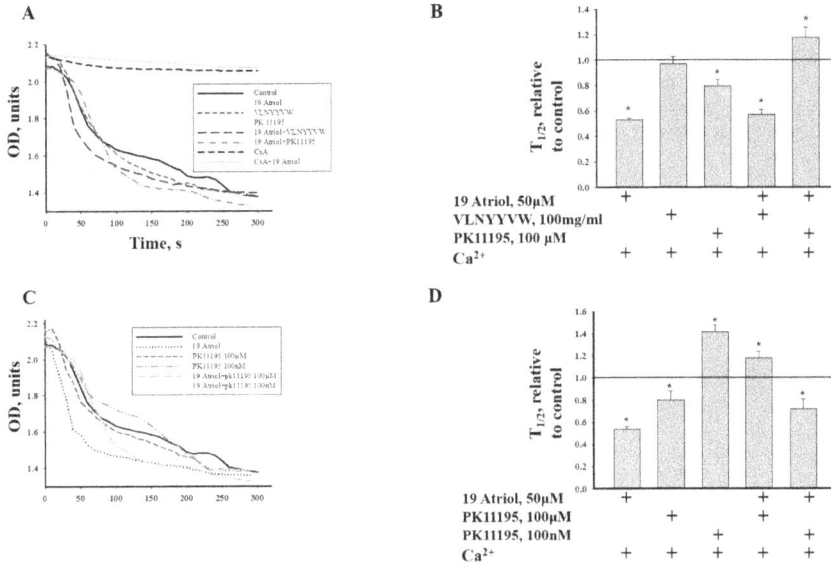

Figure 4. The effect of 19-Atriol, VLNYYVW, PK 11195, and the combined effect of 19-Atriol with VLNYYVW and PK 11195 on RLM swelling. A value of $p < 0.05$ was considered to be significant (asterisk indicates $p < 0.05$). (**A**) Curves of RLM swelling in the presence of 19-Atriol, VLNYYVW, and PK 11195 (100 μM). Trace 1 indicates the absence of swelling without Ca^{2+} addition. Trace 8 demonstrates the absence of swelling in the presence of CsA; (**C**) Curves of RLM swelling in the presence of 19-Atriol, as well as in the presence of 19-Atriol in combination with different concentrations of PK 11195 (100 μM or 100 nM); (**B,D**) Average results of the half-time ($T_{1/2}$) swelling, in comparison with the control swelling of RLM in the presence of Ca^{2+} without any other additions. Lines in (**B**) and (**D**) show swelling of RLM under control conditions.

The diagram in Figure 4B shows the average results for the half-time ($T_{1/2}$) when 50% of the maximal RLM swelling was reached. The given data demonstrate that VLNYYVW does not induce RLM swelling, while the inhibitor, 19-Atriol alone and in combination with the CRAC peptide, is able to enhance Ca^{2+}-induced swelling by two times. CsA prevented RLM swelling, supporting the notion that the effect of 19-Atriol is related to mPTP function. Strong protection of RLM swelling was found in the presence of 19-Atriol and 100 μM PK 11195, when swelling was diminished by 30% in comparison with the control. This was an unexpected result, particularly since 100 μM PK 11195 alone stimulates mPTP opening in both RLM and RBM, while 100 nM PK11195 is able to prevent pore opening. Therefore, we also compared the combined effect of 100 nM PK 11195 and 19-Atriol, and we subsequently compared this with the effect of 100 μM PK 11195 together with 19-Atriol on Ca^{2+}-induced RLM swelling. Figure 4C,D show that 100 nm PK 11195 alone suppresses RLM swelling, while 100 μM PK 11195 initiates it. However, the combination of 19-Atriol with 100 nM PK 11195 or with 100 μM PK 11195 displays the opposite effect: namely, the addition of 19-Atriol and 100 μM PK 11195 into an RLM suspension suppresses Ca^{2+}-induced RLM swelling by 40%, while the presence of 100 nM PK 11195 with 19-Atriol enhanced RLM swelling by 60%. Thus, the CRAC peptide itself does not influence RLM swelling, but the inhibitor of the CRAC domain (19-Atriol), which has a blocking effect on the CRAC peptide, is able to stimulate Ca^{2+}-induced RLM swelling. The results support the hypothesis suggesting that the CRAC domain is involved in mPTP function or regulation. The effect of 100 nM PK 11195 in combination with 19-Atriol indicates that effect of 19-Atriol is linked to TSPO, as only TSPO has nanomolar affinity to PK 11195.

2.4. Effect of the CRAC Peptide (VLNYYVW) on the Release of Apoptotic Factors (Cytochrome c, AIF, and EndoG) from RBM under mPTP Opening

It is known that the induction of mPTP initiates the release of apoptotic factors, such as cytochrome c, apoptosis-inducing factor (AIF), and endonuclease G (EndoG), from mitochondria. Earlier, we observed that in Ca^{2+}-overloaded RBM, the release of apoptotic factors (cytochrome c, AIF, and EndoG) was increased [27]. Therefore, we compare herein the levels of cytochrome c, AIF, and EndoG under a control condition and after mPTP induction. We tested the release of these factors in supernatants of RBM treated with the VLNYYVW peptide. The aliquots of non-synaptic RBM were taken from the chamber under the examined conditions, and the aquilots were subsequently centrifuged. The samples were used for electrophoresis with following Western blot. Identification of cytochrome c, AIF, and Endo G were performed using the monoclonal anti-cytochrome c antibody, the polyclonal anti-AIF antibody and the polyclonal anti-EndoG antibody (see Materials and Methods). Under the control conditions, when mPTP is closed, the release of cytochrome c was not observed in the RBM supernatant. The level of cytochrome c increased insignificantly under the same conditions in the presence of VLNYYVW alone, or in the presence of VLNYYVW together with 100 μM PK 11195 (Figure 5A).

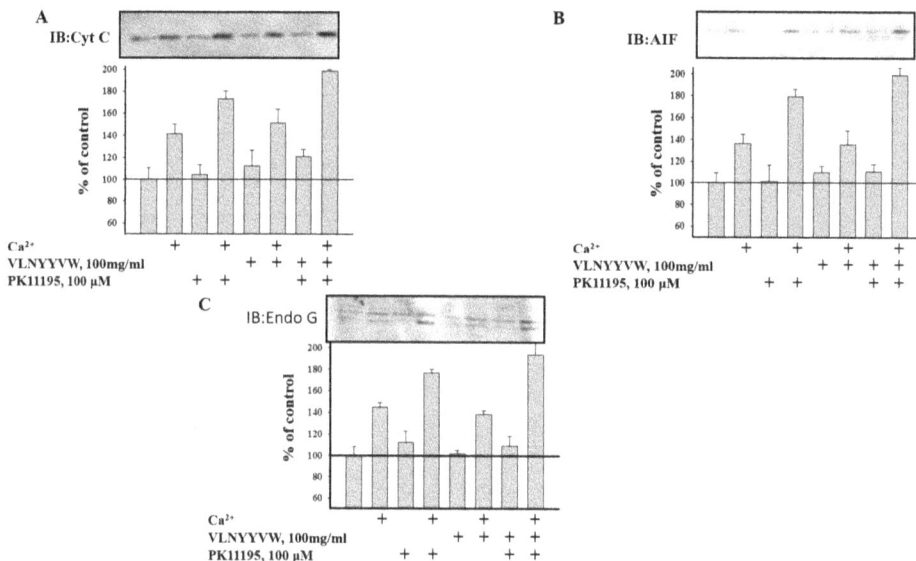

Figure 5. Release of pro-apoptotic factors cytochrome c, AIF, and EndoG from non-synaptic RBM under mPTP opening. (**A**)-Cytochrome c release; (**B**) AIF release; (**C**) Endo G release. Probes were taken in control conditions without added Ca^{2+} and in Ca^{2+}-overloaded mitochondria, as well as in the absence/presence of VLNYYVW and PK 11195, or in the combination of VLNYYVW with PK 111195. Western blots of the supernatant fraction of non-synaptic mitochondria were obtained following centrifugation. Proteins of the supernatant fractions were separated in SDS-PAGE after Western blot.

When the pore is closed, the same effect was found for AIF release, but the level of AIF release was weaker when compared with that of the cytochrome c release (Figure 5B). Furthermore, the release

Int. J. Mol. Sci. **2016**, 17, 2096

of EndoG was found not to be changed in the presence of VLNYYVW. The induction of mPTP leads to cytochrome c release from RBM. The presence of VLNYYVW does not lead to cytochrome c release, whereas 100 μM PK 11195 enhances the release of cytochrome c; however, the strongest release (by nearly two times) was observed in the presence of VLNYYVW together with 100 μM PK 11195 (Figure 5A). AIF (Figure 5B) and EndoG (Figure 5C) release were stimulated by calcium, when mPTP opened. The CRAC peptide was not able to further accelerate mPTP opening, but the CRAC peptide was able to strengthen AIF and EndoG release in cooperation with PK 11195 (Figure 5B,C). In sum, the results give reason to suppose that there is possible modulation of the VLNYYVW effect by PK 11195, indicating that their combined effect might be involved in the initiation of apoptosis, as well as in the release of apoptotic factors.

3. Discussion

The structure and activity of a membrane protein is modulated by both the interaction with its ligands and with the lipid membrane environment in which it resides. Certain drugs may modulate function by inducing physical changes in the membrane environment and by affecting the conformation and function of proteins within a membrane [28,29] that might lead to changing of the membrane permeability. The formation of cholesterol-rich domains might be used as an example. Some proteins are known to sequester to cholesterol-rich domains (raft domains) forming a CRAC motif [3,4,30–32]. In this work, for the first time, we examine the effect of the synthetic peptide-a fragment of the TSPO cholesterol-binding consensus, VLNYYVW, on induction of the mPTP, which is considered as the initial stage of apoptosis.

The initiation of apoptosis is usually preceded by the loss of the mitochondrial membrane potential ($\Delta\Psi_m$), high-amplitude swelling of the mitochondria, and apoptotic factor release, which are found during mPTP opening. Until now, the exact composition of the pore complex has not been established, but it is clear that increased permeability of the inner membrane also depends on the permeability of the outer mitochondrial membrane, mainly on VDAC [14]. The role of the C-terminal end (CRAC motif) of TSPO in mPTP has not been examined until now, therefore, the effect of the CRAC peptide (VLNYYVW) on the induction of mPTP opening in mitochondria was investigated. Here, it was shown that VLNYYVW alone does not induce the opening of mPTP in RBM (Figure 1B). However, VLNYYVW in combination with PK 11195 (100 μM) was able to accelerate pore opening in a CsA-sensitive manner. Taken together, these compounds seem to significantly stimulate calcium release from RBM, thus decreasing the threshold calcium concentration by two times and accelerating pore opening (Figure 1D). These data allowed us to hypothesize that, in the presence of PK 11195, which can modify physical properties of the membrane BLM [33,34], the CRAC peptide might promote the permeability transition. Interestingly, the peptide, VLNYYVW, was found to be able to stimulate mPTP opening in RLM (Figure 3, curve 3). In the presence of the CRAC peptide, the calcium threshold concentration needed for pore opening was decreased by 30%. That could be due to the tissue-dependent lipid composition of phospholipids, forming a mitochondrial phospholipid bilayer.

Recently, a CRAC domain ligand (19-Atriol) was identified; it is capable of inhibiting cholesterol binding at the TSPO CRAC motif [25]. In the present studies, for the first time, we used 19-Atriol to examine whether there is a relationship between 19-Atriol and mPTP. At an effective concentration of 100 μM, 19-Atriol was found to induce mPTP opening, increase calcium release and membrane depolarization, and shorten the time period prior to calcium release. Also, 100 μM of 19-Atriol caused high-amplitude RLM swelling, which was found to be Ca^{2+}-induced and CsA-sensitive, This finding suggests that there is a cause–effect relationship between 19-Atriol action and mPTP function. The experiments performed herein also revealed that the CRAC peptide itself can initiate mPTP opening in RLM. The combination of 19-Atriol with the VLNYYVW peptide caused further reduction of calcium retention in RLM (the lag-time before pore opening) by three-fold, leading to the acceleration of mPTP opening (Figure 3, curves 2, 3, and 5). 19-Atriol was shown to bind to a synthetic CRAC peptide [25], and it also inhibited binding of cholesterol at the CRAC domain in TSPO. Taken together, these results suggest that blocking of the CRAC domain, and thus suppression of cholesterol binding, might stimulate mPTP opening. The effect of 19-Atriol in combination with 100 μM of PK 11195 was found to exert an opposite effect. In this case, there was a prolongation of calcium retention by 50%, slowing down the pore opening. Thus, it is likely that PK 11195 is able to stabilize the CRAC-containing C-terminal end of TSPO, since PK 11195 is able to overcome the 19-Atriol-dependent acceleration of mPTP opening (Figure 3, curves 2 and 6). 19-Atriol-dependent mPTP opening was prevented by CsA, highlighting the relationship between the effect of 19-Atriol and mPTP function.

Additional evidence supporting this hypothesis was obtained while examining the effect of 19-Atriol in cooperation with the CRAC peptide and PK 11195 on high-amplitude Ca^{2+}-induced RLM swelling. 19-Atriol-treated RLM swelling was increased by 50% when compared with the control (Figure 4A,B). Combination of 19-Atriol with the VLNYYVW peptide was able to prevent swelling of RLM, compared to 19-Atriol-dependent swelling. The swelling of RLM, stimulated by 100 μM PK 11195, demonstrates an opposite effect in cooperation with 19-Atriol. Indeed, when together, they were able to prevent the swelling of RLM (Figure 4A,B). Additionally, the 19-Atriol-stimulated Ca^{2+}-induced swelling of RLM was CsA-sensitive. RLM swelling was caused by 19-Atriol in combination with 100 μM PK 11195. Swelling of RLM in the presence of CsA was not observed with all probes tested. These data support the relationship between 19-Atriol-dependent swelling of RLM and mPTP function. Earlier, we showed that 10–100 nM PK 11195 prevents mPTP opening [34]. If the addition of 19-Atriol with 100 μM PK 11195 results in the suppression of Ca^{2+}-induced swelling by 40%, then a combination of 100 nM PK 11195 with 19-Atriol induces RLM swelling by 60%. Given that nanomolar concentrations of PK 11195 bind specifically with TSPO, the effect of 100 nM PK 11195 with 19-Atriol seems to be TSPO-dependent and related to mPTP. It should be noted that this effect is reminiscent of the strong stimulation of mPTP opening in RBM, as a result of the combined effect of 100 nM PK 11195 with G3139 (a blocker of VDAC channels) [35]. VDAC regulates the flux of calcium ions. The VDAC channel shows very low permeability to Ca^{2+} in its normal open state. However, after VDAC closure, the permeability to Ca^{2+} can be increased by 10 times. Ca^{2+} flux into mitochondria through VDAC can lead to calcium accumulation and induction of mPTP opening. G3139 closes the VDAC channel that lead to increase of calcium permeability. Earlier, we reported that cooperation of PK 11195 together with G3139 resulted in VDAC closure and a subsequent acceleration of mPTP opening [35]. We propose that 19-Atriol, probably in complex with 100 nM PK 11195, acts in a manner that is similar to the G3139–PK 11195 combination. Since VDAC binds cholesterol, it could be a target for 19-Atriol action, which might close the VDAC channel. Blocking of cholesterol binding may lead to an increase in membrane permeability, which is sensitive to CsA. Interestingly, cholesterol is able to bind to CsA with low affinity [36]. Kinnunen et al. showed that the penetration of CsA into the lipid seems to be a specific lipid–drug interaction, which could be involved in the change of the conformation and/or orientation of CsA. The conformation/orientation of CsA in the membrane is probably sensitive to cholesterol [36]. Thus, the interaction between mPTP players and cholesterol in the mitochondrial membranes of BLM might be involved in the regulation of mPTP function. As proteins containing CRAC domains (BAX, VDAC, connexin43) were found in lipid rafts, it is possible to suppose that raft structures participate in mPTP formation/regulation. The mitochondrial cholesterol pool seems to be an important factor in the regulation of mitochondrial membrane permeabilization and cell death. It has been reported that TSPO and its ligands are implicated in mitochondrial permeability transition during apoptosis [23,37,38]. mPTP is a main checkpoint of programmed cell death in brain cells, and the induction of mPTP initiates the release of apoptotic factors, such as cytochrome c, AIF, and EndoG from mitochondria. Cytochrome c does not release from calcium-overloaded RBM in the presence of VLNYYVW. However, the CRAC peptide in combination with PK 11195 was able to stimulate cytochrome c release by two-fold as the result of this cooperation (Figure 5A). VLNYYVW alone was not able to stimulate AIF and EndoG release in calcium-overloaded RBM; however, in cooperation with PK 11195, the release of AIF and EndoG (Figure 5B,C) was accelerated. The results obtained provide reason to assume that a possible cooperative interaction between VLNYYVW and PK 11195 exists, which might lead to the release of apoptotic factors, as well as to the initiation of apoptosis. Earlier, it was shown that PK 11195-dependent induction of mPTP opening in RBM was CsA-sensitive. We hypothesize that the VLNYYVW+PK11195-dependent activation of apoptotic factors released from RBM is related to mPTP function.

Taken together, these results demonstrate that the CRAC peptide and its ligand, 19-Atriol, might mediate mPTP function and apoptosis. This effect might be modulated by PK 11195. PK 11195, incorporated into the lipid chains, is able to alter membrane fluidity, much in the same way as cholesterol [39]. Miccoli et al. [40] found an increase in mitochondrial membrane fluidity following exposure to 10 nM PK 11195 for 24 h, which was regulated via binding to TSPO. By taking these data into consideration, we suggest that PK 11195, the CRAC peptide, and 19-Atriol react with the outer membrane. If PK 111195 is able to influence mitochondrial membrane properties [39,40], then it might promote the interaction of VDAC with TSPO, both of which are mPTP regulators. Cholesterol is a critical membrane component. Many of cholesterol's effects are due to the accumulation of cholesterol in the mitochondrial membranes. Targeting mitochondrial cholesterol in a number of pathologies, such as steatohepatitis, neurodegenerative diseases, or cancer, has been reported [41,42]. The ability of the CRAC domain to bind to cholesterol allowed Lecanu et al. to show that another CRAC peptide (VLNYYVWR) had a direct action on plaque, as CRAC enabled the removal of cholesterol from plaque depot sites [43]. The authors found that the administration of the VLNYYVWR human CRAC sequence to guinea pigs fed with a high-cholesterol diet resulted in reduced circulating cholesterol levels. The CRAC peptide thus appears to be a safe prototypical drug for the treatment of dyslipidemia and atherosclerosis. Their results thus indicate that the CRAC peptide might constitute a novel and safe treatment for hypercholesterolemia and atherosclerosis [43]. However, new additional investigations should be undertaken to further understand the multifaceted function of TSPO, as well as CRAC peptides.

4. Materials and Methods

4.1. Animals

Brain and liver mitochondria were isolated from the total brain of male Wistar rats. All experiments were performed in accordance with the "Regulations for Studies with Experimental Animals" (Decree of the Russian Ministry of Health of 12 August 1997, No. 755). The protocol was approved by the Commission on Biological Safety and Ethics of the Institute of Theoretical and Experimental Biophysics, Russian Academy of Science (November 2014, protocol N45).

4.2. Isolation of Rat Liver Mitochondria (RLM)

Mitochondria were isolated from rats using a standard method [44] featuring a homogenization medium containing 210 mM mannitol, 70 mM sucrose, 1 mM EGTA, 0.05% bovine serum albumin (BSA) fraction V, and 10 mM Tris (pH 7.3). The homogenate was centrifuged at $800 \times g$ for 10 min to pellet nuclei and damaged cells. The supernatant, which contained the mitochondria, was centrifuged for 10 min at $9000 \times g$. Sedimented mitochondria were washed twice in medium containing EGTA and BSA for 10 min at $9000 \times g$ and resuspended in the same medium. The protein concentration was determined using the Bradford assay.

4.3. Isolation of Rat Brain Mitochondria (RBM)

The rats were fasted overnight before decapitation and isolation of the mitochondria. The brain was rapidly removed (within 30 s) and placed in an ice-cold solution containing 0.32 M sucrose, 1 mM EDTA, 0.5% BSA (fraction V), and 10 mM Tris-HCl (pH 7.4). All solutions used were ice-cold; all manipulations were carried out at +4 °C. The tissue was homogenized in a glass homogenizer; the ratio of brain tissue to isolation medium was 1:10 (w/v). The homogenate was centrifuged at $2000 \times g$ for 3 min. A mitochondrial pellet was obtained after centrifugation of the $2000 \times g$ supernatant at $13,500 \times g$ for 10 min. RBM was suspended in ice-cold solution containing 0.32 M sucrose, 0.1 mM EDTA, 0.05% BSA (fraction V), and 10 mM Tris-HCl (pH 7.4), and they were washed by centrifugation at $12,500 \times g$ for 10 min. The protein concentrations in the mitochondrial suspensions were 25–30 mg/mL. For the final step, the mitochondria were purified on Percoll gradient [26].

4.4. Evaluation of Mitochondrial Functions

The mitochondrial membrane potential was measured as described earlier [23] by determining the distribution of tetraphenylphosphonium (TPP$^+$) in the incubation medium with a TPP$^+$-selective electrode and Ca^{2+} transport were determined with a Ca^{2+}-sensitive electrode (Nico, Moscow, Russia). Mitochondria (1 mg protein/mL) were incubated in a medium containing 125 mM KCl, 10 mM Tris, and 2 mM K$_2$HPO$_4$, pH 7.4, at 25 °C. During the experiments, succinate (5 mM) was used as a substrate, and rotenone (5 µM) was added to the measuring medium to block Complex I dehydrogenases. The mPTP opening in RLM was induced by a threshold Ca^{2+} concentration (each addition of Ca^{2+} contained 50 µM). The RLM swelling was measured as a change in light scattering in a mitochondrial suspension at 540 nm (A$_{540}$) using a Tecan I-Control infinite 200 spectrophotometer at 25 °C (Tecan Group Ltd., Männedorf, Switzerland). The standard incubation medium for the swelling assay contained 125 mM KCl, 10 mM Tris, 2 mM KH$_2$PO$_4$, 5 mM succinate, and 5 µM rotenone. The concentration of the mitochondrial protein in each well was 0.5 mg of protein per mL. Swelling was initiated by the addition of 200 nmol of Ca^{2+} per mg of protein. The swelling process was characterized by the time needed to reach the half-maximal light-scattering signal (T$_{1/2}$).

4.5. Electrophoresis and Immunoblotting of Mitochondrial Proteins

Samples ranging from 20–30 µg of protein were mixed with Laemmli solubilization solution and boiled for 3 minutes. For immunoblotting, mitochondrial proteins solubilized in Laemmli buffer were separated under denaturing conditions on 12.5% sodium dodecyl sulfate (SDS)-polyacrylamide gel electrophoresis (PAGE) gels and transferred to nitrocellulose membranes. Precision Plus Pre-stained Standards from Bio-Rad Laboratories (Hercules, San-Diego, CA, USA) were used as markers. After overnight blocking, the membrane was incubated with the appropriate primary antibody.

4.6. Cytochrome c, AIF, and EndoG Release

Mitochondria (1 mg of protein per mL) were incubated in the medium containing 125 mM KCl, 10 mM Tris-HCl, 0.4 mM K$_2$HPO$_4$, and 5 µM rotenone, pH of 7.4, at 25 °C. Succinate (5 mM potassium succinate) was used as a mitochondrial respiratory substrate. The threshold Ca^{2+} concentration was reached by adding Ca^{2+} to the mitochondrial suspension in an open chamber with installed selective electrodes. A total of 100 nanomoles Ca^{2+} per mg of protein was sufficient to initiate mPTP opening. The release of cytochrome c, AIF, EndoG, and CNP was detected in the supernatant before and after the induction of mPTP opening. Aliquots (100 µL) were taken from the chamber and centrifuged at $10,000 \times g$ for 6 min. Then, samples of 30 µL of supernatant were taken and diluted in 10 µL of $4 \times$ Laemmli buffer. The pellet fraction was diluted in 90 µL of $1 \times$ Laemmli buffer. The samples were used for electrophoresis following Western blot. The monoclonal anti-cytochrome c antibody was used at 1:2000 dilution (# DLN 06724 from Dianova, Hamburg, Germany), the polyclonal anti-AIF antibody was used at 1:500 dilution (#PC 536 from Calbiochem®; Merck Millipore, Billerica, MA, USA), and the polyclonal anti-EndoG antibody was used at 1:500 dilution (AB3639 from Merck Millipore, Billerica, MA, USA).

4.7. Quantification and Statistical Analysis

Quantification of the band densities from Western blots was carried out using a GS800 calibrated densitometer and Gel Pro software. Films were scanned and band intensities were quantified.

For statistical analysis, relative levels of protein density were expressed as means $\pm$ SD from at least three –four different experiments. Significance was determined by using Student's *t*-test. A value of $p < 0.05$ was considered to be significant (asterisk indicates $p < 0.05$). ANOVA with Bonferoni post-hoc comparison was used with statistical significance at $p < 0.05$ for Figure 5.

5. Conclusions

In the present study we report that the peptide VLNYYVW, designed on the TSPO's CRAC domain, prevents the mPTP from opening and the release of apoptotic factors in RBM. VLNYYVW did not induce swelling in RLM. 19-Atriol, an inhibitor of the cholesterol-binding activity of TSPO's CRAC domain, alone and in combination with the peptide was able to stimulate RLM swelling in a Ca^{2+}- and CsA-sensitive manner. The TSPO specific drug ligand PK 11195 modulates the effects of the CRAC peptide and 19-Atriol on the induction of mPTP opening and apoptotic factor release. Taken together these results suggest that TSPO via its C-terminal CRAC domain participates in mPTP function/regulation and apoptosis initiation; moreover, TSPO drug ligands are pharmacologic regulators of this process.

Acknowledgments: This study was supported by grants from the Russian Foundation for Basic Research (16-04-00927 to Tamara Azarashvili) and (14-04-00625 to Olga Krestinina) the Russian Federation Government (N14.Z50.0028 to John Lemasters), the Canadian Institutes of Health Research (MOP 125983 to Vassilios Papadopoulos), and a Canada Research Chair in Biochemical Pharmacology (Vassilios Papadopoulos).

Author Contributions: Tamara Azarashvili conceived and analyzed the experiments, wrote the paper; Vladimir Akatov and Igor Beletsky—analyzed the data; Olga Krestinina, Yulia Baburina, Irina Odikova—designed and performed the experiments; John Lemasters—contributed reagents/materials/analysis tools; Vassilios Papadopoulos—contributed reagents/materials/analysis tools, wrote the paper.

Conflicts of Interest: The authors declare no conflicts of interest.

Abbreviations

TSPO   Translocator Protein
CRAC   Cholesterol recognition/interaction amino acid consensus
mPTP   Mitochondrial permeability transition pore

References

1. Colbeau, A.; Nachbaur, J.; Vignais, P.M. Enzymic characterization and lipid composition of rat liver subcellular membranes. *Biochim. Biophys. Acta* **1971**, *249*, 462–492. [CrossRef]
2. Li, H.; Yao, Z.; Degenhardt, B.; Teper, G.; Papadopoulos, V. Cholesterol binding at the cholesterol recognition/interaction amino acid consensus (CRAC) of the peripheral-type benzodiazepine receptor and inhibition of steroidogenesis by an HIV TAT-CRAC peptide. *Proc. Natl. Acad. Sci. USA* **2001**, *98*, 1267–1272. [CrossRef] [PubMed]
3. Jamin, N.; Neumann, J.M.; Ostuni, M.A.; Vu, T.K.; Yao, Z.X.; Murail, S.; Robert, J.C.; Giatzakis, C.; Papadopoulos, V.; Lacapere, J.J. Characterization of the cholesterol recognition amino acid consensus sequence of the peripheral-type benzodiazepine receptor. *Mol. Endocrinol.* **2005**, *19*, 588–594. [CrossRef] [PubMed]
4. Li, H.; Papadopoulos, V. Peripheral-type benzodiazepine receptor function in cholesterol transport. Identification of a putative cholesterol recognition/interaction amino acid sequence and consensus pattern. *Endocrinology* **1998**, *139*, 4991–4997. [CrossRef] [PubMed]
5. Murail, S.; Robert, J.C.; Coic, Y.M.; Neumann, J.M.; Ostuni, M.A.; Yao, Z.X.; Papadopoulos, V.; Jamin, N.; Lacapere, J.J. Secondary and tertiary structures of the transmembrane domains of the translocator protein tspo determined by NMR. Stabilization of the TSPO tertiary fold upon ligand binding. *Biochim. Biophys. Acta* **2008**, *1778*, 1375–1381. [CrossRef] [PubMed]
6. Bernassau, J.M.; Reversat, J.L.; Ferrara, P.; Caput, D.; Lefur, G. A 3D model of the peripheral benzodiazepine receptor and its implication in intra mitochondrial cholesterol transport. *J. Mol. Graph.* **1993**, *11*, 236–244. [CrossRef]

7. Lacapere, J.J.; Papadopoulos, V. Peripheral-type benzodiazepine receptor: Structure and function of a cholesterol-binding protein in steroid and bile acid biosynthesis. *Steroids* **2003**, *68*, 569–585. [CrossRef]

8. Giatzakis, C.; Papadopoulos, V. Differential utilization of the promoter of peripheral-type benzodiazepine receptor by steroidogenic versus nonsteroidogenic cell lines and the role of sp1 and sp3 in the regulation of basal activity. *Endocrinology* **2004**, *145*, 1113–1123. [CrossRef] [PubMed]

9. Taketani, S.; Kohno, H.; Furukawa, T.; Tokunaga, R. Involvement of peripheral-type benzodiazepine receptors in the intracellular transport of heme and porphyrins. *J. Biochem.* **1995**, *117*, 875–880. [PubMed]

10. Holt, S.A.; Le Brun, A.P.; Majkrzak, C.F.; McGillivray, D.J.; Heinrich, F.; Losche, M.; Lakey, J.H. An ion-channel-containing model membrane: Structural determination by magnetic contrast neutron reflectometry. *Soft Matter* **2009**, *5*, 2576–2586. [CrossRef] [PubMed]

11. Banati, R.B.; Newcombe, J.; Gunn, R.N.; Cagnin, A.; Turkheimer, F.; Heppner, F.; Price, G.; Wegner, F.; Giovannoni, G.; Miller, D.H.; et al. The peripheral benzodiazepine binding site in the brain in multiple sclerosis: Quantitative in vivo imaging of microglia as a measure of disease activity. *Brain* **2000**, *123*, 2321–2337. [CrossRef] [PubMed]

12. Wilms, H.; Claasen, J.; Rohl, C.; Sievers, J.; Deuschl, G.; Lucius, R. Involvement of benzodiazepine receptors in neuroinflammatory and neurodegenerative diseases: Evidence from activated microglial cells in vitro. *Neurobiol. Dis.* **2003**, *14*, 417–424. [CrossRef] [PubMed]

13. Ryu, J.K.; Choi, H.B.; McLarnon, J.G. Peripheral benzodiazepine receptor ligand pk11195 reduces microglial activation and neuronal death in quinolinic acid-injected rat striatum. *Neurobiol. Dis.* **2005**, *20*, 550–561. [CrossRef] [PubMed]

14. Lemasters, J.J.; Holmuhamedov, E. Voltage-dependent anion channel (VDAC) as mitochondrial governator—Thinking outside the box. *Biochim. Biophys. Acta* **2006**, *1762*, 181–190. [CrossRef] [PubMed]

15. Colombini, M. Vdac: The channel at the interface between mitochondria and the cytosol. *Mol. Cell. Biochem.* **2004**, *256–257*, 107–115. [CrossRef] [PubMed]

16. Campbell, A.M.; Chan, S.H. The voltage dependent anion channel affects mitochondrial cholesterol distribution and function. *Arch. Biochem. Biophys.* **2007**, *466*, 203–210. [CrossRef] [PubMed]

17. Boujrad, N.; Vidic, B.; Papadopoulos, V. Acute action of choriogonadotropin on leydig tumor cells: Changes in the topography of the mitochondrial peripheral-type benzodiazepine receptor. *Endocrinology* **1996**, *137*, 5727–5730. [PubMed]

18. Terrones, O.; Antonsson, B.; Yamaguchi, H.; Wang, H.G.; Liu, J.; Lee, R.M.; Herrmann, A.; Basanez, G. Lipidic pore formation by the concerted action of proapoptotic bax and TBID. *J. Biol. Chem.* **2004**, *279*, 30081–30091. [CrossRef] [PubMed]

19. Martinez-Abundis, E.; Correa, F.; Rodriguez, E.; Soria-Castro, E.; Rodriguez-Zavala, J.S.; Pacheco-Alvarez, D.; Zazueta, C. A CRAC-like motif in bax sequence: Relationship with protein insertion and pore activity in liposomes. *Biochim. Biophys. Acta* **2011**, *1808*, 1888–1895. [CrossRef] [PubMed]

20. Azarashvili, T.; Baburina, Y.; Grachev, D.; Krestinina, O.; Papadopoulos, V.; Lemasters, J.J.; Odinokova, I.; Reiser, G. Carbenoxolone induces permeability transition pore opening in rat mitochondria via the translocator protein tspo and connexin43. *Arch. Biochem. Biophys.* **2014**, *558*, 87–94. [CrossRef] [PubMed]

21. Epand, R.M.; Sayer, B.G.; Epand, R.F. Peptide-induced formation of cholesterol-rich domains. *Biochemistry* **2003**, *42*, 14677–14689. [CrossRef] [PubMed]

22. Miller, C.M.; Brown, A.C.; Mittal, J. Disorder in cholesterol-binding functionality of CRAC peptides: A molecular dynamics study. *J. Phys. Chem. B* **2014**, *118*, 13169–13174. [CrossRef] [PubMed]

23. Azarashvili, T.; Grachev, D.; Krestinina, O.; Evtodienko, Y.; Yurkov, I.; Papadopoulos, V.; Reiser, G. The peripheral-type benzodiazepine receptor is involved in control of Ca^{2+}-induced permeability transition pore opening in rat brain mitochondria. *Cell Calcium* **2007**, *42*, 27–39. [CrossRef] [PubMed]

24. Veenman, L.; Shandalov, Y.; Gavish, M. VDAC activation by the 18 kDa translocator protein (TSPO), implications for apoptosis. *J. Bioenerg. Biomembr.* **2008**, *40*, 199–205. [CrossRef] [PubMed]

25. Midzak, A.; Akula, N.; Lecanu, L.; Papadopoulos, V. Novel androstenetriol interacts with the mitochondrial translocator protein and controls steroidogenesis. *J. Biol. Chem.* **2011**, *286*, 9875–9887. [CrossRef] [PubMed]

26. Azarashvili, T.; Krestinina, O.; Galvita, A.; Grachev, D.; Baburina, Y.; Stricker, R.; Evtodienko, Y.; Reiser, G. Ca^{2+}-dependent permeability transition regulation in rat brain mitochondria by 2',3'-cyclic nucleotides and 2',3'-cyclic nucleotide 3'-phosphodiesterase. *Am. J. Physiol. Cell Physiol.* **2009**, *296*, C1428–C1439. [CrossRef] [PubMed]

27. Baburina, Y.; Azarashvili, T.; Grachev, D.; Krestinina, O.; Galvita, A.; Stricker, R.; Reiser, G. Mitochondrial 2′, 3′-cyclic nucleotide 3′-phosphodiesterase (CNP) interacts with mptp modulators and functional complexes (i-v) coupled with release of apoptotic factors. *Neurochem. Int.* **2015**, *90*, 46–55. [CrossRef] [PubMed]

28. Seydel, J.K.; Coats, E.A.; Cordes, H.P.; Wiese, M. Drug membrane interaction and the importance for drug transport, distribution, accumulation, efficacy and resistance. *Arch. Pharm. (Weinheim)* **1994**, *327*, 601–610. [CrossRef] [PubMed]

29. Lucio, M.; Lima, J.L.; Reis, S. Drug-membrane interactions: Significance for medicinal chemistry. *Curr. Med. Chem.* **2010**, *17*, 1795–1809. [CrossRef] [PubMed]

30. Epand, R.M. Proteins and cholesterol-rich domains. *Biochim. Biophys. Acta* **2008**, *1778*, 1576–1582. [CrossRef] [PubMed]

31. Fantini, J.; Barrantes, F.J. How cholesterol interacts with membrane proteins: An exploration of cholesterol-binding sites including CRAC, CARC, and tilted domains. *Front. Physiol.* **2013**, *4*, 31. [CrossRef] [PubMed]

32. Epand, R.M. Cholesterol and the interaction of proteins with membrane domains. *Prog. Lipid Res.* **2006**, *45*, 279–294. [CrossRef] [PubMed]

33. Hatty, C.R.; Le Brun, A.P.; Lake, V.; Clifton, L.A.; Liu, G.J.; James, M.; Banati, R.B. Investigating the interactions of the 18 kDa translocator protein and its ligand PK11195 in planar lipid bilayers. *Biochim. Biophys. Acta* **2014**, *1838*, 1019–1030. [CrossRef] [PubMed]

34. Jaremko, L.; Jaremko, M.; Giller, K.; Becker, S.; Zweckstetter, M. Structure of the mitochondrial translocator protein in complex with a diagnostic ligand. *Science* **2014**, *343*, 1363–1366. [CrossRef] [PubMed]

35. Azarashvili, T.; Krestinina, O.; Baburina, Y.; Odinokova, I.; Grachev, D.; Papadopoulos, V.; Akatov, V.; Lemasters, J.J.; Reiser, G. Combined effect of G3139 and TSPO ligands on Ca^{2+}-induced permeability transition in rat brain mitochondria. *Arch. Biochem. Biophys.* **2015**, *587*, 70–77. [CrossRef] [PubMed]

36. Soderlund, T.; Lehtonen, J.Y.; Kinnunen, P.K. Interactions of cyclosporin a with phospholipid membranes: Effect of cholesterol. *Mol. Pharmacol.* **1999**, *55*, 32–38. [PubMed]

37. Veenman, L.; Gavish, M. The role of 18 kDa mitochondrial translocator protein (TSPO) in programmed cell death, and effects of steroids on TSPO expression. *Curr. Mol. Med.* **2012**, *12*, 398–412. [CrossRef] [PubMed]

38. Azarashvili, T.; Krestinina, O.; Yurkov, I.; Evtodienko, Y.; Reiser, G. High-affinity peripheral benzodiazepine receptor ligand, PK11195, regulates protein phosphorylation in rat brain mitochondria under control of Ca^{2+}. *J. Neurochem.* **2005**, *94*, 1054–1062. [CrossRef] [PubMed]

39. Drolle, E.; Kucerka, N.; Hoopes, M.I.; Choi, Y.; Katsaras, J.; Karttunen, M.; Leonenko, Z. Effect of melatonin and cholesterol on the structure of DOPC and DPPC membranes. *Biochim. Biophys. Acta* **2013**, *1828*, 2247–2254. [CrossRef] [PubMed]

40. Miccoli, L.; Oudard, S.; Beurdeley-Thomas, A.; Dutrillaux, B.; Poupon, M.F. Effect of 1-(2-chlorophenyl)-n-methyl-n-(1-methylpropyl)-3-isoquinoline carboxamide (PK11195), a specific ligand of the peripheral benzodiazepine receptor, on the lipid fluidity of mitochondria in human glioma cells. *Biochem. Pharmacol.* **1999**, *58*, 715–721. [CrossRef]

41. Montero, J.; Mari, M.; Colell, A.; Morales, A.; Basanez, G.; Garcia-Ruiz, C.; Fernandez-Checa, J.C. Cholesterol and peroxidized cardiolipin in mitochondrial membrane properties, permeabilization and cell death. *Biochim. Biophys. Acta* **2010**, *1797*, 1217–1224. [CrossRef] [PubMed]

42. Ikonen, E. Cellular cholesterol trafficking and compartmentalization. *Nat. Rev. Mol. Cell Biol.* **2008**, *9*, 125–138. [CrossRef] [PubMed]

43. Lecanu, L.; Yao, Z.X.; McCourty, A.; Sidahmed el, K.; Orellana, M.E.; Burnier, M.N.; Papadopoulos, V. Control of hypercholesterolemia and atherosclerosis using the cholesterol recognition/interaction amino acid sequence of the translocator protein TSPO. *Steroids* **2013**, *78*, 137–146. [CrossRef] [PubMed]

44. Allshire, A.; Bernardi, P.; Saris, N.E. Manganese stimulates calcium flux through the mitochondrial uniporter. *Biochim. Biophys. Acta* **1985**, *807*, 202–209. [CrossRef]

International Journal of
Molecular Sciences

MDPI

Article

Classical and Novel TSPO Ligands for the Mitochondrial TSPO Can Modulate Nuclear Gene Expression: Implications for Mitochondrial Retrograde Signaling

Nasra Yasin [1,†], Leo Veenman [1,*,†], Sukhdev Singh [2], Maya Azrad [1], Julia Bode [3], Alex Vainshtein [1], Beatriz Caballero [1,‡], Ilan Marek [2] and Moshe Gavish [1,*]

[1] The Ruth and Bruce Rappaport Faculty of Medicine, Department of Neuroscience, Technion—Israel Institute of Technology, Haifa 32525433, Israel; nasra19@campus.technion.ac.il (N.Y.); mayabz@gmail.com (M.A.); alexanderv21184@gmail.com (A.V.); bea1979c@hotmail.com (B.C.)

[2] Faculty of Chemistry, Department of Organic Chemistry, Technion—Israel Institute of Technology, Haifa 3200003, Israel; sukhdev.giri@gmail.com (S.S.); chilanm@tx.technion.ac.il (I.M.)

[3] Schaller Research Group at the University of Heidelberg and the German Cancer Research Center (DKFZ), Im Neuenheimer Feld 581, Heidelberg 69120, Germany; j.bode@dkfz-heidelberg.de

* Correspondence: veenmanl@tx.technion.ac.il (L.V.); mgavish@tx.technion.ac.il (M.G.); Tel.: +972-4-829-5275 (L.V. & M.G.); +972-4-829-5276 (L.V. & M.G.)

† These authors contributed equally to this work.

‡ New Address: Department of Morphology and Cell Biology, Faculty of Medicine, University of Oviedo, Julian Claveria, 33008 Oviedo (Asturias), Spain.

Academic Editors: Giovanni Natile and Nunzio Denora
Received: 19 February 2017; Accepted: 27 March 2017; Published: 7 April 2017

Abstract: It is known that knockdown of the mitochondrial 18 kDa translocator protein (TSPO) as well as TSPO ligands modulate various functions, including functions related to cancer. To study the ability of TSPO to regulate gene expression regarding such functions, we applied microarray analysis of gene expression to U118MG glioblastoma cells. Within 15 min, the classical TSPO ligand PK 11195 induced changes in expression of immediate early genes and transcription factors. These changes also included gene products that are part of the canonical pathway serving to modulate general gene expression. These changes are in accord with real-time, reverse transcriptase (RT) PCR. At the time points of 15, 30, 45, and 60 min, as well as 3 and 24 h of PK 11195 exposure, the functions associated with the changes in gene expression in these glioblastoma cells covered well known TSPO functions. These functions included cell viability, proliferation, differentiation, adhesion, migration, tumorigenesis, and angiogenesis. This was corroborated microscopically for cell migration, cell accumulation, adhesion, and neuronal differentiation. Changes in gene expression at 24 h of PK 11195 exposure were related to downregulation of tumorigenesis and upregulation of programmed cell death. In the vehicle treated as well as PK 11195 exposed cell cultures, our triple labeling showed intense TSPO labeling in the mitochondria but no TSPO signal in the cell nuclei. Thus, mitochondrial TSPO appears to be part of the mitochondria-to-nucleus signaling pathway for modulation of nuclear gene expression. The novel TSPO ligand 2-Cl-MGV-1 appeared to be very specific regarding modulation of gene expression of immediate early genes and transcription factors.

Keywords: modulation of nuclear gene expression; mitochondrial 18 kDa translocator protein (TSPO); TSPO ligand; PK 11195; 2-Cl-MGV-1; retrograde mitochondrial-nuclear signaling pathway; microscopy; mitochondria; cell nucleus

1. Introduction

Various studies have shown that the 18 kDa translocator protein (TSPO) is involved in numerous functions, including glioblastoma tumorigenicity in its various aspects, e.g., cell proliferation, cell viability, etc. [1–5]. A previous common name for TSPO was peripheral-type benzodiazepine receptor (PBR) [6–8]. TSPO is primarily located in the outer mitochondrial membrane [9–11]. Thus, TSPO is a mitochondrial protein that interacts with ligands to modulate various molecular biological mechanisms. In 2006, the name translocator protein, and its acronym (TSPO), was generally adopted to reflect TSPO's participation in transport of molecules over the outer mitochondrial membrane [7,8,12]. Import into mitochondria by the TSPO includes cholesterol, protoporphyrin IX, and other tetrapyrroles [7,12,13]. Release from the mitochondria regulated by TSPO includes Ca^{2+}, ATP, and cytochrome c [5,14–16]. TSPO is involved in life essential functions, such as respiration, photosynthesis, tetrapyrrole metabolism, and programmed cell death [6,8,17,18]. Extensive studies have shown that TSPO is involved in apoptosis, gene expression, reactive oxygen species (ROS) generation, ATP production, regulation of the mitochondrial membrane potential ($\Delta\Psi m$), heme synthesis, mitochondrial cholesterol transport, neurosteroid synthesis, glutamate metabolism, cell proliferation, cell adhesion, cell migration, and cell differentiation [3,5,7,12,19,20]. In animals and humans, TSPO shows essential roles in inflammatory, immune, and stress responses, as well as in several neuropathological disorders, including neurodegeneration and brain cancer [4,7,12,21]. Based on this background, we postulated that regulation of nuclear gene expression by the TSPO can go a long way to explain TSPO's numerous functional effects [18,22].

It has been shown before that regulation of the TSPO by siRNA or application of TSPO ligands can affect gene expression in human cells in culture and bacteria [19–26]. The question we address with this study is whether TSPO is potentially directly involved in mitochondrial capability to regulate nuclear gene expression. First of all, the question became prominent how TSPO can participate in so many functions as mentioned above [1–3,13]. We hypothesized that the potential specific capability of TSPO to regulate gene expression may be the basis of TSPO's general capability to regulate numerous functions [13,18,20,22]. The other angle is that it is well known that mitochondria are able to regulate nuclear gene expression. This phenomenon is called the "retrograde mitochondrial-nuclear pathway for regulation of nuclear gene expression" [27–31]. It is primarily considered as a cellular response to stressors, as the second leg of a nuclear-mitochondrial-nuclear loop. For brevity, from here on we will refer to it as mitochondria-to-nucleus signaling.

Regarding this pathway, mitochondrial ROS generation and loss of $\Delta\psi m$ have been reported to take part in mitochondria-to-nucleus signaling [27,32]. This leads to changes in levels of ATP and NADH and the release of Ca^{2+}, which activates calcium-sensitive proteins, such as calmodulin and calcineurin. Further downstream this leads to the activation of immediate early genes and transcription factors [28,29,33,34]. This outline of the mitochondria-to-nucleus pathway mirrors well reported observations regarding TSPO's role in mitochondrial functions and mechanisms. It has been demonstrated that TSPO regulates mitochondrial ROS generation, $\Delta\psi m$ transitions, and ATP production [10,11,35,36]. This was originally studied in TSPO's function of initiating programmed cell death [3,5,37]. Furthermore, the TSPO ligand PK 11195 induces mitochondrial permeability transition pore opening, thus allowing release of Ca^{2+} from the mitochondria into the cytosol [14,38,39]. It has also been suggested that TSPO can regulate NADPH oxidase (NOX) activity [40–43]. Thus, by regulating ROS generation, $\Delta\psi m$ transitions, ATP production, Ca^{2+} release, and NADH levels, TSPO may be a key mitochondrial protein providing mitochondria the capability to regulate gene expression in the cell nucleus.

Therefore, we set out to investigate the effects of the classical TSPO ligand PK 11195 on gene expression in a time course related fashion. We studied this in cell culture of the U118MG glioblastoma cell line, as TSPO functions including gene expression have been studied by us for several years in this cell line (e.g., [5,19,20,24,37,44]). Thereby, we would be able to place the new data in a well-established context. The assumption was that the earlier the changes in gene expression occur, the more likely it is that TSPO takes part in direct control of gene expression, as opposed to secondary effects. Furthermore, we wanted to see whether genes related to transcription, post transcription, translation, and post translation would be affected. When changes in gene expression occur before changes in physiological functions, this would indicate that these changes in gene expression are most likely the causes for the changes in physiological functions. Specifically, we consider that an early regulation of transcription factors is indicative of regulation of gene expression. As we are working with glioblastoma cells (U118MG), one interest also includes how such changes in gene expression regulated by TSPO could affect tumorigenicity of these cells in this paradigm. For this purpose, we applied the TSPO ligand PK 11195 (25 μM) for several exposure times (15, 30, and 45 min, and 1, 3, and 24 h). Then, with the aid of microarray, we assayed changes in gene expression. The choice of 25 μM of PK 11195 is based on several previous studies that showed no adverse effects with this concentration, while it counteracts programmed cell death otherwise induced by various agents (e.g., [5,10,13,20,39]). Thus, in this paradigm PK 11195 at 25 μM only presents beneficial functional effects, including promotion of cell viability, and no disruptive lethal effects. In this study, apart from targeting the immediate question of regulation of nuclear gene expression by TSPO ligands, we also address the question whether modulation of gene expression may be associated with well-known TSPO functions. This was another reason to focus on PK 11195 at 25 μM, as by application of this concentration we expected to be better able to connect the present gene expression study with previous functional studies and their results that applied the same PK 11195 concentration. For these previous studies, dose response assays for application to U118MG cells were applied, which showed that 25 μM had optimal effects, without confounding side effects apparent with high doses, while lower doses in the paradigm of programmed cell death showed no effects by themselves (e.g., [5,10,13,20,44]). We also assayed whether functions predicted by the changes in gene expression could indeed also be actually observed in the present study (in particular functions that previously have not been considered part of the TSPO repertoire). Apart from the classical TSPO ligand PK 11195, we also assayed the effects of the more advanced TSPO ligand, 2-Cl-MGV-1 [13,20,45], to enhance the generality of the results.

While TSPO location in mitochondria has received a lot of attention, it is also known that TSPO can be present in perinuclear locations [7,46–48]. To investigate this further, we applied confocal microscopy to triple labeling of nuclei, mitochondria, and TSPO. Location restricted to the mitochondria would be supportive of TSPO's role in mitochondria-to-nucleus signaling for regulation of nuclear gene expression. Intranuclear location could suggest a nuclear TSPO function, different from a mitochondrial TSPO function. Other intracellular locations could suggest a more elaborate involvement of TSPO in regulation of gene expression. As an extra, the study would show whether TSPO function associated with modulation of nuclear gene expression could correlate with known TSPO functions, and might possibly also reveal hitherto unknown TSPO functions.

2. Results

The presentation of the Results is divided into 5 parts: (1) PK 11195 effects on gene expression in general; (2) Potential effects of such gene expression changes on function, uncovered by pathway analysis; (3) Microscopic correlates at cellular and intracellular levels in association with changes in gene expression due to PK 11195 exposure; (4) Actual observations of phenotypic effects of PK 11195 exposure that were predicted by pathway analysis; (5) Comparison with effects of a more advanced TSPO ligand (2-Cl-MGV-1) on gene expression. For pathway analysis, we used the "Regulator Effects" analytic (IPA®) from Qiagen to gain more insights into the affected functions associated with the changes in gene expression, as outlined in the Methods.

2.1. PK 11195 Effects on Gene Expression in General

Exposure of U118MG cells to PK 11195 (25 μM) in our paradigm induced changes in gene expression of 1.5-fold or more of various genes in a time dependent fashion. We took a 1.5-fold change as a cut off to present changes in gene expression due to application of 25 μM of PK 11195 (Table 1).

It was also taken into consideration whether groups of genes contributed to changes in functional pathways as detected by "Regulator Effects" analytic (IPA®). The complete raw data sets are filed with NCBI, names GSE77998 and "GSE85697" referenced in the section "Referenced Data Sets". "GSE77998" indicates gene expression changes after 1, 3, and 24 h of PK 11195 (25 µM) exposure in our paradigm. "GSE85697" indicates gene expression changes after 15, 30, and 45 min of PK 11195 (25 µM) exposure in our paradigm.

Table 1. Gene symbols of up-regulated genes and down-regulated genes with their expression changed due to exposure of U118MG cells to PK 11195 (25 µM) (in comparison to vehicle control in an exposure time related fashion) (cut off of 1.5): for 15, 30, 45, 60 min, 3, and 24 h.

15–45 min of PK 11195 Exposure (25 µM)					
15 min Up-regulated genes ↑		30 min Up-regulated genes ↑		45 min Up-regulated genes ↑	
FOS	↑ +2.16	FOS	↑ +3.96	EGR1	↑ +2.45
DUSP1	↑ +2.08	DUSP1	↑ +3.64	PTGS2	↑ +2.00
EGR1	↑ +1.89	EGR1	↑ +2.64	FOS	↑ +1.84
CYR61	↑ +1.82	CYR61	↑ +2.16	DUSP1	↑ +1.77
OSR1	↑ +1.71	MYC	↑ +2.03	FOSB	↑ +1.72
ANP32AP1	↑ +1.64	CXCL8	↑ +2.02	CYR61	↑ +1.69
FKBP10	↑ +1.54	PTGS2	↑ +2.01	SGK1	↑ +1.67
RNA28S5	↑ +1.54	CTGF	↑ +1.93	CXCL8	↑ +1.62
DYNC1H1	↑ +1.52	NFKBIZ	↑ +1.90	ATF3	↑ +1.58
WNK1	↑ +1.51	SGK1	↑ +1.89	ANP32AP1	↑ +1.58
15 min Down-regulated genes ↓		30 min Down-regulated genes ↓		45 min Down-regulated genes ↓	
ID3	↓ −1.83	LOC100507412	↓ −1.72	MYLIP	↓ −1.84
TUFT1	↓ −1.83	TUFT1	↓ −1.51	LOC441087	↓ −1.50
ID2	↓ −1.68	MYLIP	↓ −1.50		
PTMA	↓ −1.60	KDM3A	↓ −1.50		
ID1	↓ −1.56				
KDM3A	↓ −1.55				
NABP1	↓ −1.52				
RPL21P28	↓ −1.50				
CLK1	↓ −1.50				
1–24 h of PK 11195 Exposure (25 µM)					
1 h Up-regulated genes ↑		3 h Up-regulated genes ↑		24 h Up-regulated genes ↑	
CYR61	↑ +9.3	ID3	↑ +3.38	ASN	↑ +2.47
FOSB	↑ +4.29	ID1	↑ +2.08	SLC7A5	↑ +2.39
EGR2	↑ +3.32	GBP1	↑ +2.04	TRIB3	↑ +2.21
EGR1	↑ +3.2	SMAD6	↑ +1.89	PCK2	↑ +2.18
CTGF	↑ +3.04	ID2	↑ +1.86	LOC729779	↑ +2.12
ID3	↑ +2.62	ATOH8	↑ +1.85	PSAT1	↑ +1.97
TUFT1	↑ +2.49	TXNIP	↑ +1.84	NUPR1	↑ +1.94
ID1	↑ +2.33	NEXN	↑ +1.77	P8	↑ +1.88
SRF	↑ +2.15	SLC3A2	↑ +1.74	DDIT4	↑ +1.84
GBP1	↑ +2.12	ACTG2	↑ +1.71	FAM102A	↑ +1.82
PTGS2	↑ +2.11	FHL2	↑ +1.65	SLC1A5	↑ +1.8
TRIB1	↑ +2.09			DDIT3	↑ +1.8
ERRFI1	↑ +2.04			ATF4	↑ +1.77
ATF3	↑ +1.95			TGIF1	↑ +1.75
KLF6	↑ +1.74			SPRR2D	↑ +1.72
FOS	↑ +1.73			PHGDH	↑ +1.71
DUSP5	↑ +1.71			SLC3A2	↑ +1.69
PTGER4	↑ +1.69			PLEKHF1	↑ +1.67
SGK	↑ +1.69			FOLR3	↑ +1.65
GADD45A	↑ +1.69			BEX2	↑ +1.63
SGK1	↑ +1.69			SLC6A15	↑ +1.63
ID2	↑ +1.68			IGFBP1	↑ +1.56
FILIP1L	↑ +1.67				
1 h Down-regulated genes ↓		3 h Down-regulated genes ↓		24 h Down-regulated genes ↓	
BCL6	↓ −2.14	IL8	↓ −2.47	MYLIP	↓ −2.24
DDIT4	↓ −1.72	MYLIP	↓ −2.36	UHRF1	↓ −1.80
		SOX4	↓ −1.89	IGFBP5	↓ −1.79
				RGS4	↓ −1.72
				PDE5A	↓ −1.69
				TYMS	↓ −1.65
				ERRFI1	↓ −1.63

Int. J. Mol. Sci. **2017**, *18*, 786

As mentioned in the Introduction, we applied 25 μM of PK 11195 as this was an effective concentration determined in previous studies, in particular regarding regulation of programmed cell death, including modulation of mitochondrial membrane potential (ΔΨm), reactive oxygen species (ROS) generation, and cytochrome c release from mitochondria of U118MG cells. Importantly, this concentration also precludes confounding side effects [5]. Thereby, we assumed we would be able to associate our results with a previously established context [2,13]. First of all, we attained a general overview of gene expression changes following application of 25 μM of PK 11195, including magnitude and direction of changes in gene expression (Table 1). This provided a detailed time-course of changes in gene expression (Table 1).

After having determined changes in gene expression due to PK 11195 exposure in a time dependent fashion, we moved on to assay effects on function in general, in particular regarding induction of changes in gene expression and potentially subsequent functional effects (Table 2). Table 2 presents the gene symbols of the genes with significantly changed expression due to PK 11195 treatment, together with the general nature of the gene products (transcription factors, proteins, enzymes, and other products). Thus, as Table 1 provides a detailed time-course of changes in gene expression, Table 2 provides a detailed time course of related functional effects typically associated with the genes in question.

In short, starting at the left hand column of Table 2:

- Within 15 min, 20 genes significantly changed their expression rate in comparison to vehicle control, and 11 of them code for transcription factors, the other 9 code for proteins, enzymes, and other products.

- After 30 min (indicated in the second column) 14 genes changed significantly their expression rate from vehicle control. At this time point the number of genes coding for transcription factors is 6, the other 8 genes code for proteins, enzymes, and other products.

- After 45 min of PK 11195 exposure (indicated in the third column), 12 genes have their expression changed significantly from vehicle control, 5 of them coding for transcription factors, the other 7 genes code for proteins, enzymes, and other products.

- After 1 h (indicated in the fourth column), 25 genes significantly changed their expression rate, and 14 of them code for transcription factors, the other 11 code for proteins, enzymes, and other products.

- After 3 h (indicated in the fifth column), 14 genes changed significantly their expression rate from vehicle control. The number of genes coding for transcription factors at this time point is 6, the other 8 genes code for proteins, enzymes, and other products.

- After 24 h of PK 11195 exposure (indicated in the sixth column), 29 genes have their expression changed significantly from vehicle control. The number of genes coding for transcription factors is 6 after 24 h of PK 11195 exposure, the remaining majority of the genes (23 genes) codes for proteins, enzymes, and other products at this time point. Indeed, the biggest numbers of gene expression changes for proteins, enzymes, and other products is after 24 h.

As TSPO is well known to regulate programmed cell death, in Table 2 is also indicated with asterisks (*) which genes can be associated with programmed cell death. At every time point within the first hour of PK 11195 exposure, more than half of the genes with significantly changed gene expression are associated with programmed cell death. This number is smaller at 3 h and 24 h of PK 11195 exposure.

Table 2. Gene symbols of the genes with changed expression after15, 30, 45, 60 min, 3, and 24 h of PK 11195 exposure (of 25 μM), arranged according to their overall functions as analyzed with "Regulator Effects" analytic (IPA®). These overall functions are listed as: transcription factors, proteins, enzymes, and other products. Asterisks (*) indicate genes associated with programmed cell death. Other products include pseudogenes, ribosomal factors, etc. Further, the function of not all gene products is fully known. This time course presents 2 peaks for enhanced expression of transcription factors (at 15 min and at 1 h). Gene products including proteins, enzymes, and other products peak at 24 h.

Gene Expression Changes in U118MG Glioblastoma Cells after PK 11195 Exposure for different Time Periods					
15 min	30 min	45 min	60 min	3 h	24 h
All Types of Genes Combined					
20 Genes	14 Genes	12 Genes	25 Genes	14 Genes	29 Genes
Transcription Factors					
			ATF3 *		
			BCL6 *		
			DUSP5		
CLK1			EGR1 *		
DUSP1 *			EGR2 *		
EGR1 *			FOS *		
FOS *			FOSB *		
OSR1 *			GADD45A *		
ID1 *	DUSP1 *		ID1 *	ATOH8	ATF4 *
ID2 *	EGR1 *	ATF3 *	ID2 *	ID1 *	DDIT3 *
ID3 *	FOS *	DUSP1 *	ID3 *	ID2 *	NUPR1 *
KDM3A *	KDM3A	EGR1 *	KLF6 *	ID3 *	TGIF1
NABP1	MYC *	FOS *	SRF	SOX4	TRIB3 *
PTMA *	NFKBIZ *	FOSB *	TRIB1 *	SMAD6 *	UHRF1
(11 genes)	(6 genes)	(5 genes)	(14 genes)	(6 genes)	(6 genes)
Proteins, Enzymes, and other Products					
					ASNS *
					BEX2 *
					DDIT4 *
					ERRFI1
					FAM102A
					FOLR3
					IGFBP1 *
					IGFBP5 *
					LOC729779
					MYLIP
					PCK2
					PDE5A
			CTGF *		PHGDH
			CYR61 *		PLEKHF1
ANP32AP1			DDIT4 *		P8
CYR61 *	CTGF *		ERRFI1	ACTG	PSAT1
DYNC1H1 *	CXCL8	ANP32AP1	GBP1	FHL2 *	RGS4
FKBP10	CYR61 *	CXCL8	FILIP1L	GBP1	SLC1A5
MIR22HG	LOC100507412	CYR61 *	PTGER4	IL8	SLC3A2
RNA28S5	MYLIP	LOC441087	PTGS2 *	MYLIP	SLC6A15
RPL21P28	PTGS2 *	MYLIP	SGK	NEXN	SLC7A5
TUFT1	SGK1 *	PTGS2 *	SGK1 *	SLC3A2	SPRR2D
WNK1	TUFT1	SGK1 *	TUFT1	TXNIP *	TYMS
(9 genes)	(8 genes)	(7 genes)	(11 genes)	(8 genes)	(23 genes)

2.2. Implied Specific Functional Effects due to Gene Expression Changes Induced by Various PK 11195 Exposure Times

We used the "Regulator Effects" analytic (IPA®) to gain more insights into the potentially affected functions associated with the changes in gene expression. This showed that the products of at least some of the genes having their expression modulated after 15 min of exposure to PK 11195 are part of the canonical pathway for regulation of gene expression (Figure 1). These genes include *WNK1*, *FOS*, *SGK*, and *MYC*. Thus it appears that regulation or enabling of changes in nuclear gene expression in general indeed is one of TSPO's functions.

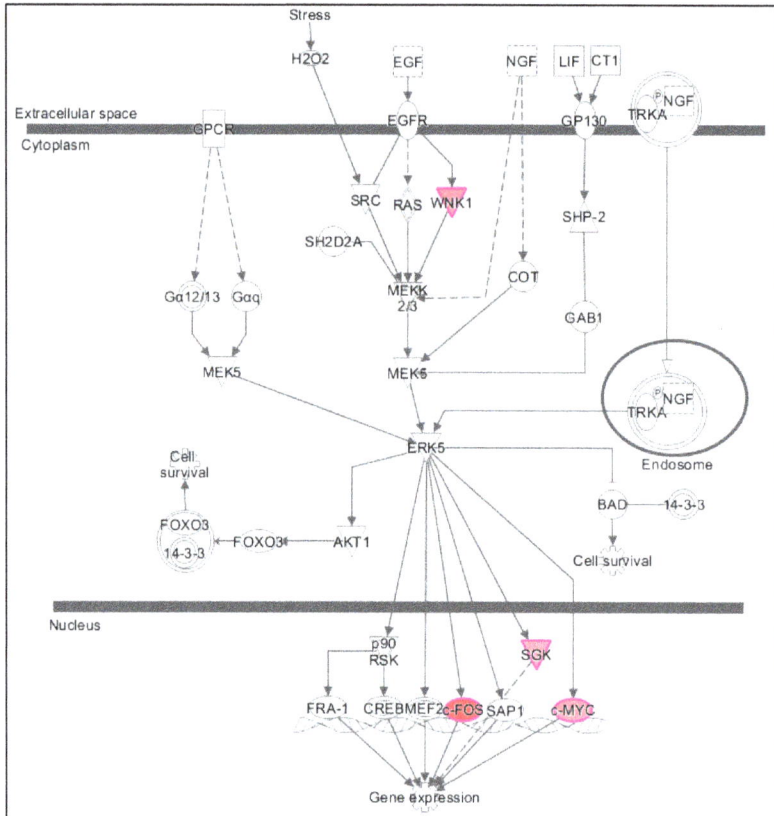

Figure 1. Specific elements of the canonical pathway for modulation of gene expression that are activated after 15 min of exposure of U118MG cells to 25 μM of PK 11195, as uncovered by "Regulator Effects" analytic (IPA®). The gene products of the genes *WNK1*, *FOS*, *SGK*, and *MYC* that are activated by the translocator protein (TSPO) ligand PK 11195 all are part of canonical pathways that converge on the final function of gene expression regulation. Furthermore, the statistically significant enhancements of expressions of the genes *WNK1*, *FOS*, *SGK*, and *MYC* all peak within one hour of exposure to PK 11195 (see Figure 2).

In Figure 2A, we established the time response curves of these 4 genes (*WNK1*, *FOS*, *SGK*, and *MYC*) that are highlighted in the canonical pathway for induction of gene expression changes presented in Figure 1, together with the 3 genes in Table 1 that showed most pronounced changes in gene expression (*FOS*, *DUSP1*, *EGR1*). *FOS*, *DUSP1*, and *EGR1* also are among the immediate early genes and transcription factors presented in Table 2. Note: the immediate early gene *FOS* is also part of the canonical pathway for induction of gene expression changes. Interestingly, the immediate early genes *FOS*, *DUSP1*, and *EGR1* consistently showed enhanced gene expression after 15, 30, and 45 min of exposure to PK 11195 (Figure 2A). Enhanced expression of *WNK1*, one of the initiators of the canonical pathway for gene expression, already peaks at 15 min. *FOS*, *MYC*, and *SGK* that are part of the intranuclear end of the canonical pathway for gene expression peak their expression at 30 min. *DUSP1* also peaks at 30 min. *EGR1* has a "small" peak at 30 min and a "high" peak at 60 min. Real time, reverse transcriptase (RT)-PCR data showed increases in expression of *FOS* and *DUSP1* compared to control by 7.5 and 3.5-fold, respectively, corroborating the microarray data. In some detail, Table 3 gives the numbers of the C_t's of *FOS* and *DUSP1*, showing that RT-PCR also gives significant differences for the expression of these genes at 30 min of exposure as microarray also showed. Figure 2B gives the fold changes provided by the $2^{-\Delta\Delta Ct}$ analysis of the *FOS* and *DUSP1* expression at 30 min measured with RT-PCR. Note regarding Figure 2B, since the samples vehicle and treated groups are not paired we are under obligation to use their averages for calculation of $2^{-\Delta\Delta Ct}$ which thus gives only one data point for each gene presented by the columns in Figure 2C, and consequently no error bars. It is for this reason that we also provide Table 3, which does give the SD's of the C_t's of *FOS* and *DUSP1* expression.

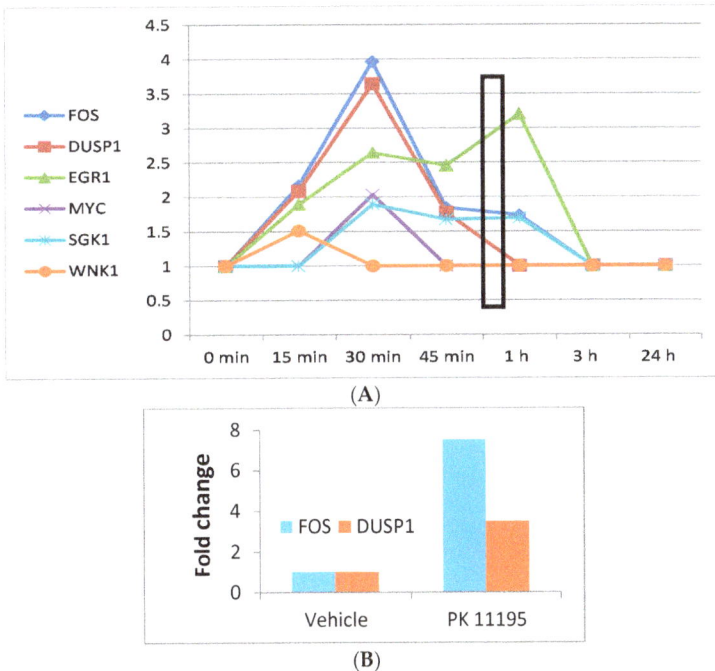

Figure 2. Effects of PK 11195 exposure on several immediate early genes of U118MG cells. (**A**) Time course of gene expression for gene products well known to take part in the initiation of modulation of gene expression assayed with microarray. These genes (*WNK1, FOS, DUSP1, EGR1, MYC, SGK1*) all present a peak of increased expression within half an hour of exposure of U118MG cells to 25 μM of PK 11195. As the data for 15, 30, and 45 min are obtained from one microarray and for 1, 3, and 24 h obtained from another microarray, a bar is placed between 45 and 60 min as a separation between the two. (Each micro array had its own untreated control as detailed in the Methods' section); (**B**) Fold change ($2^{-\Delta\Delta Ct}$) of *FOS* and *DUSP1* expression after exposure to PK 11195 is 7.5 and 3.5, respectively, compared to untreated control (vehicle).

Table 3. Real-time RT-PCR analysis (C_t) of *FOS*, *DUSP1*, and *B2M* expression in glioblastoma cells after 30 min exposure to 25 µM of PK 11195. The presentation of the C_t data as means ± SD. *** $p < 0.001$, ** $p < 0.01$, $n = 2$ (One way ANOVA, posthoc Bonferroni, multiple comparisons) shows the statistical significances of the differences between vehicle (i.e., untreated control) and PK 11195 (i.e., the treated groups) for *FOS* and *DUSP*. Each member of the biological duplicates is the average of 2 technical duplicates. Biological duplicates means cells grown in 2 different wells i.e., truly independent measurements. Technical duplicates means two measurements on the same biological sample (to achieve better accuracy); their average in the end thus is just one measurement.

C_t	Vehicle	PK 11195
B2M	22.10 ± 0.85	21.85 ± 0.21 n.s.
FOS	29.35 ± 0.07	26.20 ± 0.28 ***
DUSP1	24.90 ± 0.42	22.85 ± 0.21 **

The postulated functional effects due to the gene expression changes seen at the different time points as provided by "Regulator Effects" analytic (IPA®) are given in Table 4. As the detailed presentations of the outcomes of these analyses are very elaborate for each of the time points, they are provided in the supplementary materials. In the body of the results, we have given examples of the results of this analysis for 15 min in Figures 1–3, and for 24 h in Figures 4 and 5, as at these time points the outcome of the analysis is relatively straightforward, and insightful. For completeness, the results of the analysis for all the time points 15, 30, 45, and 60 min, and of 3 and 24 h are provided in the supplementary files, and presented summarily here in the Results section, e.g., in Table 4.

The functional effects displayed in Table 4 are provided by the downstream components ("Effects"), given by the application of "Regulator Effects" analytic (in IPA®) to our data. This gives insights into potential phenotypic and physiological effects that may result from the changes in gene expression due to exposure of U118MG cells to 25 µM of PK 11195, including the time-course of the phenotypic and physiological changes. The gene expression data for each time point, provided by the microarray assays, is given in the "Data Sets" of the figures that present the "Regulator Effect". In the body of the text, these figures are given only for 15 min and 24 h of PK 11195 exposure (Figures 3–5). The "Regulators" and "Effects" present what is known regarding upstream regulation and downstream effects of these genes. For the complete overview of the time points of 15, 30, 45 min, 1, 3, and 24 h, providing for the time course of these events, see the appendices. These appendices present "Data Sets", "Regulators", and "Effects" for each time point.

As examples in the body of the Results, the "Effects" are presented in the bottom tiers of Figures 3–5 for PK 11195 exposures of 15 min and 24 h. The "Effects" present the functional, phenotypic, and disease related effects that are known to be under control of the genes with changed expression in our study. The middle tiers of these Figures 3–5 present the genes with changed expression determined in our study (the "Data Sets") as exemplified by "Regulator Effects" analytic (IPA®). The "Regulators" presented in the top tiers of these diagrams of Figures 3–5 are factors (genes, RNAs, proteins) that reportedly can modulate expression of the genes of our "Data Sets". As mentioned, the appendices also show the assayed time points between 15 min and 24 h.

Table 4. Functional effects implied by the gene expression modulated by PK 11195. This was acquired by application of "Regulator Effects" analytic (IPA®). The time points of 15, 30, 45 min, 1, 3, and 24 h applied in this study are given, providing a time-course of functional changes. (The modulated gene expression itself is presented in Tables 1 and 2.) In the appendices more detailed presentations of the outcomes of "Regulator Effects" analytic (IPA®) are given for each time point. Functions that are upregulated are given here in bold red, functions that are down regulated are given in blue.

15 min	30 min	45 min	1 h	3 h	24 h
Gene expression modulation	Binding of protein binding site	Synthesis of DNA	Apoptosis of fibroblast cell lines	Cell death of central nervous system cells	Cell death of fibroblast cell lines
Binding of protein binding site	Synthesis of DNA	Development of neurons	Malignant solid tumor	Apoptosis of fibroblasts	Apoptosis of kidney cell lines
Transactivation of RNA	Differentiation of connective tissue cells	Formation of cellular protrusions	S phase	Apoptosis of myeloid cells	Apoptosis of epithelial cell lines
Endothelial cell development	Development of neurons	Angiogenesis	Cell cycle progression of fibroblast cell lines	Apoptosis of muscle cell lines	Abdominal cancer
Cell viability	Formation of cells	Proliferation	Development of cardiovascular system	Necrosis of epithelial tissue	Digestive system cancer
Accumulation of cells	Microtubule dynamics	Migration	Cell viability	Migration of colon cancer cell lines	Growth of digestive organ tumor
	Chemotaxis of cells	Cell growth	Formation of cellular protrusions	Cell movement of leukocyte cell lines	Growth of malignant tumor
	Cell movement of fibroblast cell lines		Growth of malignant tumor	Migration of smooth muscle cells	Epithelial cancer
	Metastasis of tumor cell lines		Proliferation of tumor cells	Migration of phagocytes	Proliferation of tumor cells
	Abdominal neoplasm		Formation of cells	Chemotaxis	
	Proliferation of lymphocytes		Development of reproductive system	Cell viability	
	Growth of tumor			Development of epithelial tissue	
	Cell viability			Proliferation of leukocyte cell lines	
	Metabolism of carbohydrate			Activation of leukocytes	
	Inflammation of body region			Inflammatory response	
				Accumulation of leukocytes	
				Proliferation of leukemia cell lines	
				Activation of tumor cell lines	

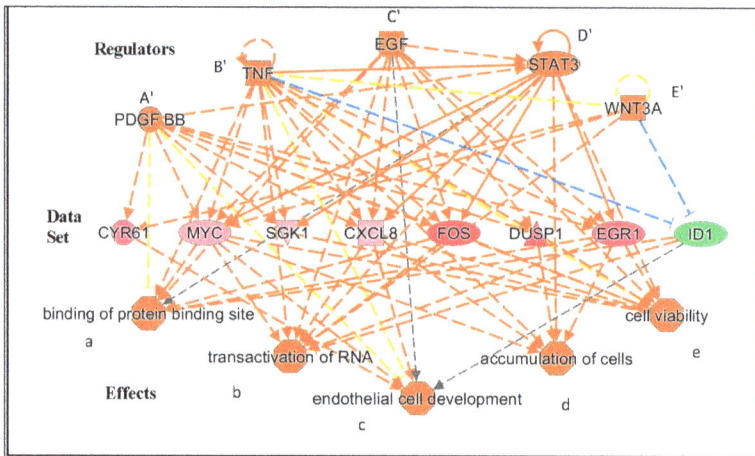

Figure 3. It presents the effects associated with changes in gene expression of 15 min of PK 11195 exposure, as determined with "Regulator Effects" analytic (in IPA®) from Qiagen as indicated in the figure (see also the Methods). The middle tier presents the genes showing significantly changed expression ("Data Set" in the middle tier). The "Effects" presented in the bottom tier indicate that, due to the significant changes in gene expression presented in the "Data Set", the following general functions can be upregulated: (1) Binding of protein binding site; (2) Transactivation of RNA; (3) Endothelial cell development; (4) Accumulation of cells; and (5) Cell viability. The top tier presents the 'Regulators' known to be associated with these "Effects" and the genes of the "Data Set". The arrows indicate the directions of the pathways, from Regulators, to Data Set, to Effect.

Figures 3–5 show that in these paradigms individual "Regulators" can modulate genes' expression from clusters of various sets of genes. A set of genes that can be modulated by one "Regulator" that affects one particular "Effect" can be considered an assembly. Such assemblies can be seen in Figures 4 and 5 (describing gene expression effects for 24 h of PK 11195 application). Interacting and overlapping assemblies can be considered super-assemblies. One such super-assembly is presented in Figure 3 (describing gene expression effects for 15 min of PK 11195 application, which actually is a relatively simple super-assembly). In the appendices, the super-assemblies induced by 15, 30, 45, and 60 min, and 3 and 24 h of PK 11195 exposures are given with more additional details. The assemblies and super-assemblies indicate to us that the functions affected are not based on the changes in expression of just one single gene, but based on several genes. This indicates that the postulations of changes in functions are robust. General terms used for these phenomena providing robustness regarding regulation are "redundancy" and "degeneracy" as exemplified in the Discussion [49–51]. We favored the application of 'Regulator Effects' analytic (IPA®), and the unadulterated presentation of its outcomes, to avoid selective bias potentially introduced by preconceived interpretations of what the effects should be. This way of gene expression analysis is particularly useful for TSPO research as it is well known that TSPO affects a broad variety of functions [1,2,13]. The functional effects are presented below, in a temporal fashion:

- Regarding functional effects at 15 min, in addition to upregulation of gene expression, in general, several functions appeared to be affected. In particular, binding of protein binding site, transactivation of RNA, cell development, cell viability, and accumulation of cells (Table 4, Figure 3). As seen in Figure 3, 15 min of PK 11195 activates a super-assembly including 5 Regulators, 8 genes, and 5 Effects. In Figures 1 and 2 is shown that PK 11195 application for 15 min affects the canonical pathway for regulation of gene expression.

- Functional effects at 30 min appeared to be more varied than at 15 min of PK 11195 exposure. For simplification, these effects can be classified as: binding of protein binding site, cell division and proliferation, cell viability, metabolism, cell differentiation, cell motility, tumorigenicity, and tissue inflammation. These functions are listed in more detail in Table 4. The super-assembly that can be distinguished at 30 min is elaborate and includes 26 Regulators, 19 genes, and 15 Effects. This super-assembly is provided in the supplementary files.

- After 45 min, cell differentiation effects appeared to be the core functional aspect of the gene expression changes, as well as angiogenesis, proliferation, migration, and cell growth (Table 4). The super-assembly seen at 45 min is relatively small, 9 Regulators, 9 genes, and 3 Effects (provided in the supplementary files). Both the 30 and 45 min of PK 11195 exposures caused gene expression changes associated with the canonical pathway for angiogenesis, also provided in the supplementary files.

- After 1 h, the functional effects in general appeared to include: upregulation of cell cycle, proliferation, cell differentiation, cell viability, and tumorigenesis, but also programmed cell death. These functions are listed in more detail in Table 4. The super-assembly seen at 1 h includes 19 Regulators, 29 genes, and 12 Effects (provided in the supplementary files).

- After 3 h, the general effect due to changes in expression of the various genes after exposure of U118MG cells to 25 µM of PK 11195 appears to imply a less tumorigenic phenotype. The majority of the 'Effects' of 3 h of PK 11195 exposure can be classified as down regulation. This down regulation relates to (1) Migration; (2) Inflammatory response; (3) Proliferation, (4) Development, including cell differentiation; (5) Cell viability; and (6) Tumorigenesis. These 'Effects' after 3 h are virtually the opposite from those seen after the shorter PK 11195 exposures. These functions are listed in more detail in Table 3. In contrast, programmed cell death is still upregulated, as was also seen after 1 h of PK 11195 exposure. These functions are listed in more detail in Table 4. The super-assembly activated by 3 h of PK 11195 exposure includes 23 Regulators, 30 genes, and 18 Effects, presented in the supplementary files.

- After 24 h of PK 11195 exposure, pathway analysis with the "Regulator Effects" analytic (IPA®) indicated that due to the significant changes in gene expression only the following general function is down-regulated: tumorigenicity (Figure 4). Several separate pathways were revealed regarding tumorigenicity, each one including just one "Regulator" and a small set of genes forming the "Data Set" (Figure 4A–C). Additional figures in the Supplementary Materials give additional, somewhat more complicated information, i.e., 2 or 3 "Regulators" together modulating 'Data Sets' of a dozen to several dozen genes (Supplementary Materials). These figures also impinge on the general theme of reduced tumorigenicity. Thus, after 24 h of PK 11195 exposure no extensive super-assembly was recognized, but several independent assemblies downregulating several aspects of tumorigenicity and upregulation of programmed cell death (Figures 4 and 5).

Regarding programmed cell death , with all PK 11195 exposure times, "Regulator Effects" analytic (IPA®) of Qiagen indicated significant changes in gene expression associated with programmed cell death (Table 2).

Taking the whole time sequence of responses to PK 11195 into account, it appears that during the first hour upregulation of tumorigenic responses prevails. At 60 min pro-aptotic gene products become relevant and remain so till at least 24 h after PK 11195 exposure. Furthermore, at 3 and 24 h of PK 11195 exposure, down regulation of tumorigenic responses becomes predominant. This response sequence may be interpreted as homeostatic.

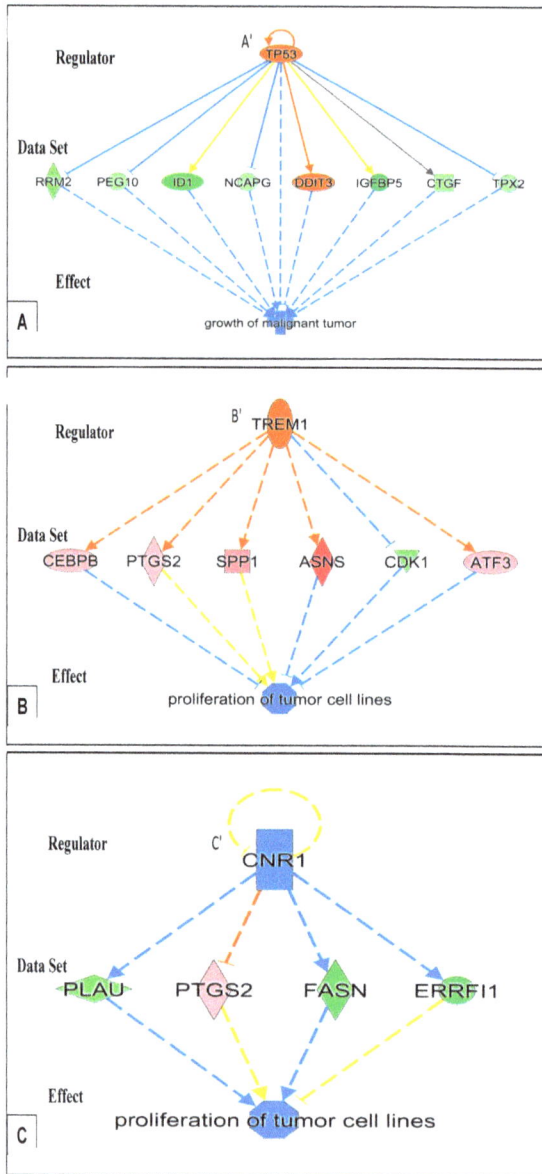

Figure 4. Potential effects on tumorigenicity due to gene expression following 24 h of exposure of U118MG cells to PK 11195 (25 μM). As determined with "Regulator Effects" analytic (IPA®) from Qiagen as indicated in the figure (see also the Methods)., in (**A–C**), individual "Regulators" (given in the upper tiers) are related to specific groups of genes with significantly changed expression ("Data Sets" given in the middle tiers), together with their particular downstream functions ("Effects" in the bottom tiers), namely, suppression of growth of malignant tumor (in (**A**) and suppression of proliferation of tumor cell lines (in (**B**,**C**). Color coding: pink/orange = upregulated, blue/green = down regulated. The configurations in seen in (**A–C**) can be considered assemblies. The arrows indicate the directions of the pathways, from Regulators, to Data Set, to Effect.

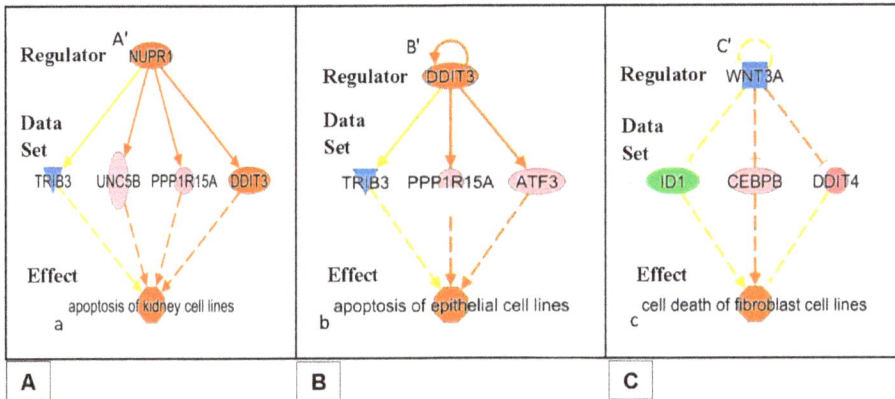

Figure 5. Potential effects on programmed cell death due to gene expression following 24 h of exposure of U118MG cells to PK 11195 (25 µM). As analyzed with Regulator Effects analytic (IPA®) from Qiagen as indicated in the figure (see also the Methods), in A,B,C, individual "Regulators" (given in the upper tiers) are related to specific groups of genes with significantly changed expression ("Data Sets" given in the middle tiers), together with their particular downstream functions ("Effects" in the bottom tiers), namely, stimulation of apoptosis of kidney cell lines (in (**A**), stimulation of apoptosis of epithelial cell lines (in (**B**)), stimulation of cell death of fibroblast cell lines (in (**C**)). Each mentioned separate set can be considered an assembly of pathways running from 1 or few Regulators via a number of genes to affect not more than 1 or 2 specific functions. Color coding: pink/orange = upregulated, blue/green = down regulated. The configurations in seen in (**A–C**) can be considered assemblies. The arrows indicate the directions of the pathways, from Regulators, to Data Set, to Effect.

2.3. Microscopic Correlates at Cellular and Intracellular Levels in Association with Changes in Gene Expression due to PK 11195 Exposure

To see whether the changes in gene expression due to PK 11195 exposures, seen at the different time points in the present study, may be associated with changes in TSPO location, confocal microscopy was applied (Figure 6). A priori, we had expected to see TSPO expression appearing in the cell nucleus after PK 11195 exposure, in association with the apparent modulation of gene expression. Actually, we did not expect to see changes in gene expression before 24 h. Thus, we expected changes in gene expression to come only after 24 h of PK 11195 exposure, potentially even associated with a shift in TSPO location. However, as mentioned above, changes in gene expression determined with microarray were already apparent within 15 min of PK 11195 application. As the gene expression assay, also our microscopic study started with the time point of 15 min, and included 30, and 45 min, 1, 3, and 24 h. A general overview of the results is given in Figure 6. For simplification, Figure 6 is restricted to the time points of 0 min exposure (vehicle control), 30, 45 min, 1, and 24 h. We found that in all instances, i.e., in vehicle control as well as at all of the time points for PK 11195 exposed cells, immunocytological TSPO labeling appears to be restricted to the cytosolic compartment of the cell, including the mitochondria (labeled with Mitotracker red). To emphasize, TSPO labeling does not double-label with DAPI labeling of the cell nuclei, as can be seen clearly in the columns named "Nucleus, TSPO" of Figures 6 and 7.

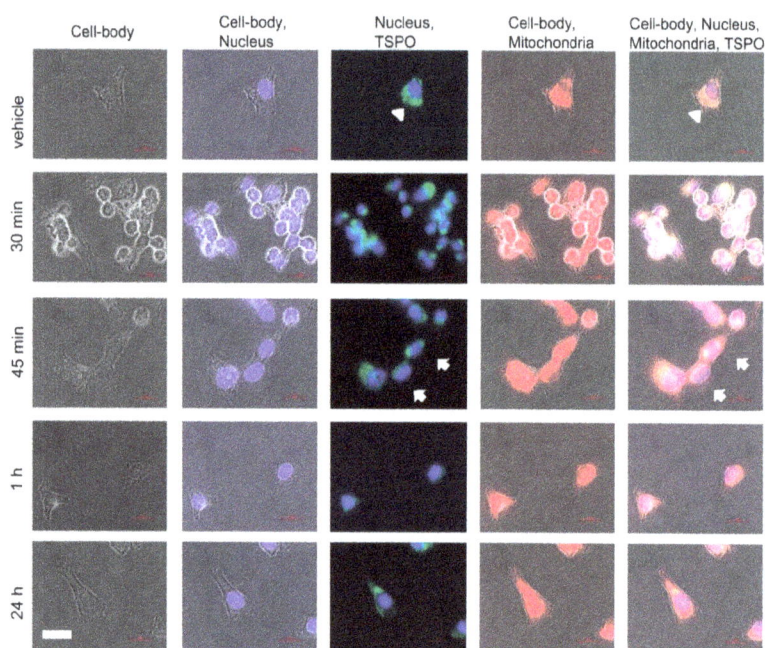

Figure 6. Phase contrast microscopic images of U118MG cells labeled for the cell nucleus (DAPI), the mitochondria (Mitotracker Red), and TSPO (immunocytological labeling). The images are indicative of morphological changes due to exposure to the TSPO ligand PK 11195, in association with intracellular TSPO location. In general, as described in the text, the main points are that TSPO does not appear in the cell nuclei, but are always co-localized with mitochondria. The arrow-heads indicate labeling of mitochondria and TSPO evenly distributed throughout the cytoplasm in the vehicle row. In the 45 min row, the arrows point at mitochondria with co-localized TSPO that are present relatively close to the cell nuclei. A more detailed description of the figure follows here: The rows present exposure times to PK 11195 (25 μM), from top to bottom: vehicle (i.e., no exposure), 30, 45 min, 1, and 24 h. In each row from column to column the same cells are shown. Phase contrast micrographs are presented in the first column. The second column shows cell nuclei (stained blue with the aid of DAPI) within the phase contrast images. The third column shows TSPO immune labeling (Alexa fluor[R] 488, i.e., green) together with the DAPI stained nuclei. Most importantly, the images in the third column show that TSPO labeling is not within the cell nuclei, but in other intracellular areas and cell organelles. In particular, there is no double-labeling of TSPO and DAPI signal. The fourth column shows mitochondrial labeling in the cells outlined by phase contrast. Here it becomes clear that the TSPO labeling in the third column covers intracellular areas occupied by mitochondria. The fifth column shows the results of all signals for the cells in question combined. This presents merged double labeling for mitochondria and TSPO (white to yellow), while phase contrast and nuclear stain is applied for orientation. The DAPI stained nuclei appear purple here due to interference. As a control for this, omission of TSPO labeling i.e., by omitting the primary anti-TSPO anti-body, results in the same nuclear stain (purple) when applying the same microscopic conditions (not shown). As a general remark, omission of TSPO antibody completely prevented the immunocytochemical labeling for TSPO (not shown). The scale bar of 20 μm in the lower left corner of the figure is for all the micrographs in this plate.

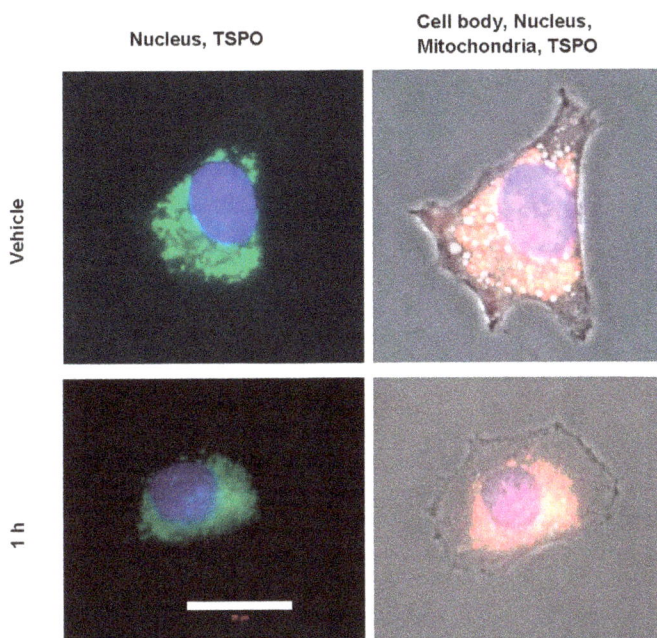

Figure 7. Phase contrast microscopic images of U118MG cells labeled for the intracellular location of TSPO. In the left hand images, cells are only viewed for labeling of the cell nucleus (DAPI) and of TSPO (immunocytological labeling). In the right hand images, cells are viewed for labeling of the cell nucleus (DAPI), TSPO (immunocytological labeling), and the mitochondria (Mitotracker Red), within the cells outlined by phase contrast. While TSPO labeling can appear in the same area of view as the cell nucleus, double labeling of TSPO and nucleus does not occur (left hand images). In the right hand images it can be seen that the TSPO labeling overlaying the nucleus double labels with mitochondria, resulting in a yellow-whitish stain. The main points are that TSPO does not appear in the cell nuclei, but are always co-localized with mitochondria. In the vehicle exposed cells (top two images), labeling of mitochondria and TSPO is evenly distributed throughout the cytoplasm, almost completely filling out the cell body. In the cells exposed for 1 h to PK 11195, labeling of mitochondria and TSPO is condensed toward the cell nucleus, with TSPO—mitochondria double labeling appearing relatively close to the nucleus. Toward the periphery of the cell body a relative broad area is devoid of mitochondrial as well as TSPO labeling. The scale bar in the lower left micrograph of the figure is of 20 μm for all the micrographs in the plate.

We think is noteworthy to mention that we encountered a potentially interesting additional observation. At every time point of PK 11195 exposure, see for example the column of Figure 6 named "Nucleus, TSPO", dense labeling of TSPO in relative proximity to the nucleus can be recognized in at least several of the PK 11195 treated cells (where this is most clear it is marked with white arrows). Such dense TSPO labeling can even become visible as "caps" adjacent to the cell nuclei. This is not apparent at all in cells not treated with PK 11195. The even spread of mitochondria labeled with TSPO throughout the cell body of vehicle control cells is indicated with white arrow heads in the top row of Figure 6. Examples of these phenomena are given as expanded images in Figure 7. For Figure 7 we chose the time points of 0 min and 1 h (i.e., with and without PK 11195 exposure), with the representations of: (1) labeling for TSPO, mitochondria, cell nuclei, and phase contrast for the cell outline to achieve a relatively complete overview of TSPO distribution inside the cell body; (2) labeling for TSPO and cell nuclei indicating the lack of TSPO labeling inside the cell nuclei, in unexposed as well as PK 11195 exposed cells. Thus, after PK 11195 exposure, TSPO labeling, as well as mitochondrial labeling, appears relatively condensed toward the nucleus, leaving more peripheral cellular regions free from this labeling , while in unexposed cells such labeling appears evenly spread over the cytosol. Note, TSPO labeling only double-labels with mitotracker red (labeling for mitochondria), never with DAPI (labeling for cell nuclei), even when TSPO labeling is in the same area of view as the cell nucleus. Figure 6 provides a more differentiated overview, including changes in the morphology of cells during the exposures to PK 11195. See a more detailed description of Figure 6 further below.

The scheme of Figure 8 presents some core observations of cell morphology presented in Figures 6 and 7. In particular, in vehicle control (0 h) TSPO labeling appears to be evenly spread throughout the cytoplasm together with the mitochondria (top of Figure 8). As illustrated in the diagram of Figure 8, commonly, in cells exposed to PK 11195, mitochondria with dense TSPO labeling can be seen relatively close to the cell nuclei (from halfway down Figure 8 to the bottom of Figure 8). Also the basic time dependent cell body shapes (round and polygonal) are diagrammatized in Figure 8. At the top of the diagram of Figure 8, the cells (not exposed to PK 11195) are polygonal. One step lower in the diagram of Figure 8, rounded cells are presented (i.e., cells exposed to PK 11195 for 15 and 30 min). From the time point of 45 min of exposure to PK 11195, the U118MG cells start to disperse from the clusters and revert to their typical polygonal and elongated shapes (bottom of the diagram of Figure 8).

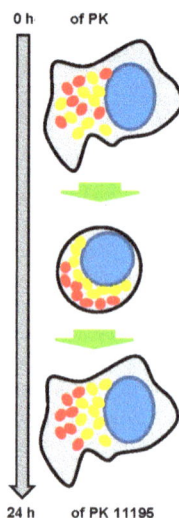

Figure 8. Scheme of observations presented in Figures 6 and 7. In vehicle control cells (i.e., 0 h of PK 11195 exposure) mitochondria with TSPO (yellow) are spread throughout the cytosol (grey). Shortly after PK 11195 exposure the cell bodies contract, become roundish, and mitochondria with TSPO appear to become more condensed toward the cell nuclei (blue). After 24 h of PK 11195 exposure the cell bodies regain their original polygonal shapes. Nonetheless, after these 24 h, mitochondria with TSPO still remain congregated in areas relatively close to the cell nucleus. Mitochondria not displaying TSPO signal are indicated with (red).

These observations are given here in more detail as a guide through the presentation of Figure 6. Going down row by row with the sequential time responses to PK 11195 exposure in Figure 6, it can be seen in the top row, or row 1 (named vehicle i.e., zero time exposure to PK 11195), that TSPO labeling is relatively evenly spread over the intracellular areas covered by mitochondria, without distinct areas of enhanced signal. This, for example, is indicated by arrow heads at the intersections of row 1 with columns 3 and 5. The cells themselves are polygonal in shape and evenly distributed over the coverslip, i.e., without clustering. In row 2 (30 min exposure to PK 11195) the cells have congregated to form clusters and frequently present roundish shapes, while the mitochondria with their TSPO appear to be condensed toward the cell nucleus (presented at the intersections with columns 3 and 5). The same is true for 15 min of PK 11195 exposure (not shown). In row 3 (45 min exposure to TSPO), the cells appear to revert to their original morphology again (as in row 1). Nonetheless, mitochondria double labeled with TSPO remain visible relatively close to the cell nuclei. This phenomenon is indicated by arrows in the images of columns 3 and 5 intersecting with row 3. Note well that this phenomenon of distinct TSPO signal present in mitochondria relatively close to the nuclei is observed at all time points of PK 11195 exposure, i.e., in Figure 6 in all rows from row 2 to row 5, but not in row 1 (which is vehicle control). In rows 4 and 5 (respectively 1 and 24 h of PK 11195 exposure), the cells continue to revert to their original morphology and at 24 h have become indistinguishable from vehicle exposed U118MG cells (row 1). The same is true for 3 h of PK 11195 exposure (not shown). The basic observations are schematized in a relatively simple manner in Figure 8, presenting that TSPO only occurs in the mitochondria in this paradigm, never in the cell nucleus.

The phenomena presented in Figures 6–8, were observed in two separate experiments, where each condition was provided in 3 wells, and where at least 10 cells, or 10 clusters of cells from each well were photographed. At present we cannot say what would be the cause, mechanism, or functional significance for the observed enhanced TSPO signal in the relative proximity of the cell nucleus in PK 11195 exposed cells.

2.4. Actual Phenotypic Effects of PK 11195 Exposure That Were Predicted by Pathway Analysis

Several functional effects implicated by the changes in gene expression described above could indeed be discerned by simple microscopic observations in our present study (Figures 6–8). For example, functional effects predicted by the changes in gene expression at 15 min included accumulation of cells, cell viability, and cell development (Table 3). Microscopically, accumulation of cells after PK 11195 exposures was very conspicuous (Figure 6). This was particularly true at 15 min and 30 min, and could still be discerned at 45 min (Figure 6). Note: accumulation of cells due to PK 11195 exposure has not been described before. Thus, we could microscopically observe the cell accumulation that was predicted by our gene expression assays. At 30 and 45 min, functional effects of gene expression changes also implicated cell migration (Table 3). The congregation and segregation of the cells under microscopic observations matched the cell migration effects implicated by the gene expression analysis (Figure 6). Possibly the microtubule dynamics presented as an 'Effect' at 30 min can be associated with changes in cell body shape observed microscopically (Figures 6 and 7). These cell soma changes may be related to other functions, such as migration, adhesion, cell division, etc., which at present cannot be decided. After 1 h, cell migration no longer did appear to be a function implicated by gene expression changes, and after 3 h such effect appeared to be reversed. Also under the microscope, signs of cell migration could no longer be discerned after one hour of PK 11195 exposure. Regulator Effects analytic (IPA®) of Qiagen also suggested cell viability effects, as uncovered at all time points. Indeed, previous experimental studies have presented PK 11195 at 25 μM to be a cell protective agent (e.g., [5]), although at that time gene expression effects were not considered.

Another major effect of gene expression changes implied at different time points was cell development and differentiation, even neuronal development and differentiation (from the time point of 15 min till the time point of 3 h) (Table 3). The suggestion that exposure to PK 11195 can lead to development and differentiation of neurons was a surprise, as U118MG cells are not considered neuronal. We proceeded to test this potentiality in PC12 cells, which can be considered neuronal progenitor cells. With previous studies we had shown that other, more advanced TSPO ligands can induce neuronal differentiation from these neural progenitor cells, including enhanced tubulin expression [46]. Indeed also in this study, PK 11195, without any additional treatment, just by itself, could very clearly induce neuronal differentiation, including development of neurites and increased expression of tubulin (Figure 9). The observed tubulin expression may be associated with microtubule dynamics also observed with our gene expression assays.

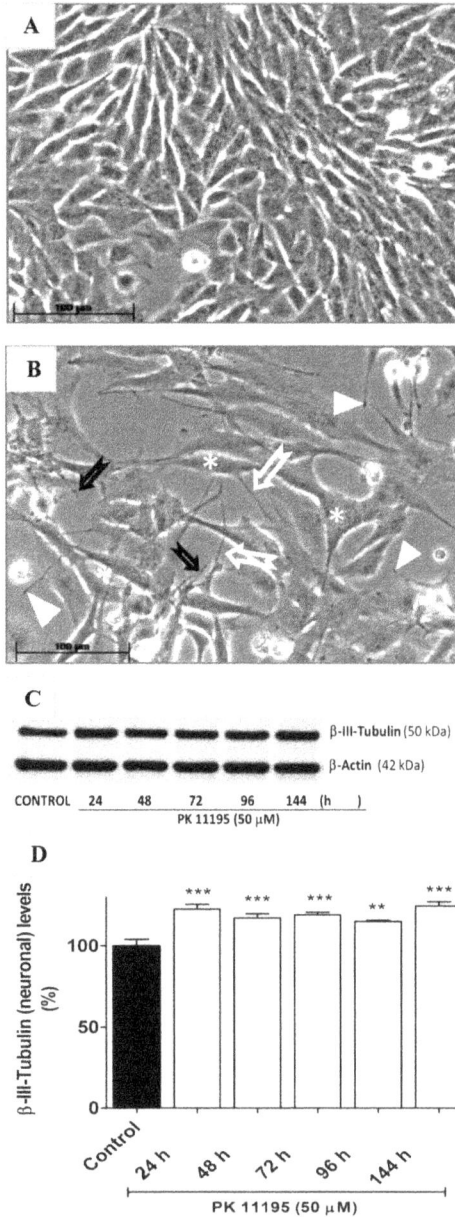

Figure 9. PK 11195 induced neuronal differentiation of rat PC12 cells. (**A**) Undifferentiated vehicle control cells; (**B**) Neuronal differentiation due to PK 11195 (50 μM), including differentiating cells (white asterisks), neurite outgrowth (white arrows), growth cones (white arrowheads), and varicosities (black arrows); (**C**) Representative western blot showing elevated β-III-tubulin expression in rat PC12 cells after exposure times of 24, 48, 72, 96, and 144 h to PK 11195 (50 μM). β-actin is the loading control; (**D**) Bar chart of Means ± SEM (*n* = 3) of the relative densities of the blot bands of β-III-Tubulin labeling in C (arbitrary units as% of control. Control = vehicle treated cells. ** *p* < 0.01, *** *p* < 0.001. (In (**A**,**B**); bars: 100 μm).

As TSPO is well known to affect programmed cell death, we paid special attention to this issue. PK 11195 exposure appeared to have a time dependent effect on gene expression related to programmed cell death, as indicated by asterisks (*) next to the genes in question in Table 2: After 15 min, 11 out of 20 genes; after 30 min, 9 out of 14 genes; after 45 min, 8 out of 12 genes; after 60 min, 17 out of 25 genes; after 3 h, 6 out of 14 genes; and after 24 h, 9 out of 29 genes. Thus, at all time points within 60 min more than half of the genes changing expression are related to programmed cell death. After 3 h it is less than half, and after 24 h less than one third (Table 2).

In general, apart from the functions discovered in the present study (e.g., accumulation of cells, neuron differentiation), the functions revealed by Regulator Effects analytic; IPA® include numerous well established TSPO functions (see Discussion and Introduction).

2.5. Effects of TSPO Ligands Other Than PK 11195 on Gene Expression

We chose PK 11195 to test potential induction of gene expression via TSPO, as PK 11195 is a classical, well studied TSPO ligand, and the results can be in a relatively facile way incorporated in the accumulated body of knowledge. Moreover, effects of 25 μm of PK 11195 are comparable to TSPO knockdown (e.g., [32,39]). However, we also wanted to determine whether more advanced TSPO ligands also present modulation of gene expression. To investigate this we applied different concentrations of a recently developed TSPO ligand 2-Cl-MGV-1 [13,20,46]. For the study of potential 2-Cl-MGV-1 effects, we used the same microarray assay method as for PK 11195. We assumed that this would give an indication of the specificity of the effects of PK 11195 as a TSPO ligand. We applied 1 h of exposure as this was an effective time period as demonstrated for changes in gene expression following PK 11195 application. We applied concentrations at the same height or higher than we applied for PK 11195 (i.e., 25, 50, and 100 μM), because the affinity of 2-Cl-MGV-1 is much lower than of PK 11195 (~240-fold) [46]. Furthermore, at 50 and 100 μM 2-Cl-MGV-1 does not show the confounding side effects that PK 11195 shows at these concentrations [5,20]. In this paradigm 2-Cl-MGV-1 significantly affected a small number of genes primarily related to gene expression regulation. Table 5 shows the genes that are regulated by 2-Cl-MGV-1, namely: *FOS*, *ZFP36*, *DUSP1*, *TUFT1*, and *ID2*, and This Table 5 shows that the concentrations of 50 and 100 μM 2-Cl-MGV-1 have robust effects on the expression of these genes. A cut off of 1.5 as applied for PK 11195, showed changes due to application of 50 and 100 μM of 2-Cl-MGV-1. Importantly, all the genes as seen in Table 5 are among the genes of which their expression is most affected by PK 11195 (see Table 1). Thus, effects on these genes by the TSPO ligands we applied in this paradigm appear to be quite specifically related to control of overall gene expression.

Table 5. Dose dependent changes in gene expression of U118MG cells due to exposure 2-Cl-MGV-1.

50 μM		100 μM	
Up-regulated genes ↑		Up-regulated genes ↑	
FOS	↑ 2.5	FOS	↑ 3.46
ZFP36	↑ 1.68	ZFP36	↑ 1.72
DUSP1	↑ 1.62	DUSP1	↑ 1.71
Down-regulated genes ↓		Down-regulated genes ↓	
TUFT1	↓ 1.64	ID2	↓ 1.68

2.6. General Observations

The TSPO ligand PK 11195, significantly and specifically, within 15 min induces changes in gene expression. These gene expression changes are in accord with actual phenotypic and functional changes. Microscopic observations imply mitochondrial TSPO in these phenomena.

3. Discussion

To answer the question whether nuclear gene expression can be modulated via the mitochondrial TSPO: (i) we first of all studied the effects of the classical TSPO ligand PK 11195 on gene expression in general, in a time dependent fashion. (ii) We then applied pathway analysis to predict potential functional implications. (iii) We also designed microscopic studies to study whether and how TSPO location could be associated with modulation of gene expression i.e., to determine whether (preferred) location of TSPO could change between mitochondria and nucleus. (iv) Microscopy was also applied to study whether functions were actually modulated as predicted by our pathway analysis. (v) Finally, comparisons with effects on gene expression by a more advanced TSPO ligand 2-Cl-MGV-1 were undertaken.

The early effect on gene expression (at 15 min of PK 11195 exposure) is an important indication provided by the present study that TSPO may be an integral part of a pathway for regulation of nuclear gene expression. PK 11195 is a specific TSPO ligand, known to affect numerous TSPO functions, similar to the effects of TSPO knockdown by genetic manipulation [14,35,37,44]. The effects seen on gene expression in the present study within this relatively short time of 15 min precede physiological changes (in particular phenotypic changes) typically caused by TSPO ligands. In general, various cellular functional changes in diverse cell cultures, including U118MG cells, typically are detected only after 24 h of TSPO ligand treatment [5,14,35,37,52]. A detailed time response study by Costa et al. [40], applying the covalent TSPO ligand irDE-MPIGA to GBM cells, showed that it takes more than 90 min for irDE-MPIGA (25 nM) to irreversibly saturate all TSPO binding sites; 3 h after ligand application $\Delta\Psi m$ collapse was observed; 6 h after ligand application externalization of phosphatidylserine was observed and cell viability was reduced. After 24 h, de-novo TSPO synthesis was observed [40]. The present study shows that modulation of gene expression in U118MG cells due to PK 11195 occurs hours before these previously reported physiological and phenotypic changes [5,14,35,37,40,52]. Thus, the changes in gene expression appear to be the cause rather than the effect of such changes. Moreover, Ingenuity pathway analysis (IPA®) also indicated that the first canonical pathway affected in our paradigm, i.e., at 15 min, was restricted to regulation of overall gene expression. Furthermore, our results showed that several key immediate early genes and transcription factor present a pronounced peak in increased expression around 30 min of PK 11195 exposure. These key immediate early genes and transcription factor are *WNK1, SGK, FOS, DUSP1,* and *EGR1*. As an additional test, also required for follow up studies, we also have applied real time RT-PCR according to standard methods to determine changes in expression of the immediate early genes *FOS* and *DUSP1*. With this RT-PCR assay, we found that relative expression of *FOS* and *DUSP1* in samples of PK 11195 treated cells was increased in comparison to vehicle treated cells. In numbers, relative concentration of *FOS* in cells treated for 30 min by PK 11195 was 7.5 compared to vehicle control, and for *DUSP1* this relative concentration was 3.5. This considerable increase in gene expression seen with real time RT-PCR basically corroborates the microarray results of 30 min exposure to PK 11195. We selected *FOS* and *DUSP1* for this purpose, as their peak expression at 30 min with microarray is one of the main indicators that the classical TSPO ligand PK 11195 can modulate nuclear gene expression, including the canonical pathway for regulation of overall gene expression, as exemplified in Figures 1 and 2. Also application of the TSPO ligand 2-Cl-MGV-1 for one hour to U118MG cells was indicative of effects on gene expression, including immediate early genes and transcription factors: *FOS, ZFP36, DUSP1, TUFT1,* and *ID2*. Thus, our experiments indicate that regulation of gene expression by different TSPO ligands appears to be a reproducible phenomenon. Subsequently over the course of 24 h of PK 11195 exposure the effects on gene expression appeared to present a very dynamic process, i.e., apart from the genes directly involved in regulation of gene expression, various genes affecting various functions over this time period were affected. Finally, after 24 h, the gene expression in this glioblastoma cell line of U118MG was regulated such that an anti-tumorigenic effect became most evident, including promotion of programmed cell death. Furthermore, changes in gene expression observed after 3 h appeared to promote effects that counteract effects determined at shorter time periods, thus being suggestive of a homeostatic effect.

Regarding the choice of the concentration of PK 11195 applied, with several previous studies we found that PK 11195 at a concentration of 25 μM could optimally prevent programmed cell death otherwise induced by various agents (ErPC3, glutamate, NO, CoCl$_2$). To emphasize, these effects 25 μM of PK 11195 are similar to the effects of TSPO knockdown. For a review of these studies see [10]. In the Kugler et al. study [5] it was determined that lower concentrations from 1 μM down to 1 nM had no effect at all on programmed cell death induced by ErPC3. Just by itself, PK 11195 concentrations of 25 μM and lower (as low as 1 nM) had no effect at all i.e., appeared neutral. Higher PK 11195 concentrations (in particular higher than 50 μM) presented increasing lethal effects on their own, an effect that is well known [5,8]. Thus, this indicated to us that 25 μM of PK 11195 presented modulation of TSPO function (reminiscent of the effect of TSPO knockdown, as mentioned) and presented the optimal concentration to a priori avoid non-effects of lower concentrations and to avoid confounding lethal effects of higher concentrations in this U118MG cell line. Furthermore, our first study to assay the effects of PK 11195 on gene expression in general showed a time dependent (from 24 to 48 h) effect of 25 μM of PK 11195 on gene expression [24], also reminiscent of the effects of TSPO knockdown on gene expression [19]. With the present study we wanted to apply shorter times of PK 11195 exposure to determine at which time points relevant changes in gene expression occur. We also wanted to optimize the application conditions for our gene expression studies. Therefore, we also took into consideration that full medium is known to be "activating" regarding gene expression [53,54]. Thus, full medium might interfere with the gene expression changes induced by PK 11195 or 2-Cl-MGV-1 and may complicate the issues at hand. This we strived to avoid. In this context, serum free culturing medium is considered to be optimal for gene expression studies as this renders cells quiescent [53,54]. Thus we applied serum free culturing medium. In short, we applied a straightforward paradigm and simple methods to clarify restricted questions.

We firstly found that changes in gene expression occur relatively early after PK 11195 application, i.e., already at 15 min. This is at least two hours before changes in physiological responses that are typically associated with TSPO function [40]. Secondly, we found that within 15 min the classical TSPO ligand PK 11195 induces considerable changes in gene expression associated with the canonical pathway for modulation of gene expression in general. We also found that not only the classical TSPO ligand PK 11195, but also the more advanced TSPO ligand 2-Cl-MGV-1 modulates expression of immediate early genes and transcription factors. These basic approaches indicate that one of TSPO's functions is to modulate nuclear gene expression.

To further gain insights whether TSPO may modulate gene expression in a fairly direct way, we applied microscopic determination of the intracellular location of TSPO. A priori, we expected that after PK 11195 exposure TSPO may be found in the cell nucleus, as a result from various physiological changes. This then would allow for fairly direct control of nuclear gene expression. However, with our double labeling studies for TSPO, mitochondria, and cell nucleus, we always found TSPO to be located in the cytosol, including the mitochondria, and never in the nucleus of U118MG cells, with and without application of PK 11195, and at each time point of PK 11195 application, from 15 min till 24 h. The mitochondrial location of the TSPO suggested to us that the mitochondria-to nucleus signaling pathway is the main venue for regulation of gene expression by TSPO. This well-known mitochondria-to-nucleus communication pathway is conserved from yeast to humans and includes mitochondrial release of Ca^{2+}, ATP, and ROS generation [27–32]. Interestingly, at all time points of PK 11195 exposure, we found that mitochondria intensely labeled for TSPO typically can be found relatively close to the cell nuclei. This never occurred with cells not exposed to TSPO. At 45 min of PK 11195 exposure, this was characterized by "caps" of mitochondrial populations expressing TSPO in the relative vicinity of the cell nuclei. We would like to believe that the relative close presence to the nucleus of mitochondria with TSPO may facilitate the regulation by TSPO of gene expression via mitochondria-to-nucleus signaling. Of course, alternative explanations are possible, such as enhanced energy requirements in particular subcellular regions [26]. In addition, it appears that within 15 min the application of PK 11195 causes the cells to contract to a round shape. This may contribute to the change of location of mitochondria toward the vicinity of the cell nucleus (see Figure 7). Or, the apparent location of the mitochondria is one of the contributing factors for cell shape change and motility. Then, already at 45 min the cells appeared to return from the morphology of clustered roundish cells back to polygonal cells evenly distributed over the culture plate. After 24 h, the cells have returned to their original appearances. Nonetheless, throughout this whole period of PK 11195 exposure, mitochondria with their TSPO remain visible close to the cell nucleus. Thus, we postulate that the morphological changes seen within 1 h of PK 11195 exposure, even starting already 15 min of PK 11195 exposure may contribute to, or at least be associated with, TSPO's ability to induce early changes in gene expression. More studies, including high power light microscopy as well as electron microscopy, are needed to elucidate the microscopic observations of the present study. We are fully aware that a restricted number of other studies also show TSPO in other locations than mitochondria, including cell nuclei [7,46,47].

The mitochondrial location of the TSPO suggested to us that the mitochondria-to-nucleus signaling pathway is the main venue for regulation of gene expression by TSPO. As mentioned, this well-known mitochondria-to-nucleus communication pathway is conserved from yeast to humans and includes mitochondrial release of Ca^{2+}, ATP, and ROS generation [27–32]. The mitochondria-to-nucleus signaling pathway, as outlined in the Introduction, includes mitochondrial ROS generation and loss of $\Delta\psi$m, leading to changes in levels of ATP and NADH, and the release of Ca^{2+}, resulting in the activation of immediate early genes and transcription factors [29–32]. As also mentioned in the Introduction, TSPO regulates mitochondrial ROS generation, $\Delta\psi$m transitions, ATP production, Ca^{2+} release, and NADPH oxidase (NOX) activity [14–16,35–37,41–44]. In studies by others, effects on free radical generation by TSPO ligands have been studied in cultured neural cells, including primary cultures of rat brain astrocytes and neurons as well as cells of the murine BV-2 microglial cell line [55]. In these studies, free radical production was measured at the time points of 2, 30, 60, and 120 min of treatment with the TSPO ligands PK 11195, Ro5-4864, and PPIX (all at 10 nM). In astrocytes, all ligands showed a significant increase in free radical production at 2 min. Thus, ROS generation induced by classical TSPO ligands, synthetic and endogenous, apparently precedes changes in gene expression. As noted, such ROS generation, may be an essential component of the mitochondria-to-nucleus signaling for modulation of nuclear gene expression [26,56,57]. Finally, the present study shows that immediate early genes which are characteristic of the mitochondrial-to-nucleus pathway (e.g., *EGR1*, *FOS*, and *MYC*), are also induced by our application of PK 11195, the classical mitochondrial TSPO ligand. Thus it appears that the well reported primary location of the TSPO in the mitochondria [7,9,11], which was also observed in the present study, as well as TSPO's well-known regulation of specific mitochondrial functions, favors a mitochondria-to-nucleus signaling pathway for TSPO's ability to regulate gene expression. As strong quantitative changes in mitochondrial ROS generation, $\Delta\psi$m transitions, ATP production, Ca^{2+} release, and NADPH oxidase (NOX) activity typically are inductive for programmed cell death, we assume that moderate or small changes may rather be related to gene expression changes. This is a subject wanting for research.

We also wanted to see whether the changes in gene expression match generally known TSPO functions (as for example reviewed in [1,2,8,13]). To study this, we applied "Regulator Effects" analytic provided by Ingenuity (info-ingenuity@qiagen.com). Indeed, the functions derived from the gene expression changes matched with well-known TSPO functions. Briefly, "Regulator Effects" analytic indicates that PK 11195 exposure time-dependently induces functional changes related to: the cell cycle; programmed cell death; proliferation; migration; development, including cell differentiation; cell viability; inflammatory and immune responses; and tumorigenesis. In the present study, our microscopic observations also presented changes related to migration, development, and differentiation, even correlating in a timely fashion with the changes in expression of the relevant genes. Worldwide, careful TSPO research over the last 40 years has shown that PK 11195 and other

TSPO ligands, as well as TSPO knockdown with genetic manipulation, modulate these same functions as seen with the gene expression analysis in this study [1,2,8,10,11,19,36,40,42,44,55,58–60].

Importantly, the pathway analysis showed that the gene expression changes presented interactive assemblies and super-assemblies. In short, in such assemblies, groups of genes provide several gene products for singular functions. In simple terms, the predictions of effects on specific functions appear to be robust. Moreover, the redundancies and degeneracies of pathways, forming the bases of the assemblies and super-assemblies, reinforce the robustness of functional gene expression effects induced by TSPO activity. It is well-known that redundancy and degeneracy in biological systems serve to stabilize them [49–51].

The experiments of the present study also provide data that functional changes predicted by observed changes in gene expression did actually occur. For example, the actual functional effects observed in the present study appear to include: stimulation of gene expression and accumulation of cells (at 15 min), activation of microtubule dynamics and cell motility (at 30 min), promotion of cell migration (at 45 min), then cell motility and accumulation of cells is reversed (at 3 h). Only at 1 and 24 h of PK 11195 expression no gene expression related to the actually observed functional effects was seen.

We were intrigued by the observed intracellular locality changes of TSPO labeling. It appears that because of some until now unknown cause and purpose, mitochondria in the relative vicinity of the cell nuclei enhance TSPO labeling in response to PK 11195 exposure. One alternative may be that mitochondria with TSPO can move from more distant areas in the cytosol to areas neighboring the cell nucleus, resulting in relative dense TSPO signal in such areas. This may potentially implicate that mitochondria not expressing TSPO are not motile in this paradigm. Such a phenomenon is not uncommon. For example in mature neurons, only one-third of axonal mitochondria are motile, the remainder thus being stationary [33]. Stationary mitochondria are considered to serve as local energy sources and buffer intracellular Ca^{2+} [33]. It is known that motility may serve to move mitochondria to the required intracellular locations for various cellular functions, such as proliferation, cell growth, cell cycle, differentiation, information transfer, apoptosis, etc. [31,56]. More studies are needed to determine whether TSPO actually may modulate mitochondrial motility.

Finally, as alluded above, also TSPO ligands other than PK 11195 can regulate gene expression. In the fifth approach of the present study this includes 2-Cl-MGV-1 which we found to modulate immediate early gene expression. As the affinity of 2-Cl-MGV-1 by design is relatively low, the concentrations of 2-Cl-MGV-1 given can be considered the equivalent of a 100 to 400 nM range of PK 11195. This is closer to the dissociation constant of PK 11195 than the 25 μM concentration used for PK 11195. Nota bene: in nature, a high affinity per se does not have to be advantageous. For example, the affinity of CO for hemoglobin is 210 higher than that of O_2. However, CO is lethal, O_2 is life giving. To further illustrate the potential general implications of our findings, other TSPO studies, applying genetic manipulation and ligands other than PK 11195 to target various specific functions, also showed modulations of gene expression [19,20,23,24,26,61–63].

In summary, our study indicates that the TSPO ligand PK 11195 can modulate gene expression in U118MG cells within 15 min. This modulation involves regulation of expression of gene products that are part of the canonical pathway for regulation of gene expression. This gene expression appears to be related to cell viability and tumorigenicity of these U118MG cells. It is likely that such modulation in gene expression occurs via mitochondria-to-nucleus signaling, probably via mechanisms including ΔΨm collapse, ROS generation, Ca^{2+} release, and ATP production (Figure 10). It is well documented that these mechanisms are under the control of mitochondrial TSPO. The modulation of gene expression by the TSPO elucidated in the present study (Figure 10) goes a long way in explaining subsequent changes in cellular and organismal functions due to application of TSPO ligands. Thus, we propose that TSPO's mitochondrial functions include modulation of nuclear gene expression via mitochondrial-nuclear signaling. This presents one way whereby TSPO can control several vital cell functions, which has major implications for the whole organism in health and disease.

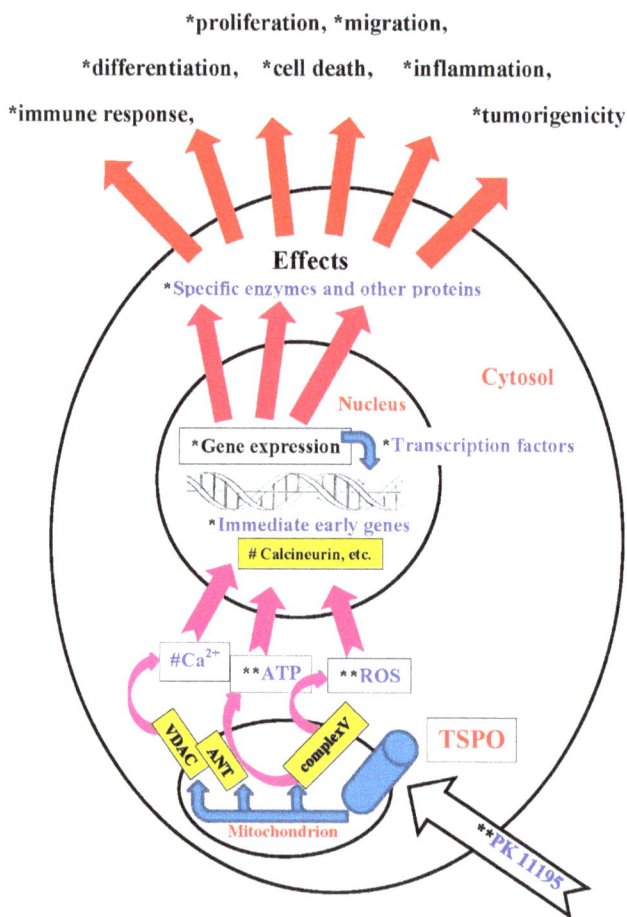

Figure 10. Regulation of gene expression by the TSPO ligand PK 11195. It is well known that the classical TSPO ligand PK 11195 can modulate mitochondrial TSPO functions such as Ca^{2+} release, ATP production, and ROS generation via modulation of mitochondrial proteins such as VDAC, ANT, and complexV (a.k.a. ATP(synth)ase). Ca^{2+} release, ATP production, and ROS are part of the mitochondrial – nuclear signaling pathway for regulation of nuclear gene expression. In this pathway, calcium sensitive proteins such as calcineurin and calmodulinKIV contribute to induction of expression of immediate early genes. The present study shows that PK 11195 exposure at first induces expression of immediate early genes, as well as other transcription factors (calcineurin, and calmodulinKIV are implicated in these effects), followed by changes in gene expression for enzymes and other proteins. Eventual potential functional implications include cell proliferation, cell migration, cell differentiation, cell death, inflammation, immune response, and tumorigenicity. While it is impossible to include all co-factors and context conditions into this diagram, it should be appreciated that by making small and big variations in the research paradigm, other final effects will become apparent regarding changes in function due to changes in gene expression, for example as a consequence of application of other TSPO ligands, TSPO knockdown, TSPO knockout, TSPO gene insertion, full medium, serum free medium, etc. The (*) and the (**) next to the phenomena modulated by PK 11195 indicate results from our present study (*) and from our previous studies (**). # indicates data from studies by others. (We consider it worthwhile for future studies to pay attention to the calcium sensitive proteins in the context of gene expression regulation via TSPO activities).

Caveats and Questions for Future Research

For most of its history, TSPO research has been challenged by oftentimes seemingly contradictory results. In this section here we only want to present a minimal sketch of this problem. For this short expose we provide a minimal number of references. For the interested reader, we refer to the numerous reviews regarding TSPO research, which directly or indirectly approach this enigma (not only referred to in this study, but also in other studies). It is well known that TSPO may contribute to various functions, sometimes appearing without any commonality [64–69]. One may find that for some researchers, gene regulation via TSPO or by TSPO ligands may be an undesired side effect. For others, it may be a desired feature, for example to render cancer cells less malignant, more differentiated, i.e., more like non-cancerous cells, or to induce regenerative responses to organ and tissue damage. These antagonistic expectancies regarding properties of TSPO and its ligands potentially may lead to widely divergent discrepancies in research approaches and interpretations of results. Another confounding aspect does relate to the now well appreciated concept that TSPO has homeostatic functions [20,23,26,38,62,63,70,71]. Assuming this to be true, many of the seemingly contradictory results can be explained as upregulatory and downregulatory characteristics of this potential homeostatic apparatus. An additional point, also redundancy and degeneracy potentially can make interpretations regarding TSPO properties difficult, as mechanisms affected by TSPO manipulations may be compensated for by other systems.It would be fascinating when somebody finds a hard-core, well defined, characteristic and/or mechanism whereby TSPO can exert its homeostatic functions.

A novel finding presented in Figure 8 should be discussed shortly. We observed two populations of mitochondria, one clearly expressing TSPO and one population of mitochondria showing no obvious presence of TSPO. To our knowledge, no quantitative studies have been performed regarding differential presence or no of TSPO in mitochondria of individual cells. Thus, our observation (striking to our minds) of mitochondria apparently with and without TSPO within one and the same cell may be a first. From another perspective, but potentially also interesting in this context, it has been observed that PK 11195 can regulate the conformation (folding or polymerization) of mitochondrial TSPO [72]. One can speculate that this regulation of TSPO protein structure may be one manner whereby immunological labeling of TSPO in mitochondria can be affected. At this point in time, one can only say that more studies are needed to resolve this matter.

Figure 10 in the manuscript simply presents the factual observations of the present study (indicated with * in Figure 10), as well as in previous studies by us (indicated with ** in Figure 10), and by others (indicated with # in Figure 10). Figure 10 also presents the core sequential associations between these observed facts. Table 6 below summarily lists these observations. Table 6 also points out potential caveats and presents questions that can be addressed by future research. Regarding each observation presented in the Table 6, it is obvious that a widely branched tree of assumptions, postulations, questions, and research approaches will emerge in relation to each observation, not to mention each potential caveat and questions associated with these factual observations. It is left to the TSPO enthusiast, and the general interested reader, which approaches based on their experience, expertise, and interests they feel should be pursued. It is obvious that with the very restricted number of TSPO studies available till now that the final word defining TSPO and its related mechanisms definitely has not been said. The authors here attempted to provide an aid for choice of questions for future research. It has been well recognized that TSPO presents a target that can be utilized for treatment of various diseases. In particular its characteristic as a receptor make it per definition a target for the development of ligands designed cure such diseases [22,45,71,73].

Table 6. Caveats and future questions regarding this and other TSPO research.

Observations	Caveats or Questions	Answer or Future Studies
PK 11195 applications affect gene expression and the related functions.	Is this context dependent?	The application of PK 11195 can be further refined (dose, time window, co-factors, etc.).
TSPO ligands other than PK 11195 affect gene expression and related functions.	Is this context dependent?	If desired, the application of ligands other than PK 11195 can be further refined (dose, time window, co-factors, etc.).
TSPO knockdown affects gene expression and related functions.	Is this context dependent?	If desired, the application of TSPO knockdown and knockout can be further refined (transient, stable, time course of effects, co-factors, etc.).
Interaction between PK 11195 and TSPO. PK 11195 can interact with TSPO to finally affect gene expression.	PK 11195 can also bind to other receptors. This can also occur via other receptors.	Knockout all the other receptors binding to PK 11195 (Difficult in practice). TSPO can be knocked down to address this question (This has been done). It can be checked whether other TSPO ligands can affect gene expression (This has been done).
Interactions between TSPO and PK 11195 affect mitochondrial to nuclear signaling, for example via the initiating steps of ROS regeneration, and Ca^{2+} and ATP release.	Will different concentrations of ATP and Ca^{2+} and levels of ROS relate to different changes in gene expression?	ATP and Ca^{2+} levels and ROS generation can be measured after TSPO manipulations. Gene expression can be measured. Protein expression of calcium sensitive proteins can be measured.
Mitochondrial to nuclear signaling apparently induced by TSPO and its ligands implicates various calcium sensitive proteins.	Which calcium sensitive proteins can be activated due to TSPO modulations?	One can measure calcium binding proteins after TSPO manipulations.
Immediate early genes and other transcription factors are activated as a consequence of TSPO knockdown and TSPO ligand applications.	TSPO ligands can act via other receptors than TSPO. TSPO knockdown can possibly be compensated by various cellular mechanisms.	One should not rely on one method to induce effects of TSPO manipulation on gene expression. The time interval between measurements after TSPO manipulation, such as knockdown or knockout, or application of TSPO ligands, or other agents affecting TSPO activity should be as short as possible. In this way, compensatory events are precluded.
Immediate early genes induce changes in expression of other genes.	How is it 'decided' which genes will be modulated? One molecule such as PK 11195 by itself cannot determine which complex of genes will change expression.	TSPO modulation has to be combined with co-factors or different contexts. The assumption is that these additional variations, together with TSPO modulation determine the patterns in changes of gene expression. For example switching from minimal to maximal cell culture medium results in major changes in numbers of gene expression changes.
The final changes in gene expression correlate with functional changes.	Does this always occur?	Modulation of TSPO activity, modulation of gene expression, and modulation of function always have to be considered in association with each other and subjected to combined studies (an approach taken in the present study).
Modulation of gene expression typically has functional effects.	How important is control of TSPO and its ligands of gene expression?	We think it is very important, at least in cell culture and also in animal models we have seen major phenotypic changes.

4. Experimental Section

4.1. Materials

Cell culture materials were obtained from Biological Industries (Beit Ha'emek, Israel). PK 11195 was from PerkinElmer (Hopkinton, MA, USA). Stock solution of PK 11195 was 10^{-2} M in 100% ethanol, kept at -20 °C. RNeasy Mini kit Qiagen and RNase–free DNase Set were from ELDAN (Petach Tikva, Israel). Human HT-12v4.0 Expression BeadChip Kit-24 samples were obtained via Danyel BIOTECH (Rehovot, Israel) from Illumina (San Diego, CA). DAPI Fluoromount-G^R was from Southern Biotech (Birmingham, AL, USA), MitoTracker^R Red CMXRos from Cell Signaling (Rehovot, Israel), anti TSPO primary antibodies from Abcam (Cambridge, UK), and secondary antibody (AffiniPure, Alexa fluor^R 488-conjugated, Donkey Anti Rabbit IgG(H + L)) from Jackson ImmunoResearch (Philadelphia, PA, USA).

Cell Culture

Cells of the human glioblastoma cell line U118MG were allowed to multiply under sterile conditions, at 37 °C, under humidified air with 5% CO_2, in full medium, i.e., MEM-EAGLE supplemented with 10% fetal bovine serum (FBS), 2% glutamine, and 0.05% gentamycin, as described previously [2]. For the gene expression studies and the preferred approach of serum deprived medium was applied i.e., 0.5% FBS instead of 10% FBS, as this renders cells quiescent [53,54].

To study the effects of PK 11195 on neuronal differentiation, rat pheochromocytoma PC12 cells of flat, polygonal and attached phenotype, were generously provided by Ilana Gozes of the Tel Aviv University. This strain of PC12 cells has been studied extensively for differentiation effects of various types of TSPO ligands [20,22]. The PC12 cells were cultured in DMEM supplemented with 8% fetal calf serum (FCS), 8% heat-inactivated horse serum (HS), 1% glutamine and 0.1% penicillin (100,000 units/mL) + streptomycin (100 mg/mL) solution at 37 °C, 5% CO_2, 90% relative humidity.

4.2. Exposure to PK 11195 and 2-Cl-MGV-1

U118MG cells (1.255×10^6) were seeded in Petri dishes (i.e., 28.5×10^3 cells/cm^2) and allowed to proliferate for 3 days in full medium. Experiments for gene expression assayed with microarray typically consisted of three experimental groups and a vehicle control group ($n = 3$ for each group). Serum deprived medium was applied for 24 h, in the experimental groups ending with the inclusion of a choice from various TSPO ligand exposures. PK 11195 (25 μM final concentration) exposures were for 15, 30, 45 min, 1, 3, or 24 h (the vehicle control group is without PK 11195 inclusion). To verify whether other TSPO ligands could elicit gene expression effect, we also applied 25, 50, and 100 μM of 2-Cl-MGV-1 for one hour, according to the same paradigm as applied for PK 11195. 2-Cl-MGV-1 is a recently developed TSPO ligand with glial protective and neuronal differentiation effects [13,20,22,45].

Exposure to PK 11195 and 2-Cl-MGV-1 implies serum deprived medium with 1% alcohol (vehicle), PK 11195 (25 μM final concentration), or 2-Cl-MGV-1 (25, 50, and 100 μM) for the required time periods. Previously we had found that 25 μM of PK 11195 is the optimal concentration for various types of experiments with U118MG cells (e.g., [5,24,35,44]). The cells were collected by trypsinization, washed by centrifugation in phosphate buffered saline (PBS) ($400\times$ g, 5 min), and lysed (RLT lysis buffer provided with the RNeasy Mini Kit diluted with β-mercaptoethanol (1:100)), according to the manufacturer's instructions. Lysates were stored at -70 °C.

4.3. RNA Extraction

Lysates from 3.85×10^6 cells were taken for RNA extraction, according to the instructions of the RNeasy$^®$ Mini Kit. Amount of obtained RNA was determined from a 1.5 μL sample by NanoDrop™ (Thermo, Rockland, DE, USA). For microarray assay, RNA was diluted to 50 ng/μL. RNA quality was verified in 1 μL of each sample [19,24].

4.4. Gene Expression Assay

100 ng of total RNA was amplified into biotinylated cRNA by in vitro transcription using the TargetAmp Nano labeling kit for Illumina BeadChips (Epicentre, an Illumina Company). The biotinylated cRNA was purified, fragmented, and hybridized to a HumanHT-12v 4.0 Expression BeadChip, according to the instructions of the Direct Hybridization assay (Illumina). The hybridized chip was stained with streptavidin-Cy3 (Amersham^{TM}, GE Healthcare, Little Chalfont, UK) and scanned with an Illumina HiScan. The scanned images were imported into GenomeStudio (Illumina) for extraction and quality control. Then, the data were imported into JMP® Genomics software version 6.0 (SAS Institute, Cary, NC, USA), and statistical analysis was applied. The raw data are at the NCBI data bank (File: GSE77998): http://www.ncbi.nlm.nih.gov/geo/query/acc.cgi?acc=GSE77998. And the second file (File: GSE85697) at: http://www.ncbi.nlm.nih.gov/geo/query/acc.cgi?acc=GSE85697.

4.5. Pathway Analysis

The raw data obtained with the microarray chip for gene expression assay were transformed to log2 and filtered out for transcripts with expression at background level. Background signal was provided by probes that contained scrambled DNA that does not correspond to any human DNA sequence. Transcripts were also filtered out when they only showed differences of <5% between the samples from the different time points (vehicle control = 0 h, and PK 11195 treatments for 15, 30, 45 min, 1, 3, and 24 h). Basic gene expression data analysis was performed using JMP® Genomics version 6.0 (SAS, Cary, NC, USA). Principal component analysis and variance component analysis were performed and followed by one-way analysis of variance (ANOVA). For selection of individual genes with substantially changed expression, we applied a cut off of a minimal difference of 2-fold as well as an adjusted *p*-value of ≤0.05 for multiple comparisons [74]. Furthermore, Ingenuity Pathway Analysis (IPA®) software of QIAGEN [75] was applied. In particular, the "Regulator Effects" analytic (IPA®) was applied to obtain insights in the known downstream "Effects" (functional, phenotypic, disease related) related to genes with significant changes in expression ("Data Set"). Thus, the Data Set comprises genes with changed expression in our study. This software also shows the upstream "Regulators" (genes, RNAs, or proteins) that are known to be able to affect the genes of the "Data Set" in relation to their known functional, phenotypic, and disease related effects.

4.6. Real-Time RT-PCR

U118MG cells (1.255×10^6) were seeded in Petri dishes (i.e., 28.5×10^3 cells/cm^2) for treatments, and allowed to proliferate for 3 days in full medium, just as for the microarray assay described above. Then, as described above, serum deprived medium was applied for 24 h, prior to PK 11195 exposures. We prepared for real-time reverse transcriptase (RT)-PCR according to procedures described previously [19,24]. Two biological duplicates were applied, each sample averaged from technical duplicates. Amount of obtained RNA was determined from a 1.5 µL sample by NanoDrop^{TM} (Thermo, Rockland, DE, USA), as described above. RNA (1 µg) was reverse transcribed using the Verso^{TM} cDNA Synthesis Kit (Thermo Fisher Scientific, Waltham, MA, USA). The resulting cDNA was used as a template for TaqMan qPCR, using *DUSP*, *FOS*, and *B2M* specific primers and probes (Primer Design), and ABsolute^{TM} Blue qPCR ROX Mix (Thermo Fisher Scientific, Waltham, MA, USA). *B2M* was used as a normalizing gene. qPCR results were analyzed using Rotor-Gene Q 2.3.1.49 (QIAGEN) software.

4.7. Microscopic Studies

For triple labeling studies, cells were seeded in 6 well plates (each well holding a square microscopic cover glass) at the exact same cell density and growing conditions as applied for mRNA extraction i.e., 133×10^3 cells per well. Stock solution of PK 11195 was diluted (to 25×10^{-4} M) in 100% ethanol. MitoTracker^R Red was dissolved in 100% DMSO (1 µM). For experiments, MitoTracker^R Red solution was diluted to a concentration of 100 nM in serum deprived medium (DMSO final concentration 0.01%). The treatments were 25 µM final concentration of PK 11195. Final concentration of the vehicle was 1% of ethanol.

The time periods of PK 11195 exposure for the microscopic studies were 0, 15, 30, and 45 min, and at 1, 3, and 24 h, ending together with the application of serum deprived medium for 24 h as described above. At the last hour of this 24 h period, MitoTracker[R] Red of 100 nM was applied. Then the cells were fixated with 500 µL of 4% paraformaldehyde per well (2 × 20 min), and washed by PBS (2 × 5 min), followed by 0.5% Triton in PBS-500 µL per well for 10 min. Then blocking solution (5% bovine serum albumin (BSA) in PBS with 0.1% Triton) was applied for 1 h, followed by primary antibody against TSPO, at a final concentration of 1:100 in blocking solution TSPO (500 µL for each well), overnight at 4 °C. In one set, the primary antibody against TSPO was omitted. Next day, the cover glasses with the cells in the wells were washed with PBS (with 0.1% Triton, 2 × 5 min), and then the secondary antibody in blocking solution (1:100) for TSPO labeling was applied for 1 h at room temperature. Then the cover glasses with the cells were washed with PBS (2 × 5 min), and DDW (2 × 5 min), and coverslipped onto microscopic slides with DAPI Fluoromount-G[R] and dried. Then these cells prepared for nuclear staining with DAPI, mitochondrial labeling with MitoTracker[R], and TSPO immunofluorescent labeling with Alexa fluor[R] 488, were observed with the aid of an Axio Observer z1 inverted microscope (Zeiss, Oberkochen, Germany), applying ZEN (Zeiss) for capture and image analysis.

For microscopic studies of neuronal differentiation induce by PK 11195, exposure for 48 h to 50 µM of PK 11195 was applied. Changes in cell morphology of differentiating PC12 cells include neurite sprouting, which is the hallmark of neuronal differentiation [20,22]. We used a Zeiss Axio observer inverted microscope, a Colibri led illumination light source, and a high resolution B/W CCD camera (Zeiss HS). Cell images were captured with the aid of Zeiss Axiovision 4.8 software for data acquisition.

4.8. Western Blot Analysis of Tubulin Expression in Relation to Neuronal Differentiation

Protein levels of cell pellets were measured according to Bradford [76] using bovine serum albumin as a standard. From PC12 cells treated with PK 11195 as described above, collected samples of with equal amounts of protein (20 µg protein/lane) were prepared in 1× sample buffer [0.125 M 2-amino-2-(hydroxymethyl) propane-1,3-diol (Tris)-HCl, pH 6.8, glycerol (20% v/v), SDS (sodium dodecyl sulfate) (0.1% w/v), 0.14 M β-mercaptoethanol, and bromophenol blue (0.005% w/v)]. The 12% SDS-polyacrylamide gels were run and analyzed as described previously [77]. To detect immunoreactivity related to our proteins of choice, we applied the appropriate primary antibodies: Monoclonal Anti-β-tubulin III (neuronal) Clone 2G10 from mouse (T8578, Sigma, Rehovot, Israel) at 0.5 µg/mL, as advised by the datasheet instructions. Monoclonal Anti-β-Actin antibody produced in mouse (A5441, Sigma, Rehovot, Israel) diluted 1:5000 was applied as a loading reference. The primary antibodies were labeled with IgG secondary antibody linked to horseradish peroxidase (anti-mouse IgG diluted 1:5000 as required; GE Healthcare, Buckinghamshire, UK). Binding of antibodies to their antigens was detected with the EZ-ECL-detection reagent. Labeling was captured on X-Omat blue XB-1 Kodak scientific Imaging Film. Band intensity was quantified by using Quantity one 1D-analysis software (Bio-Rad, Hercules, CA, USA).

5. Conclusions

- Our study indicates that the classical TSPO ligand PK 11195 can modulate gene expression in U118MG cells.
- Robust and significant changes in gene expression can already be seen within 15 min and appear to be associated with cell morphological changes within the same time frame.
- At least at 15 min of PK 11195 exposure, expression of several elements of the canonical pathway for regulation of gene expression in U118MG cells is enhanced.

- After 24 h of exposure to PK 11195, changes in gene expression appear to be related to cell viability and tumorigenicity of these U118MG cells.
- This modulation in gene expression most likely occurs via mitochondria-to-nucleus signaling, probably via mechanisms including $\Delta\Psi$m collapse, ROS generation, Ca^{2+} release, and ATP production (Figure 9). It is well documented by previous studies that $\Delta\Psi$m collapse, ROS generation, Ca^{2+} release, and ATP production are under the control of mitochondrial TSPO.
- Thus, TSPO does not just modulate local mitochondrial functions, it also modulates nuclear gene expression.
- Phenotypic changes predicted by the changes in gene expression did actually occur, e.g., cell migration, cell accumulation, cell differentiation, and others.
- The novel TSPO ligand 2-Cl-MGV-1 also specifically modulated gene expression of immediate early genes.
- The modulation of gene expression by the TSPO elucidated in the present study goes a long way in explaining subsequent changes in cellular and organismal functions due to application of TSPO ligands (Figure 9).
- Thus, modulation of nuclear gene expression via the mitochondrial TSPO can induce several vital cell functions, which has major implications for the whole organism in health and disease.
- We believe that our study provides more understanding in the overall biological function of TSPO.

Supplementary Materials: Supplementary materials can be found at www.mdpi.com/1422-0067/18/4/786/s1.

Acknowledgments: We thank Liat Linde and Nili Avidan of the Genomics Core Facility of the Rappaport Family Medical Research Institute, and Edith Suss-Toby and Ortal Schwartz, of the Bioimaging Center of the Biomedical Core Facility of the Bruce Rappaport Faculty of Medicine, all of the Technion, Israel Institute of Technology for their expert assistance and advice, in particular regarding the assays for gene expression. This work is supported in part by a joint grant from the Center for Absorption in Science of the Ministry of Immigrant Absorption and the Committee for Planning and Budgeting of the Council for Higher Education under the framework of the KAMEA program (Leo Veenman). The Israel Science Foundation is thankfully acknowledged for their support for this research (Leo Veenman and Moshe Gavish). Furthermore, we wish to acknowledge the anonymous reviewers and editors for their invaluable suggestions of improvements, and their unwavering encouragement to continue with this project.

Author Contributions: Leo Veenman and Nasra Yasin wrote the paper, including preparation of the presentations; Nasra Yasin, Bea Caballero, Maya Azrad, and Alex Vainshtein performed the experiments, and wrote the parts related to the experiments; Sukhdev Singh synthesized 2-Cl-MGV-1; Moshe Gavish and Ilan Marek contributed reagents/materials/analysis tools; Nasra Yasin and Leo Veenman analyzed the data; Leo Veenman and Julia Bode conceived and designed the experiments.

Conflicts of Interest: The authors declare no conflict of interest.

References

1. Gavish, M.; Bachman, I.; Shoukrun, R.; Katz, Y.; Veenman, L.; Weisinger, G.; Weizman, A. Enigma of the peripheral benzodiazepine receptor. *Pharmacol. Rev.* **1999**, *51*, 629–650. [PubMed]
2. Veenman, L.; Gavish, M. The peripheral-type benzodiazepine receptor and the cardiovascular system. Implications for drug development. *Pharmacol. Ther.* **2006**, *110*, 503–524. [CrossRef] [PubMed]
3. Veenman, L.; Gavish, M. The role of 18 kDa mitochondrial translocator protein (TSPO) in programmed cell death, and effects of steroids on TSPO expression. *Curr. Mol. Med.* **2012**, *12*, 398–412. [CrossRef] [PubMed]
4. Starosta-Rubinstein, S.; Ciliax, B.J.; Penney, J.B.; McKeever, P.; Young, A.B. Imaging of a glioma using peripheral benzodiazepine receptor ligands. *Proc. Natl. Acad. Sci. USA* **1987**, *84*, 891–895. [CrossRef] [PubMed]
5. Kugler, W.; Veenman, L.; Shandalov, Y.; Leschiner, S.; Spanier, I.; Lakomek, M.; Gavish, M. Ligands of the mitochondrial 18 kDa translocator protein attenuate apoptosis of human glioblastoma cells exposed to erucylphosphohomocholine. *Cell. Oncol.* **2008**, *30*, 435–450. [PubMed]
6. Yeliseev, A.A.; Kaplan, S. TspO of rhodobacter sphaeroides. A structural and functional model for the mammalian peripheral benzodiazepine receptor. *J. Biol. Chem.* **2000**, *275*, 5657–5667. [PubMed]

7. Papadopoulos, V.; Baraldi, M.; Guilarte, T.R.; Knudsen, T.B.; Lacapere, J.J.; Lindemann, P.; Norenberg, M.D.; Nutt, D.; Weizman, A.; Zhang, M.R.; et al. Translocator protein (18 kDa): New nomenclature for the peripheral-type benzodiazepine receptor based on its structure and molecular function. *Trends Pharmacol. Sci.* **2006**, *27*, 402–409. [CrossRef] [PubMed]

8. Veenman, L.; Papadopoulos, V.; Gavish, M. Channel-like functions of the 18-kDa translocator protein (TSPO): Regulation of apoptosis and steroidogenesis as part of the host-defense response. *Curr. Pharm. Des.* **2007**, *13*, 2385–2405. [CrossRef] [PubMed]

9. McEnery, M.W.; Snowman, A.M.; Trifiletti, R.R.; Snyder, S.H. Isolation of the mitochondrial benzodiazepine receptor: Association with the voltage-dependent anion channel and the adenine nucleotide carrier. *Proc. Natl. Acad. Sci. USA* **1992**, *89*, 3170–3174. [CrossRef] [PubMed]

10. Caballero, B.; Veenman, L.; Gavish, M. Role of mitochondrial translocator protein (18 kDa) on mitochondrial-related cell death processes. *Recent Pat. Endocr. Metab. Immune Drug Discov.* **2013**, *7*, 86–101. [CrossRef] [PubMed]

11. Veenman, L.; Shandalov, Y.; Gavish, M. VDAC activation by the 18 kDa translocator protein (TSPO), implications for apoptosis. *J. Bioenerg. Biomembr.* **2008**, *40*, 199–205. [CrossRef] [PubMed]

12. Zeno, S.; Veenman, L.; Katz, Y.; Bode, J.; Gavish, M.; Zaaroor, M. The 18 kDa mitochondrial translocator protein (TSPO) prevents accumulation of protoporphyrin IX. Involvement of reactive oxygen species (ROS). *Curr. Mol. Med.* **2012**, *12*, 494–501. [CrossRef] [PubMed]

13. Veenman, L.; Vainshtein, A.; Yasin, N.; Azrad, M.; Gavish, M. Tetrapyrroles as endogenous TSPO ligands in eukaryotes and prokaryotes: Comparisons with synthetic ligands. *Int. J. Mol. Sci.* **2016**, *17*, 880. [CrossRef] [PubMed]

14. Wu, Y.; Kazumura, K.; Maruyama, W.; Osawa, T.; Naoi, M. Rasagiline and selegiline suppress calcium efflux from mitochondria by PK11195-induced opening of mitochondrial permeability transition pore: A novel anti-apoptotic function for neuroprotection. *J. Neural Transm.* **2015**, *122*, 1399–1407. [CrossRef] [PubMed]

15. Azarashvili, T.; Grachev, D.; Krestinina, O.; Evtodienko, Y.; Yurkov, I.; Papadopoulos, V.; Reiser, G. The peripheral-type benzodiazepine receptor is involved in control of Ca^{2+}-induced permeability transition pore opening in rat brain mitochondria. *Cell Calcium* **2007**, *42*, 27–39. [CrossRef] [PubMed]

16. Rosenberg, N.; Rosenberg, O.; Weizman, A.; Veenman, L.; Gavish, M. In vitro effect of FGIN-1-27, a ligand to 18 kDa mitochondrial translocator protein, in human osteoblast-like cells. *J. Bioenerg. Biomembr.* **2014**, *46*, 197–204. [CrossRef] [PubMed]

17. Balsemão-Pires, E.; Jaillais, Y.; Olson, B.J.; Andrade, L.R.; Umen, J.G.; Chory, J.; Sachetto-Martins, G. The Arabidopsis translocator protein (AtTSPO) is regulated at multiple levels in response to salt stress and perturbations in tetrapyrrole metabolism. *BMC Plant Biol.* **2011**, *11*, 108. [CrossRef] [PubMed]

18. Vanhee, C.; Zapotoczny, G.; Masquelier, D.; Ghislain, M.; Batoko, H. The Arabidopsis multistress regulator TSPO is a heme binding membrane protein and a potential scavenger of porphyrins via an autophagy-dependent degradation mechanism. *Plant Cell* **2011**, *23*, 785–805. [CrossRef] [PubMed]

19. Veenman, L.; Bode, J.; Gaitner, M.; Caballero, B.; Pe'er, Y.; Zeno, S.; Kietz, S.; Kugler, W.; Lakomek, M.; Gavish, M. Effects of 18-kDa translocator protein knockdown on gene expression of glutamate receptors, transporters, and metabolism, and on cell viability affected by glutamate. *Pharmacogenet. Genom.* **2012**, *22*, 606–619. [CrossRef] [PubMed]

20. Vainshtein, A.; Veenman, L.; Shterenberg, A.; Singh, S.; Masarwa, A.; Dutta, B.; Island, B.; Tsoglin, E.; Levin, E.; Leschiner, S.; et al. Quinazoline based tricyclic compounds that regulate programmed cell death, induce neuronal differentiation, and are curative in animal models for excitotoxicity and hereditary brain disease. *Cell Death Discov.* **2015**, *1*, 15027. [CrossRef] [PubMed]

21. Karlstetter, M.; Nothdurfter, C.; Aslanidis, A.; Moeller, K.; Horn, F.; Scholz, R.; Neumann, H.; Weber, B.H.; Rupprecht, R.; Langmann, T. Translocator protein (18 kDa) (TSPO) is expressed in reactive retinal microglia and modulates microglial inflammation and phagocytosis. *J. Neuroinflamm.* **2014**, *11*, 3. [CrossRef] [PubMed]

22. Veenman, L.; Vainsthein, A.; Gavish, M. TSPO as a target for treatments of diseases, including neuropathological disorders. *Cell Death Disease* **2015**, *6*, e1911. [CrossRef] [PubMed]

23. Maaser, K.; Sutter, A.P.; Krahn, A.; Höpfner, M.; Grabowski, P.; Scherübl, H. Cell cycle-related signaling pathways modulated by peripheral benzodiazepine receptor ligands in colorectal cancer cells. *Biochem. Biophys. Res. Commun.* **2004**, *324*, 878–886. [CrossRef] [PubMed]

24. Bode, J.; Veenman, L.; Vainshtein, A.; Kugler, W.; Rosenberg, N.; Gavish, M. Modulation of gene expression associated with the cell cycle and tumorigenicity of glioblastoma cells by the 18 kDa Translocator Protein (TSPO). *Austin J. Pharmacol. Ther.* **2014**, *2*, 1053.

25. Yeliseev, A.A.; Kaplan, S. A sensory transducer homologous to the mammalian peripheral-type benzodiazepine receptor regulates photosynthetic membrane complex formation in Rhodobacter sphaeroides 2.4.1. *J. Biol. Chem.* **1995**, *270*, 21167–21175. [CrossRef] [PubMed]

26. Liu, G.J.; Middleton, R.J.; Kam, W.W.; Chin, D.Y.; Hatty, C.R.; Chan, R.H.; Banati, R.B. Functional gains in energy and cell metabolism after TSPO gene insertion. *Cell Cycle* **2017**, *19*. [CrossRef] [PubMed]

27. Parikh, V.S.; Morgan, M.M.; Scott, R.; Clements, L.S.; Butow, R.A. The mitochondrial genotype can influence nuclear gene expression in yeast. *Science* **1987**, *235*, 576–580. [CrossRef] [PubMed]

28. Ždralević, M.; Guaragnella, N.; Giannattasio, S. Yeast as a tool to study mitochondrial retrograde pathway en route to cell stress response. *Methods Mol. Biol.* **2015**, *1265*, 321–331. [PubMed]

29. Cagin, U.; Enriquez, J.A. The complex crosstalk between mitochondria and the nucleus: What goes in between? *Int. J. Biochem. Cell Biol.* **2015**, *63*, 10–15. [CrossRef] [PubMed]

30. Butow, R.A.; Avadhani, N.G. Mitochondrial signaling: The retrograde response. *Mol Cell* **2004**, *14*, 1–15. [CrossRef]

31. Biswas, G.; Srinivasan, S.; Anandatheerthavarada, H.K.; Avadhani, N.G. Dioxin-mediated tumor progression through activation of mitochondria-to-nucleus stress signaling. *Proc. Natl. Acad. Sci. USA* **2008**, *105*, 186–191. [CrossRef] [PubMed]

32. Gomes, A.P.; Price, N.L.; Ling, A.J.; Moslehi, J.J.; Montgomery, M.K.; Rajman, L.; White, J.P.; Teodoro, J.S.; Wrann, C.D.; Hubbard, B.P.; et al. Declining NAD$^+$ induces a pseudohypoxic state disrupting nuclear-mitochondrial communication during aging. *Cell* **2013**, *155*, 1624–1638. [CrossRef] [PubMed]

33. Lin, M.Y.; Sheng, Z.H. Regulation of mitochondrial transport in neurons. *Exp. Cell Res.* **2015**, *334*, 35–44. [CrossRef] [PubMed]

34. Yi, M.; Weaver, D.; Hajnóczky, G. Control of mitochondrial motility and distribution by the calcium signal: A homeostatic circuit. *J. Cell Biol.* **2004**, *167*, 661–672. [CrossRef] [PubMed]

35. Shargorodsky, L.; Veenman, L.; Caballero, B.; Pe'er, Y.; Leschiner, S.; Bode, J.; Gavish, M. The nitric oxide donor sodium nitroprusside requires the 18 kDa translocator protein to induce cell death. *Apoptosis* **2012**, *17*, 647–665. [CrossRef] [PubMed]

36. Veenman, L.; Alten, J.; Linnemannstöns, K.; Shandalov, Y.; Zeno, S.; Lakomek, M.; Gavish, M.; Kugler, W. Potential involvement of F0F1-ATP(synth)ase and reactive oxygen species in apoptosis induction by the antineoplastic agent erucylphosphohomocholine in glioblastoma cell line: A mechanism for induction of apoptosis via the 18 kDa mitochondrial translocator protein. *Apoptosis* **2010**, *15*, 753–768. [PubMed]

37. Veenman, L.; Gavish, M.; Kugler, W. Apoptosis induction by erucylphosphohomocholine via the 18 kDa mitochondrial translocator protein: Implications for cancer treatment. *Anticancer Agents Med. Chem.* **2014**, *14*, 559–577. [CrossRef] [PubMed]

38. Hong, S.H.; Choi, H.B.; Kim, S.U.; McLarnon, J.G. Mitochondrial ligand inhibits store-operated calcium influx and COX-2 production in human microglia. *J. Neurosci. Res.* **2006**, *83*, 1293–1298. [CrossRef] [PubMed]

39. Azarashvili, T.; Krestinina, O.; Baburina, Y.; Odinokova, I.; Grachev, D.; Papadopoulos, V.; Akatov, V.; Lemasters, J.J.; Reiser, G. Combined effect of G3139 and TSPO ligands on Ca^{2+}-induced permeability transition in rat brain mitochondria. *Arch. Biochem. Biophys.* **2015**, *587*, 70–77. [CrossRef] [PubMed]

40. Costa, B.; da Pozzo, E.; Giacomelli, C.; Taliani, S.; Bendinelli, S.; Barresi, E.; da Settimo, F.; Martini, C. TSPO ligand residence time influences human glioblastoma multiforme cell death/life balance. *Apoptosis* **2015**, *20*, 383–398. [CrossRef] [PubMed]

41. Xiao, J.; Liang, D.; Zhang, H.; Liu, Y.; Li, F.; Chen, Y.H. 4′-Chlorodiazepam, a translocator protein (18 kDa) antagonist, improves cardiac functional recovery during postischemia reperfusion in rats. *Exp. Biol. Med.* **2010**, *235*, 478–486. [CrossRef] [PubMed]

42. Lehtonen, M.T.; Akita, M.; Frank, W.; Reski, R.; Valkonen, J.P. Involvement of a class III peroxidase and the mitochondrial protein TSPO in oxidative burst upon treatment of moss plants with a fungal elicitor. *Mol. Plant Microbe Interact.* **2012**, *25*, 363–371. [CrossRef] [PubMed]

43. Guilarte, T.R.; Loth, M.K.; Guariglia, S.R. TSPO finds NOX2 in microglia for redox homeostasis. *Trends Pharmacol. Sci.* **2016**. [CrossRef] [PubMed]

44. Zeno, S.; Zaaroor, M.; Leschiner, S.; Veenman, L.; Gavish, M. CoCl$_2$ induces apoptosis via the 18 kDa translocator protein in U118MG human glioblastoma cells. *Biochemistry* **2009**, *48*, 4652–4661. [CrossRef] [PubMed]

45. Kim, T.; Pae, A.N. Translocator protein (TSPO) ligands for the diagnosis or treatment of neurodegenerative diseases: A patent review (2010–2015; part 1). *Expert Opin. Ther. Pat.* **2016**, *26*, 1325–1351. [CrossRef] [PubMed]

46. Mukherjee, S.; Das, S.K. Translocator protein (TSPO) in breast cancer. *Curr. Mol. Med.* **2012**, *12*, 443–445. [CrossRef] [PubMed]

47. Ruksha, T.; Aksenenko, M.; Papadopoulos, V. Role of translocator protein in melanoma growth and progression. *Arch. Dermatol. Res.* **2012**, *304*, 839–845. [CrossRef] [PubMed]

48. Bonsack, F., 4th; Alleyne, C.H., Jr.; Sukumari-Ramesh, S. Augmented expression of TSPO after intracerebral hemorrhage: A role in inflammation? *J. Neuroinflamm.* **2016**, *13*, 151. [CrossRef] [PubMed]

49. Randles, M.; Lamb, D.E.; Odat, E.; Taleb-Bendiab, A. Distributed redundancy and robustness in complex systems. *J. Comput. Syst. Sci.* **2011**, *77*, 293–304. [CrossRef]

50. Macia, J.; Solé, R.V. Distributed robustness in cellular networks: Insights from synthetic evolved circuits. *J. R. Soc. Interface* **2009**, *6*, 393–400. [CrossRef] [PubMed]

51. Feala, J.D.; Cortes, J.; Duxbury, P.M.; McCulloch, A.D.; Piermarocchi, C.; Paternostro, G. Statistical properties and robustness of biological controller-target networks. *PLoS ONE* **2012**, *7*, e29374. [CrossRef] [PubMed]

52. Rosenberg, N.; Rosenberg, O.; Weizman, A.; Leschiner, S.; Sakoury, Y.; Fares, F.; Soudry, M.; Weisinger, G.; Veenman, L.; Gavish, M. In vitro mitochondrial effects of PK 11195, a synthetic translocator protein 18 kDa (TSPO) ligand, in human osteoblast-like cells. *J. Bioenerg. Biomembr.* **2011**, *43*, 739–746. [CrossRef] [PubMed]

53. Lagutina, I.S.; Mezina, M.N.; Prokof'ev, M.I.; Zakharchenko, V.I.; Galat, V.I. A study of factors affecting the efficiency of electrofusion of enucleated eggs and cell-donors of nuclei. *Ontogenez* **2002**, *33*, 100–116. [PubMed]

54. Tullai, J.W.; Schaffer, M.E.; Mullenbrock, S.; Sholder, G.; Kasif, S.; Cooper, G.M. Immediate-early and delayed primary response genes are distinct in function and genomic architecture. *J. Biol. Chem.* **2007**, *282*, 23981–23995. [CrossRef] [PubMed]

55. Jayakumar, A.R.; Panickar, K.S.; Norenberg, M.D. Effects on free radical generation by ligands of the peripheral benzodiazepine receptor in cultured neural cells. *J. Neurochem.* **2002**, *83*, 1226–1234. [CrossRef] [PubMed]

56. Chen, C.; Chen, Y.; Guan, M.X. A peep into mitochondrial disorder: Multifaceted from mitochondrial DNA mutations to nuclear gene modulation. *Protein Cell* **2015**, *6*, 862–870. [CrossRef] [PubMed]

57. Arnould, T.; Vankoningsloo, S.; Renard, P.; Houbion, A.; Ninane, N.; Demazy, C.; Remacle, J.; Raes, M. CREB activation induced by mitochondrial dysfunction is a new signaling pathway that impairs cell proliferation. *EMBO J.* **2002**, *21*, 53–63. [CrossRef] [PubMed]

58. Choi, J.; Ifuku, M.; Noda, M.; Guilarte, T.R. Translocator protein (18 kDa)/peripheral benzodiazepine receptor specific ligands induce microglia functions consistent with an activated state. *Glia* **2011**, *59*, 219–230. [CrossRef] [PubMed]

59. Levin, E.; Premkumar, A.; Veenman, L.; Kugler, W.; Leschiner, S.; Spanier, I.; Weisinger, G.; Lakomek, M.; Weizman, A.; Snyder, S.H.; et al. The peripheral typebenzodiazepine receptor and tumorigenicity: Isoquinoline binding protein (IBP) antisense knockdown in the C6 glioma cell line. *Biochemistry* **2005**, *44*, 9924–9935. [CrossRef] [PubMed]

60. Bode, J.; Veenman, L.; Caballero, B.; Lakomek, M.; Kugler, W.; Gavish, M. The 18 kDa translocator protein influences angiogenesis, as well as aggressiveness, adhesion, migration, and proliferation of glioblastoma cells. *Pharmacogenet. Genom.* **2012**, *22*, 538–550. [CrossRef] [PubMed]

61. Wu, M.; Kalyanasundaram, A.; Zhu, J. Structural and biomechanical basis of mitochondrial movement in eukaryotic cells. *Int. J. Nanomed.* **2013**, *8*, 4033–4042.

62. Zhang, Z.W.; Zhang, G.C.; Zhu, F.; Zhang, D.W.; Yuan, S. The roles of tetrapyrroles in plastid retrograde signaling and tolerance to environmental stresses. *Planta* **2015**, *242*, 1263–1276. [CrossRef] [PubMed]

63. Wang, H.; Zhai, K.; Xue, Y.; Yang, J.; Yang, Q.; Fu, Y.; Hu, Y.; Liu, F.; Wang, W.; Cui, L.; et al. Global deletion of TSPO does not affect the viability and gene expression profile. *PLoS ONE* **2016**, *11*, e0167307. [CrossRef] [PubMed]

64. Middleton, R.J.; Liu, G.J.; Banati, R.B. Guwiyang Wurra—"Fire Mouse": A global gene knockout model for TSPO/PBR drug development, loss-of-function and mechanisms of compensation studies. *Biochem. Soc. Trans.* **2015**, *43*, 553–558. [CrossRef] [PubMed]

65. Gut, P.; Zweckstetter, M.; Banati, R.B. Lost in translocation: The functions of the 18-kD translocator protein. *Trends Endocrinol. Metab.* **2015**, *26*, 349–356. [CrossRef] [PubMed]

66. Midzak, A.; Zirkin, B.; Papadopoulos, V. Translocator protein: Pharmacology and steroidogenesis. *Biochem. Soc. Trans.* **2015**, *43*, 572–578. [CrossRef] [PubMed]

67. Li, F.; Liu, J.; Liu, N.; Kuhn, L.A.; Garavito, R.M.; Ferguson-Miller, S. Translocator protein 18 kDa (TSPO): An old protein with new functions? *Biochemistry* **2016**, *55*, 2821–2831. [CrossRef] [PubMed]

68. Gatliff, J.; Campanella, M. TSPO: Kaleidoscopic 18-kDa amid biochemical pharmacology, control and targeting of mitochondria. *Biochem. J.* **2016**, *473*, 107–121. [CrossRef] [PubMed]

69. Selvaraj, V.; Tu, L.N. Current status and future perspectives: TSPO in steroid neuroendocrinology. *J. Endocrinol.* **2016**, *231*, R1–R30. [CrossRef] [PubMed]

70. Batoko, H.; Veljanovski, V.; Jurkiewicz, P. Enigmatic Translocator protein (TSPO) and cellular stress regulation. *Trends Biochem. Sci.* **2015**, *40*, 497–503. [CrossRef] [PubMed]

71. Da Pozzo, E.; Giacomelli, C.; Barresi, E.; Costa, B.; Taliani, S.; Passetti Fda, S.; Martini, C. Targeting the 18-kDa translocator protein: Recent perspectives for neuroprotection. *Biochem. Soc. Trans.* **2015**, *43*, 559–565. [CrossRef] [PubMed]

72. Issop, L.; Ostuni, M.A.; Lee, S.; Laforge, M.; Péranzi, G.; Rustin, P.; Benoist, J.F.; Estaquier, J.; Papadopoulos, V.; Lacapère, J.J. Translocator protein-mediated stabilization of mitochondrial architecture during inflammation stress in colonic cells. *PLoS ONE* **2016**, *11*, e0152919. [CrossRef] [PubMed]

73. Kim, T.; Pae, A.N. Translocator protein (TSPO) ligands for the diagnosis or treatment of neurodegenerative diseases: A patent review (2010–2015; part 2). *Expert Opin. Ther. Pat.* **2016**, *26*, 1353–1366. [CrossRef] [PubMed]

74. Benjamini, Y.; Hochberg, Y. Controlling the false discovery rate: A practical and powerful approach to multiple testing. *J. R. Stat. Soc. Ser. B* **1995**, *57*, 289–300.

75. Krämer, A.; Green, J.; Pollard, J.; Tugendreich, S. Causal analysis approaches in ingenuity pathway analysis. *Bioinform.* **2014**, *30*, 523–530. [CrossRef] [PubMed]

76. Bradford, M.M. A rapid and sensitive method for the quantitation of microgram quantities of protein utilizing the principle of protein-dye binding. *Anal. Biochem.* **1976**, *72*, 248–254. [CrossRef]

77. Caballero, B.; Veenman, L.; Bode, J.; Leschiner, S.; Gavish, M. Concentration-dependent bimodal effect of specific 18 kDa translocator protein (TSPO) ligands on cell death processes induced by ammonium chloride: Potential implications for neuropathological effects due to hyperammonemia. *CNS Neurol. Disord. Drug Targets* **2014**, *13*, 574–592. [CrossRef] [PubMed]

MDPI AG

St. Alban-Anlage 66

4052 Basel, Switzerland

Tel. +41 61 683 77 34

Fax +41 61 302 89 18

http://www.mdpi.com

International Journal of Molecular Sciences Editorial Office

E-mail: ijms@mdpi.com

http://www.mdpi.com/journal/ijms

www.ingramcontent.com/pod-product-compliance
Lightning Source LLC
Chambersburg PA
CBHW051859210326
41597CB00033B/5958